Analytical Chemistry

Theoretical and Metrological Fundamentals

K. Danzer

Analytical Chemistry

Theoretical and Metrological Fundamentals

With 138 Figures and 31 Tables

 Springer

K. Danzer
Am Friedensberg 4
07745 Jena
Germany
e-mail: claus.danzer@jetzweb.de

Library of Congress Control Number 2006932265

ISBN-10 3-540-35988-5 Springer Berlin Heidelberg New York
ISBN-13 978-3-540-35988-3 Springer Berlin Heidelberg New York
DOI 10.1007/b103950

Springer is a part of Springer Science+Business Media
springer.com
© Springer-Verlag Berlin Heidelberg 2007

Cover-Design: WMXDesign GmbH, Heidelberg
Typesetting: PTP-Berlin Protago-TEX-Production GmbH, Germany
Production: LE-TEX Jelonek, Schmidt & Vöckler GbR, Leipzig, Germany
Printed on acid-free paper   2/3141/YL - 5 4 3 2 1 0

Preface

Analytical chemistry can look back on a long history. In the course of its development, analytical chemistry has contributed essentially to the progress of diverse fields of science and technology. Significant progress in natural philosophy, chemistry and other disciplines has been founded on results and discoveries in analytical chemistry. Well known examples include Archimedes' principle, iatrochemistry, emergence and overcoming of phlogiston theory, basic chemical laws on stoichiometry and mass proportions as well as the discoveries of elements and of nuclear fission.

As chemistry split into inorganic, organic and physical chemistry in the middle of the nineteenth century, the question occasionally arose as to whether analytical chemistry was likewise an autonomous field of chemistry. At the beginning of the twentieth century, analytical chemistry was established by Wilhelm Ostwald and other protagonists on the basis of chemical and physicochemical laws and principles. In addition, analytical procedures and instruments spread into other chemical fields. However, in the 1950s and 1960s a new generation of analysts opened their mind to ideas coming from other scientific fields dealing with measurements from other points of view such as information science, metrology, technometrics, etc.

As a result, efforts made towards the acceptance of analytical chemistry as an autonomous field of chemistry increased. Analysts in USA and Europe founded organisations and forums that focussed ideas and developments in progressive directions. In this way, not only were national and international societies formed but also regular conferences like Pittcon and Euroanalysis. In this connection the so-called Lindau circle should be mentioned where analysts from Germany, Austria and Switzerland made public relevant fundamentals of analytical chemistry derived from statistics, information theory, system theory, signal theory, game theory, decision theory, metrology, etc. In parallel with this, such modern principles have been successfully applied in selected case studies, mainly by scientists from USA, UK, and Canada. From this development, real theoretical and metrological fundamentals of analytical chemistry have crystallized. However, the acceptance of analytical chemistry as an automomous branch of chemistry has not been realised. Gradually, however, progress in analytical chemistry culminated in the 1990s in the proclamation of a new scientific discipline, *analytical science*. Protagonists of these activities included Michael Widmer and Klaus Doerffel.

Today, analytical chemistry is still a discipline within chemistry. Although characterized as an auxiliary science, analytical chemistry continues to develop and grow just as before. It is no detraction being characterized as an auxiliary discipline as mathematics shows. However, it is likely that efforts to make analytical chemistry a more independent science will be repeated in the future from time to time.

Analytical chemistry possesses today a sound basis of chemical, physical, methodical, metrological, and theoretical fundamentals. The first of these are usually taken as the basis of classical textbooks on analytical chemistry. The others are found in diverse publications in the field of analytical chemistry and chemometrics. It is essential to state that chemometrics *is not* the theoretical basis of analytical chemistry but it contributes significantly to it. Frequently, analytical chemistry is considered to be a measuring science in chemistry. Therefore, its object is the generation, evaluation, interpretation, and validation of measuring signals as well as the characterization of their uncertainty. With this aim, the analyst needs knowledge of the general analytical process, statistics, optimization, calibration, chemometrical data analysis, and performance characteristics.

In this book the attempt is made to summarize all the components that can be considered as building blocks of a theory of analytical chemistry. The "building" constructed in this way is a provisional one. It is incomplete and, therefore, extension, reconstruction and rebuilding have to be expected in the future.

A large number of mathematical formulas will be found in the book. This may be regarded as a disadvantage particularly because some of them are not readily to apply in daily analytical practice. However, great scientists, explicitly Emanuel Kant, said that a scientific branch contains only so much of science as it applies mathematics. Consequently, all the relationships which can be described mathematically should be so described. It is true despite Werner Heisenbergs statement: Although natural processes can be described by means of simple laws which can be precisely formulated, these laws, on the other hand, cannot directly be applied to actions in practice.

Wherever possible, official definitions of IUPAC, ISO and other international organizations have been used, in particular in the "Glossary of Analytical Terms" compiled at the end of the book. However, uniformity could not be achieved in every case. In a few instances, special comments and proposals (characterized as such) have been added. Although progress in the field of harmonization of nomenclature and definitions has been considerable, some things still remain to be done.

Some of the contents of various sections have been published previously and are, of course quoted verbatim. In this connection the author is grateful to Klaus Doerffel, Karel Eckschlager, Günter Ehrlich, Andrzej Parczewski, Karol Florian, Mikulas Matherny, Günter Marx, Dieter Molch, Eberhard Than, Ludwig Küchler and others for long-standing collaboration resulting in mutually complementary papers and books. A chapter on statistics was previously published in another book (*Accreditation and Quality Assurance in Analytical Chemistry* edited by Helmut Günzler at Springer). Gratefully I have adapted essential parts of the translation of Gaida Lapitajs here in this book.

Stimulations and ideas have arisen from discussions with William Horwitz, Duncan Thorburn Burns, Alan Townshend, Koos van Staden, and other members of diverse IUPAC Commissions and Task Groups as well as officers of the Analytical Chemistry Division. From a 20-year membership of various IUPAC bodies I have gained a variety of feedback and experience which has proved to be useful in the writing of this book. In particular, I owe Lloyd Currie and Mattias Otto a great debt of gratitude. They have permitted me to use essential parts of common publications on calibration quoted in Chap. 6. Many of the results stem from work carried out with colleagues, collaborators, postdocs and graduate students. All of them are deserving of my thanks but, due to the large numbers involved, this cannot be done here. Some representative colleagues include Jürgen Einax, Hartmut Hobert, Werner Schrön, Manfred Reichenbächer, Reiner Singer, Dietrich Wienke, Christoph Fischbacher, Kai-Uwe Jagemann, Michael Wagner, Katrin Venth, Raimund Horn, Gabriela Thiel, Demetrio de la Calle García, and Balint Berente.

The author gratefully acknowledges the cooperation with and the support of the editorial staff of Springer. Particularly I have to thank Peter Enders, Birgit Kollmar-Thoni, and – last but not least – John Kirby, the English copy editor, who has essentially improved the English. Many thanks for that.

No book is free of errors and this one will be no exception. Therefore, the author would be grateful to readers who point out errors and mistakes and suggest any improvement.

Jena, July 2006 Klaus Danzer

Contents

Symbols

A	Matrix of sensitivity factors in multicomponent analysis (e.g., absorbance factors)	(6.66)
a	Regression coefficient (calibration coefficient): intercept (general and regression in case of errors in both variables)	
a_x	Regression coefficient a for the estimation of y from x	(6.9)
a_{x_w}	Regression coefficient a in case of weighted calibration	(6.37)
a_y	Regression coefficient a for the estimation of x from y	(6.10)
$acc(x)$, $acc(\overline{x})$	Accuracy of an analytical result x and of a mean $\overline{x}$, respectively	(7.10) (7.11)
antilg x	$= 10^x$, antilogarithm of x	
b	Regression coefficient (calibration coefficient): slope (general and in case of regression having errors in both variables)	
b_x	Regression coefficient b for the estimation of y from x	(6.8)
b_{x_w}	Regression coefficient b in case of weighted calibration	(6.36)
b_y	Regression coefficient b for the estimation of x from y	(6.10)
$bias(x)$, $bias(y)$	Bias (systematic deviation) of an individual test value x (or y) from the (conventional) true value x_{true} (or y_{true})	(4.1)

$bias(\overline{x})$, $bias(\overline{y})$	Bias of an average test value $\overline{x}$ (or $\overline{y}$) from the (conventional) true value x_{true} (or y_{true})	
$cnf(\overline{x})$, $cnf(\overline{y})$	Total (two-sided) confidence intervall of a mean	(4.17)
$cond(A)$	Condition number of the matrix A	(6.80)
$cov(p_i, p_j)$	Covariance of the quantities p_i and p_j	
CB	Confidence band (of a calibration line)	(6.26a)
CR	Concordance rate (proportion of correct test results in screening)	(4.53a)
d_{crit}	Critical diameter (in microprobe analysis)	
d_{ij}	Distance (in multivariate data)	(8.12)
df	Discriminant function	(8.16)
dv	Discriminant variable	
E	Information efficiency	
E_{in}	Input energy	
E_{out}	Output energy	
E	Residual matrix (in multivariate data analysis)	
e, e_x, e_y	Error term (general, in x, and in y, respectively), in concrete terms, e_y may be the deviation of an individual value from the mean or the residual of a mathematical model, see next line (e_x analogous)	
e_y	$= y_i - \overline{y}$, deviation of an individual measured value from the mean (also d_y)	
e_y	$= y_i - \hat{y}$, deviation of an individual measured value from the estimate of the corresponding mathematical model	
$\hat{F}$	Test statistic (estimate) of the F test	(4.37)

$F_{1-\alpha,\nu_1,\nu_2}$	Quantile of the F distribution at the level of significance $1-\alpha$ and for the degrees of freedom ν_1 and ν_2	
$\hat{F}_{\max}$	Test statistic (estimate) of HARTLEY's test	(4.37')
FNR	False negative rate (of screening tests)	(4.51)
FPR	False positive rate (of screening tests)	(4.50)
$f_{det}(\dots)$	Deterministic function of ...	
$f_{emp}(\dots)$	Empirical function of ...	
$\hat{G}$	Test statistic (estimate) of GRUBB's outlier test	(4.36)
$\hat{G}_{\max}$	Test statistic (estimate) of COCHRAN's test	(4.38)
$G_{\alpha,n}$	Significance limits of GRUBB's outlier test for a risk of error α and n individual measurements	
$g_{accept}(n)$	Acceptance function in attribute testing	
$g_{reject}(n)$	Rejection function in attribute testing	
$g_{accept}(n,s,\overline{\alpha})$	Acceptance function in variable testing	
$g_{accept}(n,s,\overline{\beta})$	Rejection function in variable testing	
H'	a posteriori information entropy	
H	a priori information entropy	
H_0	Null hypothesis	
H_A	Alternative hypothesis	
$hom(A)$	Quantity characterizing the homogeneity of an analyte A in a sample	(2.9)
I	Information content	(9.1)
I_{Aj}	Specific strength of an influence factor j on the signal of analyte A	
$I(p,p_0)$	Divergence measure of information	(9.9)
J	Information flow	(9.40)

L	Loading matrix (of principal component analysis and factor analysis)	(6.44) (8.19)
LCL	Lower control limit (in quality control)	
LWL	Lower warning limit (in quality control)	
l_x, l_y, l_z	Dimensions (lengths) in x-, y- and z-direction, respectively	
lb a	Binary logarithm of a	
M	Information amount	
$M(n)$	Information amount of multicomponent analysis (of n components)	(9.21)
$mad\{y_i\}$	Median absolute deviation (around the median), i.e., the median of the differences between an individual measured value and the median of the series: $mad\{y_i\} = med\{\|y_i - med\{y_i\}\|\}$	
$med\{y_i\}$	Median of a set of measured values	(4.22)
N	Number of repeated measurements for the estimation of x by means of a calibration function	
N_{pp}	Noise amplitude, peak-to-peak noise	
$N(\mu, \sigma^2)$	Normal distribution with mean μ and variance σ^2	
$\overline{N(t)}$	Time average of noise	
NPR	Negative prediction rate (in screening tests)	(4.55)
n	Number of individual measurements (also in establishing a calibration model), number of objects (in a data set)	
$n(t)$	Noise component of a signal function in the time domain	
$net_i(t), net^{(i)}$	Net function (propagation function) in a neural network (layer i)	(6.117)
$O_i(t), O^{(i)}$	Output of a layer i of a neural network	(6.119)

P	Two-sided level of significance of statistical tests ($P = 1 - \alpha$)	
$\overline{P}$	One-sided level of significance of statistical tests ($\overline{P} = 1 - \overline{\alpha}$)	
$P(A)$	Probability of an event A (ratio of the number of times the event occurs to the total number of trials)	
$P(A)$	Probability that the analyte A is present in the test sample	
$P(\overline{A})$	Probability of the opposite of the event A (ratio of the number of times the event does not occur to the total number, $P(A) + P(\overline{A}) = 1$)	
$P(\overline{A})$	Probability that the analyte A is present in the test sample	
$P(A|B)$	Conditional probability: probability of an event B on the condition that another event A occurs	
$P(T^+|A)$	Probability that the analyte A is present in the test sample if a test result T is positive	
$\boldsymbol{P}$	Score matrix (of principal component analysis)	(6.44)
PB	Prediction band (of a calibration line)	(6.26b)
PPR	Positive prediction rate (in screening tests)	(4.54)
PV	Prevalency of screening tests: probability that the analyte A is present in a given number of samples	(4.52)
p	Number of primary samples (in sampling); Number of calibration points	
p_i, p_j, p_{ij}	Influence parameters	(4.25)
$prec(x)$	Precision of an analytical result x	(7.8)
$prd(x), prd(a)$	Prediction interval of a quantity x (and a, respectively)	
q	Number of subsamples (in sampling)	(2.5)
$\hat{q}_R$	Test statistic (estimate) of the DAVID test	(4.33)

Q	Categorial variable characterizing chemical entities (species) being investigated, e.g., elements, isotopes, ions, compounds	
$\hat{Q}$	Test statistic (estimate) of DIXON's outlier test	(4.35)
$Q(z)$	Evaluation relationship (of quantities characterizing qualitative properties) in form of a function, atlas, or table	
$Q = f^{-1}(z)$	Evaluation function of quantities characterizing qualitative properties, viz type of species, Q, in dependence of signal position, z	
R	Range ($R = y_{max} - y_{min}$)	
R	Resolution power	
$\mathbf{R}$	Correlation matrix	(6.4)
R_C	Risk for customers (in quality assurance)	(4.56b)
R_M	Redundancy	(9.31a)
R_M	Risk for manufacturers (in quality assurance)	(4.56a)
R_t	Temporal resolution power	(7.57)
R_z	Analytical resolution power	(7.53)
$RMSP$	Root mean standard error of prediction	(6.128)
RR	Recovery rate	
$R\{q\}$	Range of a quantity q	
r	Number of measurements at a subsample	(2.4)
r_M	Relative redundancy	(9.31b)
r_{xy}	Correlation coefficient between the (random) variables x and y	(6.3)
$rob(A/B\ldots; f_1\ldots)$	Robustness of a procedure to determine an analyte A with regard to disturbing components ($B,\ldots$) and factors ($f_1,\ldots$)	(7.31)

$rug(A/B \ldots; f_1 \ldots; u_1 \ldots)$	Ruggedness of a procedure to determine an analyte A with regard to disturbing components $(B, \ldots)$, factors $(f_1, \ldots)$, and unknowns $(u_1, \ldots)$	(7.33)
S	Sensitivity, derivative of the measured quantity (response) y with respect to the analytical quantity x	
S_{total}	Total multicomponent sensitivity	(7.18)
$\mathbf{S}$	Covariance matrix	(6.5)
	Sensitivity matrix	(7.17)
S_{AA}	Sensitivity of the response of the analyte A with respect to the amount x_A of the analyte A (also S_A)	(7.12)
S_{AB}	Cross sensitivity (partial sensitivity) of the response of the analyte A with respect to the amount x_B of the component B	
S_{Ai}	Cross sensitivity (partial sensitivity) of the response of the analyte A with respect to the amount x_i of species i $(i = B, C, \ldots N)$	(3.11)
S_{xx}, S_{yy}	Sum of squared deviations (of x and y, respectively)	(6.3′)
S_{xy}	Sum of crossed deviations (of x and y)	(6.3′)
$S_\bullet, S_{\bullet i}$	Sum over a variable index (e.g., at constant i); the dot replaces the index over that the summation is carried out	
$S(x)$	Sensitivity function	(6.64)
s	Estimate of the standard deviation (SD of the sample)	(4.12)
s^2	Estimate of the variance (variance of the sample)	(4.10)
s_N	Standard deviation of noise	(7.5)
$s_{p_i p_j}, s(p_i, p_j)$	Covariance of the quantities p_i and p_j	(4.27)
$s_{y.x}$	Residual standard deviation (of the calibration); estimate of the error of the calibration model	(6.19)

$sel(A, B, \ldots N)$	Selectivity of a multicomponent analysis with regard to the analytes $A, B, \ldots N$	(7.24)
$snr(\overline{y})$	Signal-to-noise ratio of an average signal intensity $\overline{y}$ (related to N_{pp})	(7.6)
$spec(A/B, \ldots, N)$	Specificity of the determination of an analyte A with regard to the accompanying components $B, \ldots N$	(7.26)
S/N	Signal-to-noise ratio of an average signal intensity $\overline{y}$ or a net intensity $\overline{y}_{net}$ (related to s_y)	(7.1)
$(S/N)_c$	Critical signal-to-noise ratio	(7.51)
SS	Sum of squares	
$ssr(y)$	Sum of squares of residuals (of y)	(6.12)
t_W	Test statistic (estimate) of the generalized t test (WELCH test)	(4.43)
TE	Test efficiency of screening tests	(4.53b)
TNR	True negative rate (of screening tests)	(4.49)
TPR	True positive rate (of screening tests)	(4.48)
$\hat{t}$	Test statistic (estimate) of STUDENT's t test	(4.41)
$t_{1-\alpha,\nu}$	Quantile of the t-distribution at the level of significance $1 - \alpha$ and for ν degrees of freedom	
UCL	Upper control limit (in quality control)	
UWL	Upper warning limit (in quality control)	
$u(x)$	Combined uncertainty of an analytical value x	(4.31)
$u(x(p_1, p_2, \ldots))$	Uncertainty of an analytical value x, combined from the uncertainties of the parameters $p_1, p_2, \ldots$	
$u(y)$	Combined uncertainty of a measured value y	
$u(y(p_1, p_2, \ldots))$	Uncertainty of a measured value y, combined from the uncertainties of the parameters $p_1, p_2, \ldots$	(4.25)

$U(x)$	Extended combined uncertainty (limit) of an analytical value x	
$U(y)$	Extended combined uncertainty (limit) of a measured value y	(4.29)
$unc(\overline{x})$	Uncertainty interval of a mean $\overline{x}$	(4.32)
$unc(\overline{y})$	Uncertainty interval of a mean $\overline{y}$	(4.30)
w_{y_i}, w_i	Weight coefficient (in weighted calibration)	(6.34)
x	Analytical value: analyte amount, e.g., content, concentration	
x_{LD}	Limit of detection	(7.44)
x_{LQ}	Limit of quantitation	(7.48)
x_Q	Analytical value (amount) of a species Q	
x_{test}	Analytical value of a test sample	
x_{true}	(Conventional) true value of a (certified) reference sample	
$\overline{x}$	Arithmetic mean of x (of the sample)	
$\hat{x}$	Estimate of x	
$x(Q)$	Sample composition (function)	
$x = f(Q)$	Analytical function in the sample domain, sample composition function	
$Y(\omega)$	Signal function in the frequency domain, FOURIER transform of $y(t)$	
y	Measured value, response (e.g., intensity of a signal)	
y_c	Critical value	(7.41)
y_{z_i}	Measured value (e.g. intensity) of a signal at position z_i	
$\overline{y}$	Arithmetic mean of y (of the sample)	
$\overline{\overline{y}}$	Total mean of several measured series	
$\overline{y}_{BL}$	Blank	
$\overline{y}_{geom}$	Geometric mean of y (of the sample)	
$\hat{y}$	Estimate of y	

y^*	Outlier-suspected value among the measured values	
$y(t)$	Signal function in the time domain	
$y_{true}(t)$	Component of the signal function in the time domain that is considered being true (influenced by noise)	
$y(z)$	Signal record (spectrum, chromatogram, etc.)	
$y = f(x)$	Calibration function (in the stricter sense, i.e. of quantities characterizing quantitative properties, e.g., signal intensity vs analyte amount)	
$y = f(z)$	Signal function (measurement function, analytical function in the signal domain)	
z	Signal position: measuring quantity that depends on a qualitative property of the measurand. Therefore, analytes may be identified by characteristic signal positions. The z-scale may be directly or reciprocally proportional to an energy quantity or time	
z	z-scores, standardized analytical values	(8.9)
$z(Q)$	Signal assignment function (table, atlas etc.)	
$z = f(Q)$	Calibration function of quantities characterizing qualitative properties, viz signal position as a function of the type of species	
α	Risk of the error of the first kind (two-sided)	
α	Regression coefficient (intercept)	(4.2)
α_i	Coefficient characterizing influences	(5.3)
$\overline{\alpha}$	Risk of the error of the first kind (one-sided)	
β	Risk of the error of the second kind (two-sided)	

β	Regression coefficient (slope)	(4.3)
β_j	Coefficient characterizing influences	(5.4)
$\overline{\beta}$	Risk of the error of the second kind (one-sided)	
$\hat{\chi}^2$	Test statistic (estimate) of the χ^2 test	
$\chi^2_{1-\alpha,v}$	Quantile of the χ^2 distribution at the significance level $1-\alpha$ and for v degrees of freedom	
$\Delta\overline{x}_{cnf}$	Confidence limit of a mean $\overline{x}$	
$\Delta\overline{x}_{prd}$	Prediction limit of a mean $\overline{x}$	
$\Delta\overline{y}_{cnf}$	Confidence limit of a mean $\overline{y}$	
Δ^2	Test statistic (estimate) for VON NEUMANN's trend test	(4.34)
$\mathscr{F}\{x(t)\}$	FOURIER transform of the time function $x(t)$ into the frequency function $X(\omega)$	
$\mathscr{F}^{-1}\{X(\omega)\}$	FOURIER backtransform of the frequency function $X(\omega)$ into the time function $x(t)$	
γ	Exponent characterizing nonlinear errors	(4.5)
λ	Eigenvalue (characteristic root) of a matrix	
$\mathcal{M}$	Multiplet splitting according to relevant rules	
μ_x	Mean of x of the population	
μ_y	Mean of y of the population	
v	Number of the statistical degrees of freedom	
v	Dispersion factor	(4.20)
σ	Standard deviation (of the population)	
σ^2	Variance (of the population)	(4.8)
Ξ	KAISER's selectivity	(7.21)

Ψ_A	KAISER's specificity with regard to an analyte A	(7.22)
$\Psi_{xx}(\tau)$	Autocorrelation function (ACF) of a function $x(t)$ with time lag τ	(2.11)
$(\alpha\beta)_{ij}$	Coefficient characterizing interactions of influences α and β	(5.4)
$(\alpha\beta\gamma)_{ijk}$	Coefficient characterizing interactions of influences α, β, and γ	(5.5)
$\{x_i\}$	Set (series, sequence) of analytical results whose terms are $x_1, x_2, \ldots, x_i, \ldots$	
$\{y_i\}$	Set of measured values (observations of a measurement series) whose terms are $y_1, y_2, \ldots, y_i, \ldots$	
$\hat{a}, \hat{b}, \ldots$	Estimates of the quantities $a, b, \ldots$	

In general, variables are expressed by italics, vectors by bold small letters, and matrices by bold capital letters.

Abbreviations and Acronyms

This list of abbreviations contains both acronyms which are generally used in analytical chemistry and such applied in the book. In addition to terms from analytical methods, essential statistical and chemometrical terms as well as acronyms of institutions and organizations are included. Terms of very particular interest are explained on that spot.

2D	Two-Dimensional (e.g., 2D-NMR)
3D	Three-Dimensional
AA	Activation Analysis
AAS	Atomic Absorption Spectrometry
ACF	Autocorrelation Function
ACV	Analytical value at critical (measuring) value
AEM	Analytical Electron Microscopy
AES	Auger Electron Spectroscopy
AES	(Atomic Emission Spectrometry) → OAES, OES
AFM	Atomic Force Microscopy
AFS	Atomic Fluorescence Spectrometry
ANN	Artificial Neural Networks
ANOVA	Analysis Of Variance
AOAC	Association of Official Analytical Chemists
ARM	Atomic Resolution Microscopy
ARUPS	Angle Resolved Ultraviolet Photoelectron Spectrometry
ASV	Anodic Stripping Voltammetry
ATR	Attenuated Total Reflectance
BASIC	Programming language: Beginners All-purpose Symbolic
BCA	Beckman Glucose Analyzer
BIPM	Bureau International des Ponds et Mesures
CA	Chemical Analysis
CARS	Coherent Anti-Stokes Raman Spectrometry
CCC	Counter Current Chromatography
CCD	Charge-Coupled Device
CE	Capillary Electrophoresis

C–E–R	Calibration-Evaluation-Recovery (Function)
CFA	Continuous Flow Analysis
CGC	Capillary Gas Chromatography
CMP	Chemical Measurement Process
CPAA	Charged Particle Activation Analysis
CRM	Certified Reference Material
CV	Critical value
CV-AAS	Cold Vapour Atomic Absorption Spectrometry
DCM	Dielectric Constant Measurement (Dielcometry)
DENDRAL	Expert system: Dendritic Algorithm
DIN	Deutsches Institut für Normung
DSC	Differential Scanning Calorimetry
DTA	Differential Thermal Analysis
DTG	Differential Thermogravimetry
EBV	Errors in Both Variables (Model, Procedure)
ECA	Electrochemical Analysis
ECD	Electron Capture Detector
ED	Electron Diffraction
EDX	Energy-Dispersive X-Ray (Spectrometry)
ED-XFA	Energy-Dispersive X-Ray Fluorescence Analysis
EDL	Electrodeless Discharge Lamp
EELS	Electron Energy Loss Spectrometry
EM	Electron Microscopy
EMP	Electron Microprobe
EN	European Norm
EPA	Environmental Protection Agency (USA)
EPH	Electrophoresis
EPMA	Electron Probe Microanalysis
ESAC	Expert Systems in Analytical Chemistry
ESCA	Electron Spectroscopy for Chemical Analysis
ESD	Estimated Standard Deviation
ESR	Electron Spin Resonance (Spectroscopy)
ETA	Electrothermal Atomizer
ETA-AAS	Electrothermal Atomizing Atomic Absorption Spectrometry
EXAFS	Extended X-Ray Absorption Fine Structure (Spectrometry)
FAB	Fast Atom Bombardment
FANES	Furnace Atomization Non-Thermal Emission Spectrometry
FD	Field Desorption
FEM	Field Electron Microscopy

FFT	Fast Fourier Transform
FIA	Flow Injection Analysis
FI-AP	Field Ion Atom Probe
FID	Flame-Ionization Detector
FIM	Field Ion Microscopy
fn	false negative (decisions in screening tests)
fp	false positive (decisions in screening tests)
FNAA	Fast Neutron Activation Analysis
FORTRAN	Programming language: Formula Translation
FT	Fourier Transform
FT-ICR-MS	Fourier Transform Ion Cyclotron Resonance Mass Spectrometry
FT-IR, FTIR	Fourier Transform Infrared (Spectrometry)
FT-MS	Fourier Transform Mass Spectrometry
FT-NMR	Fourier Transform Nuclear Magnetic Resonance Spectrometry
FWHM	Full Width at Half Maximum
GA	Genetic Algorithm
GC	Gas Chromatography
GC-IR	Gas Chromatography Infrared Spectrometry Coupling
GC-MS	Gas Chromatography Mass Spectrometry Coupling
GDL	Glow Discharge Lamp
GDMS	Glow Discharge Mass Spectrometry
GDOS	Glow Discharge Optical Spectrometry
GF-AAS	Graphite Furnace Atomic Absorption Spectrometry
GLC	Gas Liquid Chromatography
GLS	Gaussian least squares (regression)
GLP	Good Laboratory Practice
GMP	Good Manufacturing Practice
HCL	Hollow Cathode Lamp
HEED	High Energy Electron Diffraction
HEIS	High Energy Ion Scattering
HG-AAS	Hydride Generation Atomic Absorption Spectrometry
HPLC	High Performance Liquid Chromatography
HPTLC	High Performance Thin-Layer Chromatography
HR	High Resolution
HRMS	High Resolution Mass Spectrometry
IC	Ion Chromatography
ICP	Inductively Coupled Plasma (Spectrometry)

ICP-MS	Inductively Coupled Plasma Mass Spectrometry
ICP-OES	Inductively Coupled Plasma Optical Emission Spectrometry
ICR	Ion Cyclotron Resonance
ID	Ion Diffraction
IDMS	Isotope Dilution Mass Spectrometry
IEC	Ion Exchange Chromatography
IEC	International Electrotechnical Commission
IFCC	International Federation of Clinical Chemistry
IM	Ion Microscopy
IMS	Ion Mobility Spectrometry
INAA	Instrumental Neutron Activation Analysis
INDOR	Internuclear Double Resonance
INS	Inelastic Neutron Scattering
IR, IRS	Infrared Spectroscopy
IRM	Infrared Microscopy
ISE	Ion-Selective Electrodes
IRS	Internal Reflectance Spectroscopy
ISO	International Organisation for Standardization
ISS	Ion Surface Scattering, Ion Scattering Spectrometry
IU	Insulin unit
IUPAC	International Union of Pure and Applied Chemistry
IUPAP	International Union of Pure and Applied Physics
LAMMS	Laser (Ablation) Micro Mass Spectrometry
LASER	Light Amplification by Stimulated Emission of Radiation
LC	Liquid Chromatography
LC-MS	Liquid Chromatography Mass Spectrometry Coupling
LD	Limit of Detection
LDMS	Laser Desorption Mass Spectrometry
LEED	Low Energy Electron Diffraction
LEIS	Low Energy Ion Scattering
LEMS	Laser Excited Mass Spectrometry
LIDAR	Light Detection And Ranging (analogous to RADAR)
LIMS	Laboratory Information Management Systems
LISP	List Processing
LLC	Liquid Liquid Chromatography
LM	Light Microscopy
LMA	Laser Microspectral Analysis
LM-OES	Laser Micro Optical Emission Spectroscopy
LOD	Limit Of Detection

LOS, LoS	Level of significance
LQ	Limit of quantitation
LRMA	Laser Raman Micro Analysis
MALDI	Matrix-Assisted Laser Desorption and Ionization
MALDI-TOF	MALDI Time-Of-Flight (Mass Spectrometry)
MEIS	Medium Energy Ion Scattering
MES	Mössbauer Effect Spectroscopy
MIP	Microwave-Induced Plasma
MLD	Measured value at Limit of Detection
MLQ	Measured value at Limit of Quantitation
MORD	Magneto Optical Rotary Dispersion
MÖS	Mössbauer Spectrometry
MS	Mass Spectrometry
MS-MS	Tandem Mass Spectrometry
MS^n	n-fold Tandem Mass Spectrometry
M-OES	Micro(spark) Optical Emission Spectroscopy
MWS	Microwave Spectroscopy
NAA	Neutron Activation Analysis
NBS	National Bureau of Standards, USA (today: NIST)
ND	Neutron Diffraction
NEXAFS	Near-Edge X-Ray Absorption Fine Structure (Spectroscopy)
NIR, NIRS	Near Infrared (Spectrometry)
NIST	National Institute of Standards and Technology, USA
NMR	Nuclear Magnetic Resonance (Spectroscopy)
NOE	Nuclear Overhauser Effect
NQR	Nuclear Quadrupole Resonance
OAES	Optical Atomic Emission Spectroscopy
OCS	Out-of-control situations (in quality control)
OES	Optical Emission Spectroscopy
OIML	Organisation Internationale de Métrologie Légale
OLS	Ordinary least squares (regression)
ORD	Optical Rotary Dispersion
PARC	Pattern Recognition
PAS	Photo Acoustic Spectroscopy
PASCAL	Programming language called by BLAISE PASCAL (1623–1662)
PC	Paper Chromatography
PCR	Principal component regression
PES	Photoelectron Spectrometry
PIXE	Particle Induced X-Ray Emission

PLS	Partial least squares (regression)
PMT	Photomultiplier Tube
ppm	part per million (10^{-6} corresponding to 10^{-4} per cent)
ppb	part per billion (10^{-9} corresponding to 10^{-7} per cent)
ppt	part per trillion (10^{-12} corresponding to 10^{-10} per cent)
ppq	part per quatrillion (10^{-15} corresponding to 10^{-13} per cent)
PROLOG	Programming language: <u>Pro</u>gramming in <u>Log</u>ic
QCC	Quality control charts
QMS	Quadropol Mass Spectrometer
R&D	Research and Development
RADAR	Radiowave Detection And Ranging
RBS	Rutherford Backscattering Spectrometry
REELS	Reflection Electron Energy Loss Spectrometry
REM	Reflection Electron Microscopy
RHEED	Reflection High Energy Electron Diffraction
RF	Refractometry
RIMS	Resonance Ionization Mass Spectrometry
RM	Reference Material
RMSP	Root mean standard error of prediction
RPC	Reversed Phase Chromatography
RPLC	Reversed Phase Liquid Chromatography
RRS	Resonance Raman Scattering
RS	Raman Spectroscopy
SAM	Scanning Auger Microscopy
SAM	Standard Addition Method
SEC	Size Exclusion Chromatography
SEM	Scanning Electron Microscopy
SERS	Surface Enhanced Raman Scattering
SFC	Supercritical Fluid Chromatography
SI	Système International (d'Unités)
SIMS	Secondary Ion Mass Spectrometry
SNMS	Sputtered Neutrals Mass Spectrometry
SNR	Signal-to-Noise Ratio
SOP	Standard Operating Procedure
SPE	Solid Phase Extraction
SPME	Solid Phase Micro Extraction
SRM	Standard Reference Material
SSMS	Spark Source Mass Spectrometry
STEM	Scanning Transmission Electron Microscopy

STM	Scanning Tunneling Microscopy
SV	Standard value
TA	Thermal Analysis
TCD	Thermal Conductivity Detector
TEELS	Transmission Electron Energy Loss Spectrometry
TEM	Transmission Electron Microscopy
TG	Thermogravimetry
TGA	Thermogravimetric Analysis
THEED	Transmission High Energy Electron Diffraction
TIC	Total Ion Chromatogram
TID	Thermoionic Detector
TLC	Thin-Layer Chromatography
TMA	Thermomechanical Analysis
tn	true negative (decisions in screening tests)
TOF-(MS)	Time-Of-Flight (Mass Spectrometry)
tp	true positive (decisions in screening tests)
TXRF	Total Reflection X-ray Fluorescence (Spectrometry)
UPS	Ultraviolet Photoelectron Spectrometry
UV	Ultraviolet (radiation)
UV-VIS	Ultraviolet-Visible (Spectrometry)
VIS	Visible (radiation)
WD-XFA	Wavelength-Dispersive X-Ray Fluorescence (Spectrometry)
XAS	X-Ray Absorption Spectrometry
XD	X-Ray Diffraction
XFA	X-Ray Fluorescence Analysis
XFS	X-Ray Fluorescence Spectrometry
XPS	X-Ray Photoelectron Spectrometry
XRD	X-Ray Diffraction
XRF	X-Ray Fluorescence (Spectrometry)
ZAF	Z (stands for atomic number) Absorption Fluorescence (Correction) (XFA)
ZAAS	Zeeman Atomic Absorption Spectrometry

1 Object of Analytical Chemistry

Analytical chemistry is one of the oldest scientific disciplines. Its history can be traced back to the ancient Egyptians, about four to five thousand years ago (SZABADVARY [1966], MALISSA [1987], YORDANOV [1987]). But notwithstanding its long history, analytical chemistry is always an essential factor in the development of modern scientific and industrialised society.

The development of chemistry itself has progressed significantly by analytical findings over several centuries. Fundamental knowledge of general chemistry is based on analytical studies, the laws of simple and multiple proportions as well as the law of mass action. Most of the chemical elements have been discovered by the application of analytical chemistry, at first by means of chemical methods, but in the last 150 years mainly by physical methods. Especially spectacular were the spectroscopic discoveries of rubidium and caesium by BUNSEN and KIRCHHOFF, indium by REICH and RICHTER, helium by JANSSEN, LOCKYER, and FRANKLAND, and rhenium by NODDACK and TACKE. Also, nuclear fission became evident as HAHN and STRASSMANN carefully analyzed the products of neutron-bombarded uranium.

In recent times, analytical chemistry has stimulated not only chemistry but many fields of science, technology and society. Conversely, analytical chemistry itself has always been heavily influenced by fields like nuclear engineering, materials science, environmental protection, biology, and medicine. Figure 1.1 shows by which challenges analytical chemistry has been stimulated to improved performances within the last half century.

WILHELM OSTWALD [1894], who published the first comprehensive textbook on analytical chemistry, emphasized in it the service function of analytical chemistry. This fact has not changed until now. Interactions with all the fields of application have always had a promoting influence on analytical chemistry.

1.1
Definition of Analytical Chemistry

In the second half of the twentieth century, analytical chemistry was defined as the *chemical discipline that gains information on the chemical composition and structure of substances, particularly on the type of species, their amount,*

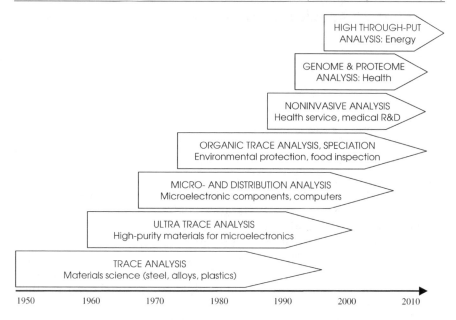

Fig. 1.1. Economic and social challenges and stimuli for the development of some important branches of analytical chemistry

possible temporal and spatial changes, and structural relationships between the constituents (see, e.g. ECKSCHLAGER and DANZER [1994]; DANZER et al. [1976]). Sometimes the development of methods and instruments is included or is central to such definitions, e.g. that of WPAC [1993]: *"Analytical chemistry is found to be a scientific discipline that develops and applies methods, instruments, and strategies to obtain information on the composition and nature of matter in space and time"* (KELLNER et al. [1998]).

However, now and then analytical chemists feel uneasy with such kinds of definitions which do not reflect completely the identity and independence of analytical chemistry. Chemists of other branches (inorganic, organic, and physical chemists) as well as physicists and bioscientists also obtain information on inanimate or living matter using and developing high-performance analytical instruments just as analytical chemists do.

Consequently, there exists a wide range of diverse definitions. One of the most appropriate is that of REILLEY [1965]: *"Analytical chemistry is what analytical chemists do"* which is, however, not really helpful.

The first definition that is focused directly on the role of analytical signals was given by PUNGOR who characterizes analytical chemistry as *"a science of signal production and interpretation"* (VERESS et al. [1987], LEWENSTAM and ZYTKOW [1987]). ZOLOTOV [1984] characterized chemical, physicochemical and physical methods of analytical chemistry as follows: *"All of them, however, have the same feature: it is the dependence of signal on analyte concentration. The important task of analytical chemistry is therefore the discovery and implantation of these dependencies into analytical procedures"* (LEWENSTAM and ZYTKOW [1987]).

To overcome the unsatisfactory situation in the understanding the meaning of analytical chemistry at the end of the last century, an international competition was organized in 1992 by noted European analytical chemists and the Fresenius Journal of Analytical Chemistry to characterize analytical chemistry as an autonomous field of science by a topical and proper definition. The title of this competition was *"Analytical Chemistry – today's definition and interpretation"* and 11 out of 21 contributions were published in Fresenius J Anal Chem (FRESENIUS and MALISSA [1992]; CAMMANN [1992]; VALCARCEL [1992]; ZUCKERMAN [1992]; ZHOU NAN [1992]; KOCH [1992]; PEREZ-BUSTAMANTE [1992]; ORTNER [1992]; DANZER [1992]; GREEN [1992]; STULIK and ZYKA [1992]; KUZNETSOV [1992]).

The first prize winner – CAMMANN [1992] – defined analytical chemistry *"as the self-reliant chemical sub-discipline which develops and delivers appropriate methods and tools to gain information on the composition and structure of matter, especially concerning type, number, energetic state and geometrical arrangement of atoms and molecules in general or within any given sample volume. ... In analytical chemistry, special techniques are used to transform measured chemical signals, derived mostly from specific interaction between matter and energy, into information and ordered knowledge"*.

Other remarkable aspects expressed in the published contributions of the competition are the following: *Analytical chemistry ...*

- *"... is the chemical metrological science"* (VALCARCEL [1992])
- *"... can also be considered applied physical chemistry"*[1] (CAMMANN [1992])
- *"... is a science devoted to the analytical cognition of substances: their properties, composition, structure and state, steric and inner relations, behaviour in chemical reaction systems at all levels of the chemical organisation of matter"* (ZUCKERMAN [1992])
- *"... is a branch of science which comprises the theory and practice of acquiring information about chemical characteristics of any matter of system, present in particular state, from its bulk or from a specified region ... "* (ZHOU NAN [1992])
- *"... is a multidiscipline, comprising various fields of chemistry with special understanding of physics, mathematics, computer science, and engineering"* (KOCH [1992])
- *"... uses chemical, physicochemical, and physical or even biological methods for analytical signal production, followed by problem- and matter-related signal processing and signal interpretation in order to provide*

---

[1] Note of the author: this interpretation is not reversible: physical chemistry is not theoretical analytical chemistry

reliable (quality assured) qualitative, quantitative and/or structural information about a sample" (KOCH [1992])

- *"...is the branch of chemistry, the aim of which is the handling of any type of matter in order to separate, identify, quantify and speciate its components by extracting the pertinent information of analytical interest contained in a representative sample"* (PEREZ-BUSTAMANTE [1992])

- *"...is based upon the symbiotic knowledge and application of chemistry, physics and applied mathematics, to establish a congruent trinomium involving: chemical species – instrumentation and metrology – data handling and processing (chemometrics)"* (PEREZ-BUSTAMANTE [1992])

- *"...is the science of the creative derivation of information using proper methods to answer the following four basic questions which also resemble the relevant fundamental fields of analytical chemistry: What and how much? (Bulk analysis), How structured? (Structure analysis), How bound? (Speciation), How distributed? (Topochemical analysis)"* (ORTNER [1992])

- *"...embraces that domain of the physical sciences that allows the interaction of molecules and matter to be understood and their composition to be determined."* (GREEN [1992])

All these definitions express essential aspects of analytical chemistry and the analytical work. Some others – with originality – could be added, such as that from MURRAY [1991] who characterized analytical chemistry briefly and aptly as *"the science of chemical measurements"*.

Different opinions can be found about the status of analytical chemistry as being a branch of chemistry independent from other chemical disciplines or being a physical discipline (GREEN [1992]), or even being an autonomous science, occasionally called Analytics or Analytical Sciences. On the other hand, wide agreement can be stated about the aim to obtain information on matter via representative samples and the inclusion of structural information. Remarks on the general importance of analytical signals can be repeatedly found.

Considering the recent development of analytical chemistry and the significance of analytical signals for which reasons will be given in Chap. 3, the following object characterization is proposed:

> *Analytical Chemistry is the science of chemical measurement. Its object is the generation, treatment and evaluation of signals from which information is obtained on the composition and structure of matter.*

Principle goals are the identification and recognition of sample constituents with regard to their type (elements, isotopes, ions, modifications, complexes, functional groups, compounds, viz. simple, macromolecular or biomolecular, and occasionally sum parameters, too), their amount (absolute or relative: content, concentration), and binding state (oxidation state, binding form and type: inorganic, organic, or complex bound).

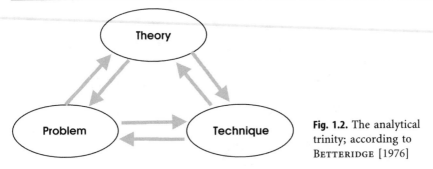

Fig. 1.2. The analytical trinity; according to BETTERIDGE [1976]

The opinions about the status of analytical chemistry – being an independent or an auxiliary science – have varied over many decades. However, it is not really a desirable to be independent among the scientific or chemical disciplines and is not a disadvantage to be thought of as an auxiliary science. In some respects, various respectable sciences like mathematics, physics, and biology have a service function, too. Therefore, the question about independence is relative and of secondary importance.

In contrast to classical analysis, the concept of modern analytical chemistry has changed in so far as the problem that has to be solved is included in the analytical process. The analytical chemist is considered as a *"problem solver"* (LUCCHESI [1980]) and the concept is represented in the form of the *"analytical trinity"* (BETTERIDGE [1976]) as shown in Fig. 1.2.

Nowadays, analytical chemistry has a large variety of methods, techniques and apparatus at its disposal and is able to play its "instruments" with high virtuosity. Therefore, the wide range of performance which analytical chemistry can achieve is extremely varied and extends from simple binary decisions (qualitative analysis) to quantitative analysis at the ultratrace level, from structure elucidation and species identification to studies of the dynamics and the topology of multispecies systems by means of temporally and spatially high-resolving techniques.

1.2
Repertoire of Analytical Chemistry

The continual progress of analytical chemistry is attributed to the increasing demands from science and technology as well as from society.

Performance parameters have rapidly and drastically improved by the high demands of the main focuses of development. In Fig. 1.3 it is shown how the efficiency of analytical methods has successively improved in the last 50 years.

The progress was caused by the development of completely new methods, techniques, and principles (e.g. microprobe techniques) as well as the introduction of new system components into common instruments (e.g. flash

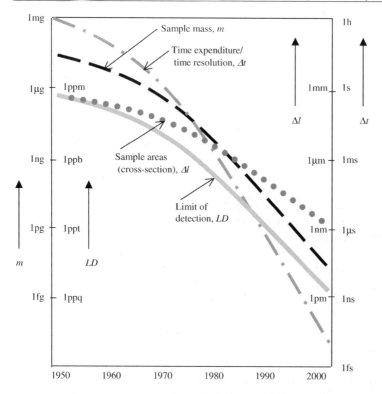

Fig. 1.3. Technical progress in analytical chemistry with regard to sample mass m, limit of detection LD, sample areas (cross-section) Δl, and time expenditure/time resolution Δt

excitation, capillary columns) and methods (e.g. enrichment techniques, hyphenated methods). Additionally, the use of chemometric procedures has considerably improved the power and range of analytical methods.

Analytical investigations usually concern samples which are temporally and locally invariant. This kind of analysis is denoted as *bulk analysis (average analysis)*. On the other hand, analytical investigations can particularly be directed to characterize temporal or local dependences of the composition or structure of samples. One has to perform *dynamic analysis* or *process analysis* on the one hand and *distribution analysis, local analysis, micro analysis,* and *nano analysis* on the other.

According to the demands of the analysis, analytical chemistry can be classified into analysis of major components (*major component analysis, precision analysis, investigation of stoichiometry*), minor components, and trace components (*trace analysis, ultra trace analysis*). On the other hand, analytical problems are differentiated according to the number of analytes involved. Accordingly, single component and *multicomponent analysis* are distinguished.

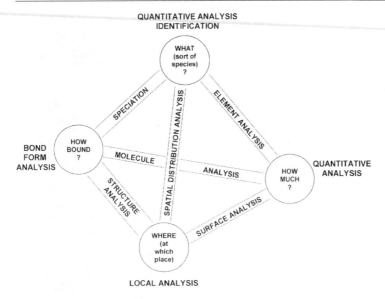

Fig. 1.4. The tetrahedron of the analytical repertoire

In principle, structure analysis can be considered as distribution analysis in atomic dimensions. However, from the practical point of view it makes sense to deal separately with structure analysis and to differentiate between *molecular structure analysis* and *crystal structure analysis*. Further structure investigations concern near-orders in solids and liquids (e.g. glass).

The relations between the questions that are answered by analytical chemistry are shown in Fig. 1.4. The tetrahedron represents the basic analytical repertoire in a simplified way. It can be seen that all the analytical treatments are connected with each other.

In Fig. 1.5 the repertoire of analytical chemistry is classified more in detail according to element and structure analysis and according to the extent of quantification.

The different ways of species analysis – qualitative and quantitative – are well known. However, in structure analysis, they can also be differentiated between qualitative and quantitative ways according to the type and amount of information obtained (ECKSCHLAGER and DANZER [1994]). Identification of a sample or a given constituent may have an intermediate position between species and structure analysis. In any case, identification is not the same as qualitative analysis. The latter *is the process of determining if a particular analyte is present in a sample* (PRICHARD et al. [2001]). Qualitative analysis seeks to answer the question of whether certain components are present in a sample or not. On the other hand, identification is the process of finding out what unknown substance(s) is or are present (ECKSCHLAGER and DANZER [1994]). In Sects 9.1 and 9.3 it will be shown that there is a

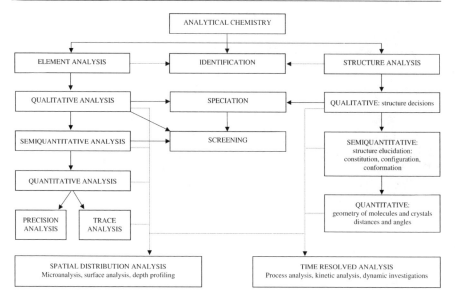

Fig. 1.5. Survey of element and structure analytical standard procedures

clear difference between the information contents of qualitative analysis and identification.

Another aspect of modern analytical chemistry is the possibility of *multi-component analysis*. Especially spectroscopic and chromatographic methods are able to detect and determine a large number of species simultaneously. Therefore, such methods like ICP-OES, ICP-MS, TXRF, and chromatography are the work-horses in today's analytical chemistry.

Notwithstanding the formal classification given in Fig. 1.5 there is no fundamental difference between qualitative and quantitative analysis. In each case a specific signal is generated which may be evaluated to meet any component of the following logical sequence:

Detection → Assurance → Semiquantifying → Quantifying

This chain of information can be broken and applied at each point according to the special demands. By the way, even the detection limit involves a numerical estimation.

A rough gradation of analyte amounts has been done by DE GRAMONT [1922] who investigated 82 elements in minerals, ores and alloys by means of atomic spectroscopy using so-called "raies ultimes" (last lines, ultimate lines, i.e. such lines which disappear at definite concentrations).

The application of combinatorial principles in chemical synthesis, particularly in the search for active substances, requires analytical methods with high throughput (VON DEM BUSSCHE-HÜNNEFELD et al. [1997]). *Screening techniques* can be used to analyse a large number of test samples in a short

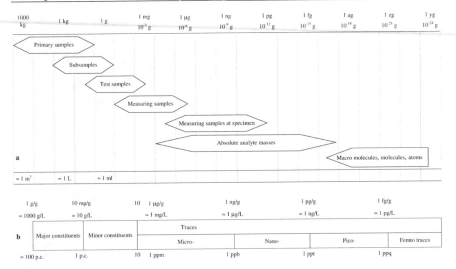

Fig. 1.6. Absolute masses of samples and analytes (**a**) and relative analyte amounts (**b**) relevant in analytical investigations

time. Methods of screening must be inexpensive with regard to both time and cost. Naturally they must be reliable enough that the risk of erroneous decisions (false positive and false decisions, respectively) is low.

Structure elucidation of a compound follows another logical sequence:

Constituents → Gross composition → Sum formula → Constitution → Configuration → Conformation → Quantitation: Intramolecular Dimensions

This strengthens the case for treating structure analysis as a particular field of analytical chemistry despite the fact that, from the philosophical point of view, structure analysis can be considered as distribution analysis (topochemical analysis) of species in atomic dimensions. Structure analysis of solids follows a similar scheme like that given above. The characteristics of molecules are then linked with those of crystals and elementary cells.

A quantification of the repertoire of analytical chemistry is shown in Fig. 1.6. The field of operation covers over 30 orders of magnitude and more when the amount of lots is included. On the other hand, the relative amounts (contents, concentrations) with which the analyst has to do covers 20 orders and more because single atom detection has become reality now.

The fields of application of analytical chemistry extend from research to service, diagnosis, and process control, from science to technology and society, from chemistry to biology, health services, production, environmental protection, criminalogy and law as well as from chemical synthesis to materials sciences and engineering, microelectronics, and space flight. In brief, analytical chemistry plays an important role in every field of our life.

Today, analytical chemistry has such a wide variety of methods and techniques at its disposal that the search for general fundamentals seems to be very difficult. But independent from the concrete chemical, physical and technical basis on which analytical methods work, all the methods do have one principle in common, namely the extraction of information from samples by the *generation, processing, calibration,* and *evaluation of signals* according to the logical steps of the *analytical process.*

References

Betteridge D (1976) *Tell me the old old story or the analytical trinity.* Anal Chem 48:1034A

Cammann, K (1992) *Analytical chemistry – today's definition and interpretation.* Fresenius J Anal Chem 343:812

Danzer, K (1992) *Analytical chemistry – today's definition and interpretation.* Fresenius J Anal Chem 343:827

Danzer, K, Than, E, Molch, D (1976) *Analytik – Systematischer Überblick.* Akademische Verlagsgesellschaft Geest & Portig, Leipzig

de Gramont, MA (1922) *Tableau des raies de grande sensibilité des éléments destine aux recherches analytiques.* C R Hebd Séances Acad Sci 171:1106

Eckschlager, K, Danzer, K (1994) *Information theory in analytical chemistry.* Wiley, New York, Chichester, Brisbane

Fresenius, W, Malissa, H (1992) *Competition "analytical chemistry – today's definition and interpretation".* Fresenius J Anal Chem 343:809

Green, JD (1992) *Analytical chemistry – today's definition and interpretation.* Fresenius Z Anal Chem 343:829

Kellner, R (1994) *Education of analytical chemists in Europe.* Anal Chem 66:98A

Kellner, R, Mermet, J-M, Otto, M, Widmer, HM [eds] (1998) *Analytical chemistry. The approved text to the FECS curriculum analytical chemistry.* Wiley-VCH, Weinheim, New York, Chichester

Koch, KH (1992) *Analytical chemistry – today's definition and interpretation.* Fresenius J Anal Chem 343:821

Kusnetsov, VI (1992) *Analytical chemistry – today's definition and interpretation.* Fresenius J Anal Chem 343:834

Lewenstam, A, Zytkow, JM (1987) *Is analytical chemistry an autonomous field of science?* Fresenius Z Anal Chem 326:308

Lucchesi, CA (1980) *The analytical chemist as problem solver.* Internat Lab 11/12:67

Malissa, H (1987) *Analytical chemistry: kitchenmaid or lady of science – deduction versus induction.* Fresenius Z Anal Chem 326:324

Murray, RW (1991) *Analytical chemistry: the science of chemical measurement.* Anal Chem 63: 271A

Ortner, HM (1992) *Analytical chemistry – today's definition and interpretation.* Fresenius J Anal Chem 343:825

Oswald, W (1894) *Die wissenschaftlichen Grundlagen der Analytischen Chemie.* Verlag Wilhelm Engelmann, Leipzig

Perez-Bustamante, JA (1992) *Analytical chemistry – today's definition and interpretation.* Fresenius J Anal Chem 343:823

Prichard, E, Green, J, Houlgate, P, Miller, J, Newman, E, Phillips, G, Rowley, A (2001) *Analytical measurement terminology – handbook of terms used in quality assurance of analytical measurement.* LGC, Teddington, Royal Society of Chemistry, Cambridge

Reilley, CN (1965) AS National Meeting, Fisher Award Address

Stulik, K, Zyka, J (1992) *Analytical chemistry – today's definition and interpretation.* Fresenius J Anal Chem 343:832

Szabadvary, F (1966) *History of analytical chemistry.* Pergamon Press, Oxford

Valcarcel, M (1992) *Analytical chemistry – today's definition and interpretation.* Fresenius J Anal Chem 343:814

Veress, GE, Vass, I, Pungor, E (1987) *Analytical chemical methods as systems producing chemical information by inference.* Fresenius Z Anal Chem 326:317

von dem Bussche-Hünnefeld C, Balkenhohl F, Lansky A, Zechel C (1997) *Combinatorial chemistry.* Fresenius J Anal Chem 359:3

WPAC (1993) *Edinburgh definition of analytical chemistry;* see Kellner, R (1994)

Yordanov, N (1987) *The position of analytical chemistry from the viewpoint of moderate scientism – a critical outlook.* Fresenius Z Anal Chem 326:303

Zhou Nan (1992) *Analytical chemistry – today's definition and interpretation.* Fresenius J Anal Chem 343:812

Zolotov YA (1984) *Khimiczeskaja enciklopedia.* Nauka, Moscow

Zuckerman, AM (1992) *Analytical chemistry – today's definition and interpretation.* Fresenius J Anal Chem 343:817

2 The Analytical Process

Analytical chemistry is a problem-solving science. Independent from the concrete analytical method, the course of action, called *analytical process*, is always very similar. The analytical process starts with the analytical question on the subject of investigation and forms a closed chain to the answer to the problem. Using a proper sampling technique a test sample is taken that is adequately prepared and then measured. The measured data are evaluated on the basis of a correct calibration and then interpreted with regard to the object under study.

The analytical process in the broader sense is represented in Fig. 2.1. Frequently – in a stricter sense – only the lower part (grey background) from sample through to information (or essential parts from it) are regarded as representing the analytical process.

The lower part of the analytical process (grey) is – mostly merged as a black box – sometimes known as the *chemical measurement process* (CMP); see CURRIE [1985, 1995, 1999].

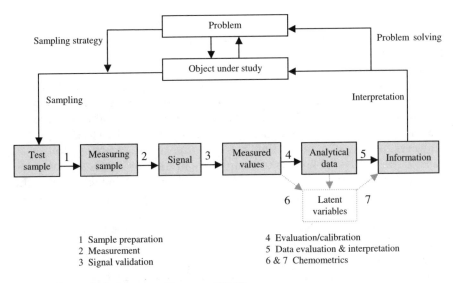

1 Sample preparation
2 Measurement
3 Signal validation

4 Evaluation/calibration
5 Data evaluation & interpretation
6 & 7 Chemometrics

Fig. 2.1. The analytical process (DANZER [2004])

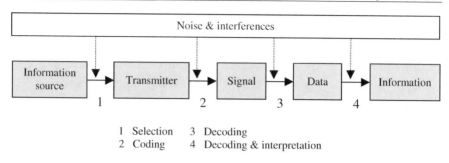

1 Selection   3 Decoding
2 Coding   4 Decoding & interpretation

Fig. 2.2. General principle of processing information

Different from former representations of the analytical process (DANZER et al. [1976], LUCCHESI [1980], TYSON [1989], KELLNER et al. [1998]), signal generation is inserted as a separate step. This is because at first signals are produced from measuring samples and then measured values are taken from the signals. For this, signals must be selected from a signal function. After validation, parameters are chosen, the values of which are subject to evaluation (e.g., signal position, differences of positions, signal intensity measured in form of peak height or peak area).

In addition to the measured values and the analytical values (e.g. content, concentration), latent variables are included in the scheme. Latent variables can be obtained from measured values or from analytical values by means of mathematical operations (e.g. addition, subtraction, eigenanalysis). By means of latent variables and their typical pattern (represented in chemometric displays) special information can be obtained, e.g. on quality, genuineness, authenticity, homogeneity, origin of products, and health of patients.

The analytical process in the stricter sense or chemical measurement process, respectively, has a conspicuous similarity with the general information process which is shown in Fig. 2.2.

Interference plays an analogous role in the course of the analytical process as well as noise, which is mainly manifested by random deviations.

The scheme given in Fig. 2.1 represents normal analytical procedures in *off-line analysis*. It contains all the steps that must be considered in principle. However, there could be reasons to reduce the course of action. Not in all cases is the analyst able to take samples by himself or check the sampling procedure. Sometimes he or she must accept a situation in which he or she has to receive a given sample (e.g., in extreme cases, extraterrestrial samples, autopsy matter).

In some cases the object under study has to be continuously monitored. Then *on-line analytical methods* are applied by which the system can be directly measured. The analytical process then runs without sampling and sample preparation, as can be seen in Fig. 2.3a. The analytical process is shortened even more in the case of *in-line analysis* where measurement and

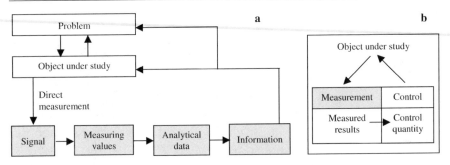

Fig. 2.3. The analytical process in case of on-line analysis (**a**) and in-line analysis (**b**)

control equipment are coupled with and installed in the object under study (Fig. 2.3b).

Other forms of analytical investigations may also be applied, such as at-line analysis (discontinuous direct analysis of objects, e.g., noninvasive blood glucose monitoring) and trans-line analysis (remote sensing, teleanalysis).

In the following, the stages of the analytical process will be dealt with in some detail, viz. sampling principles, sample preparation, principles of analytical measurement, and analytical evaluation. Because of their significance, the stages signal generation, calibration, statistical evaluation, and data interpretation will be treated in separate chapters.

2.1
Principles of Sampling

Sampling is the most important step in the course of the analytical process. It is essential for obtaining reliable analytical results that the samples are taken from the lot in a representative way. Therefore, sampling strategy and design, sampling technique and sample size must be harmonized in an optimum way. There exist general guidelines for sampling (HORWITZ [1990], GY [1982, 1992], KRAFT [1980], WEGSCHEIDER [1996]) as well as special instructions for various fields of application (e.g. TAGGART [1945], KEITH [1988], MARKERT [1994] BARNARD [1995]).

The general demand on sampling is that it has to be carried out in a representative way, to be precise, representative with regard to both the properties of the material and the analytical problem. This twofold representativeness of the sample means that sampling strategies are different in case of average analysis (bulk analysis) and testing homogeneity within a lot of *n* batches. Figure 2.4 shows two possible sampling schemes for answering both questions.

Sampling is a crucial step of the analytical process, particularly in cases where there are large differences between the material under investigation and the test sample (laboratory sample) with regard to both amounts and properties, especially grain size, fluctuations of quality and inhomogeneities.

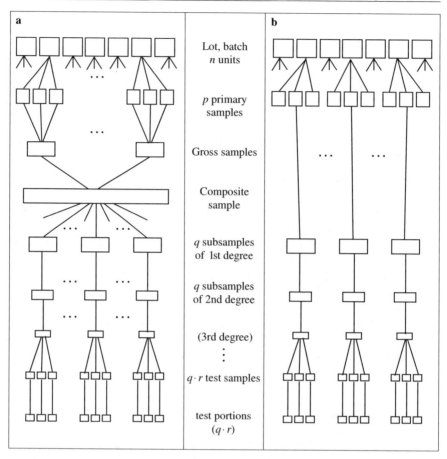

Fig. 2.4. Sampling schemes for bulk analysis (**a**) and for testing homogeneity of several batches (**b**), according to DANZER [1995a] and following HORWITZ [1990]

A material is regarded as being inhomogeneous if the chemical composition is not the same in all of its parts, i.e.

– In spatial dimensions or in partial volumes of compact solids, or
– In portions of bulk materials (grained and powdered substances) and
– In time sequences of natural or technical processes.

The technical terms homogeneity and inhomogeneity defined in analytical chemistry must be distinguished from the physicochemical concept of homogeneity and heterogeneity (DANZER and EHRLICH [1984]). Whereas the thermodynamical definition refers to morphology and takes one-phase- or multi-phase states of matter as the criterion, the analytical-chemical definition is based on the concentration function

$$x = f(l) \tag{2.1}$$

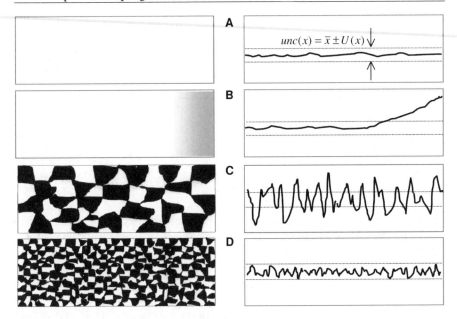

Fig. 2.5. Illustration of the terms homogeneity, heterogeneity and inhomogeneity from the physicochemical and analytical viewpoint

and is statistically determined. Instead of the distance l applied in Eq. (2.1), other spatial dimensions of the sample (area, surface or volume, respectively) or even the time t may be of relevance.

In contrast to the physicochemical categorical properties of homogeneity and heterogeneity, the transition from homogeneity to inhomogeneity in analytical chemistry (the *degree* of homogeneity) may be characterized by a variable, see, e.g., Eq. (2.9).

Figure 2.5 illustrates the state of affairs, and shows that heterogeneous material may be characterized by an inhomogeneous (C) or homogeneous (D) concentration function dependent on the relation between the total variation of concentration and the uncertainty of measurement on the one hand and the sample amount (or microprobe diameter in case of distribution-analytical investigations) on the other.

The sample A is homogeneous both from the physicochemical and the analytical-chemical point of view because the variation of the concentration is within the uncertainty of the analytical measurement $unc(x)$. In contrast, sample B is homogeneous from the physico-chemical but not from the analytical-chemical viewpoint because the systematical change of the concentration exceeds $unc(x)$. Whereas sample C is heterogeneous from the physico-chemical point of view and inhomogenous from the analytical-chemical viewpoint (the concentration deviations plainly exceed $unc(x)$), sample D is heterogeneous from the physico-chemical standpoint. The an-

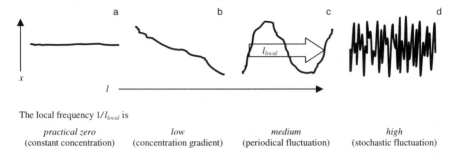

The local frequency $1/l_{local}$ is

practical zero   low   medium   high
(constant concentration)   (concentration gradient)   (periodical fluctuation)   (stochastic fluctuation)

Fig. 2.6. Characteristic concentration functions and local frequencies

alytical decision about homogeneity depends on the diameter of the micro-probe or the size of the portion of the test sample. If the sample size is large enough the material may be characterized to be homogeneous as illustrated in Fig. 2.5D.

In analogy to the time frequency, f, the spatial concentration behaviour may be characterized by the *local frequency*, $f_{local} = 1/l_{local}$. In Fig. 2.6, four types of spatial concentration functions are shown. These types and combinations of them can characterize all the variations of concentration in analytical practice both in one- and more-dimensional cases.

In addition to the local frequency, the amplitude of the concentration fluctuation is a crucial parameter for the characterization of inhomogeneity (see Fig. 2.5, right hand side). It is essential for sampling problems that inhomogeneities, as real concentration differences Δx between parts of sample portions or areas, respectively, can only be ensured if Δx is significantly greater than the uncertainty $U(x)$ of the analytical method. The criterion for testing homogeneity is the F-test:

$$F = \frac{\sigma_{total}^2}{\sigma_{anal}^2} \tag{2.2}$$

which takes as basis the addition of variances according to the model

$$\sigma_{total}^2 = \sigma_{anal}^2 + \sigma_{inhom}^2 \quad . \tag{2.3a}$$

The inhomogeneity variance σ_{inhom}^2 determines the sampling error σ_{samp}^2 which can be estimated experimentally and which is crucial for representative sampling[1]

$$\sigma_{total}^2 = \sigma_{anal}^2 + \sigma_{samp}^2 \tag{2.3b}$$

The relationship between the estimated variances s^2 has to consider the number of primary samples, p, subsamples, q, and test samples, $q \cdot r$

---

[1]   The inhomogeneity- and the sampling variance are adequate only with a given risk of error α: $\sigma_{inhom}^2 \overset{\alpha}{=} \sigma_{samp}^2$

$$s_{\bar{x}}^2 = \frac{s_{anal}^2}{q \cdot r} + \frac{s_{samp}^2}{p \cdot q} \quad . \tag{2.4}$$

In the case of sampling according to the scheme given in Fig. 2.4a, the critical number of samples required at least for reliable analytical results can be derived from Eq. (2.4) (DANZER [1995b]):

$$q = \frac{p \, s_{anal}^2 + r \, s_{samp}^2}{r \, p \, s_{\bar{x}}^2} \tag{2.5}$$

General relationships such as Eq. (2.6) which has been created by KRAFT [1980]:

$$q = \left(\frac{t \cdot s}{U} \right)^2 \tag{2.6}$$

lack information what s and U precisely mean (U is described as the precision of sampling plus analysis and relates to the combined uncertainty as defined in Sect. 4.2 according to ISO [1993] in certain respect.

The critical weight of samples can be derived from the general condition of representativeness of sampling expressed by the null hypothesis H_0: $\sigma_{total}^2 = \sigma_{anal}^2$ which is tested by means of FISHER's F-test

$$\hat{F} = \frac{s_{total}^2}{s_{anal}^2} \quad . \tag{2.7}$$

The general principles of testing chemical homogeneity of solids are given e.g. by MALISSA [1973], COCHRAN [1977], and DANZER et al. [1979]. The terms of variation σ_{total}^2 and σ_{anal}^2 can be separated by analysis of variance (Sect. 5.1.1). According to DANZER and KÜCHLER [1977] there exists an exponential dependence between the total variance and the reciprocal sample mass

$$s_{total}^2 = s_{anal}^2 \cdot \exp \left(\frac{a}{m} \right) \tag{2.8a}$$

where a is material-specific parameter. Developing Eq. (2.8a) into an arithmetical progression, the following relationship results:

$$s_{total}^2 = s_{anal}^2 \left(1 + \frac{a}{m} \right) \tag{2.8b}$$

when the higher powers of m are neglected.

The exponentially decrease of the total variance with increasing sample mass is shown in Fig. 2.7. It can be seen that the uncertainty of sampling, s_{samp}^2, decreases and becomes statistically insignificant when the sample amount m exceeds the critical sample mass. Instead of m_{crit} the proportional critical sample volume v_{crit} may also be considered, represented, e.g. by a critical microprobe diameter d_{crit}. Results of homogeneity investigations of alloys, ores, and lamellar eutectics by EPMA (Electron Microprobe Analysis), which correspond to the curve of Fig. 2.7, have been presented by DANZER and KÜCHLER [1977].

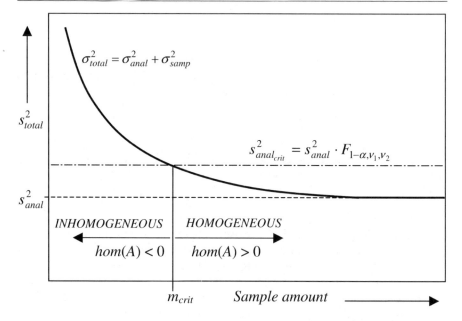

Fig. 2.7. Dependence of the total variance on the sample amount characterized by mass; m_{crit} is the *critical sample mass*. The statement of homogeneity or inhomogeneity is always related to a given analyte A

In the course of time, several measures have been proposed to characterize homogeneity of solids, such as homogeneity constants, factors etc. (GY [1982], SINGER and DANZER [1984], KURFÜRST [1991], INGAMELLS and SWITZER [1973], INGAMELLS [1974, 1976]). However, unfortunately, quantities like this have a special background with regard to the material and, therefore, do not have any general decision power. Taking the homogeneity number of SINGER and DANZER [1984] as starting point, a useful measure of homogeneity can be developed.

The state of affairs with regard to the distribution of a particular species A in bulk material can be characterized by the quantity *homogeneity* defined by

$$hom(A) = 1 - \frac{\hat{F}}{F_{1-\alpha,\nu_1,\nu_2}} \quad . \tag{2.9}$$

This homogeneity value will become positive when the null hypothesis is not rejected by FISHER's F-test ($\hat{F} \leq F_{1-\alpha,\nu_1,\nu_2}$) and will be closer to 1 the more homogeneous the material is. If inhomogeneity is statistically proved by the test statistic $\hat{F} > F_{1-\alpha,\nu_1,\nu_2}$, the homogeneity value becomes negative. In the limiting case $\hat{F} = F_{1-\alpha,\nu_1,\nu_2}$, $hom(A)$ becomes zero.

In the case that inhomogeneities are not attributed to random variations but to systematic changes as, e.g. shown in Fig. 2.5B, then other strategies of sampling and evaluation must be used. Numerous different sampling designs have been proposed. Surveys can be found, e.g. in COCHRAN [1977], GILBERT [1987], KEITH [1988] NOTHBAUM et al. [1994], and EINAX et al.

[1997]. Diverse systematical strategies of sampling require other statistical evaluation procedures than simple analysis of variance (ANOVA) as applied in case of stochastic sampling. In case of regular arrangements of sampling points, the statistical evaluation can be carried out in several ways (DANZER [1995b], EINAX et al. [1997]):

(1) *Two-way layout ANOVA* according to DANZER and MARX [1979]. In this way it is possible to obtain not only information on homogeneity/inhomogeneity, but also on certain preferred directions of inhomogeneities.

(2) *Regression analysis* by means of linear or quasi-linear models (PARCZEWSKI [1981], SINGER and DANZER [1984], PARCZEWSKI et al. [1986]).

(3) *Gradient analysis* (PARCZEWSKI [1981], SINGER and DANZER [1984], PARCZEWSKI et al [1986], PARCZEWSKI and DANZER [1993]) which is based on two-dimensional regression models and adds pictorial information to statistical decisions. Figure 3.12 shows such a graphical representation of an element distribution on a surface.

(4) *Pattern recognition methods* (PARC) can be used if no a priori information on the type of inhomogeneities (stochastic or systematic) is available. By reversing the aim of PARC (classifying objects into classes), homogeneity is proved by the impossibility to form point classes having significantly different concentrations (DANZER and SINGER [1985]).

(5) *Trigonometric functions* have been used by INCZEDY [1982] to describe inhomogeneities quantitatively.

(6) *Autocorrelation and time series analysis* have been successfully applied in testing spatial inhomogeneities (EHRLICH and KLUGE [1989], DOERFFEL et al. [1990]). This techniques are generalized in the *theory of stochastic processes* (BOHACEK [1977a, b]) which is widely used in chemical process analysis and about them.

The main goal of time-series analysis (BOX and JENKINS [1976], CHATFIELD [1989], METZLER and NICKEL [1986]) apart from process analysis is time-dependent sampling. In both cases fluctuations in time $x(t)$ matter and can be considered as a simple stochastic process or as time series.

Process sampling and analysis will be carried out with several aims (KATEMAN [1990]):

- *Control*, i.e. sampling, analysis and reconstruction of the process
- *Monitoring* for process safety and warnings
- *Description*, i.e. mathematical decomposition of the variability by time series analysis

A time series $x(t)$ can be decomposed according to HARTUNG et al. [1991] into *even (smooth)* components, $x_{even}(t)$, *seasonal* components, $x_{seas}(t)$, and *irregular* components, $x_{irreg}(t)$:

$$x(t) = x_{even}(t) + x_{seas}(t) + x_{irreg}(t) \quad . \tag{2.10}$$

Seasonal process components are of particular interest for environmental studies and production processes. The different components of $x(t)$ can be estimated by smoothing and filtering methods.

The correlation of values within a time series plays an important role in sampling. It can be characterized by means of the autocorrelation function (DOERFFEL and WUNDRACK [1986], CHATFIELD [1989], HARTUNG et al. [1991])

$$\Psi_{xx}(\tau) = \lim_{T \to \infty} \frac{1}{2T} \int_{-T}^{+T} x(t)\, x(t + \tau)\, dt \tag{2.11}$$

where $x(t)$ is the value of the time series at the time t, τ the time lag (time difference between two values), $x(t + \tau)$ the value for lag τ in relation to t, and T the total time. The autocorrelation function corresponds to the time average $\overline{x(t)\, x(t + \tau)}$ of the function $x(t)$. For stochastical time series the autocorrelation function (ACF) is represented by a typical curve as shown in Fig. 2.8.

The correlation time (time constant) corresponds to the lag at which the maximum value $\Psi_{xx}(0)$ is decreased by the factor $1/e$ and the correlation becomes practically zero. Correlation time is an important quantity in sam-

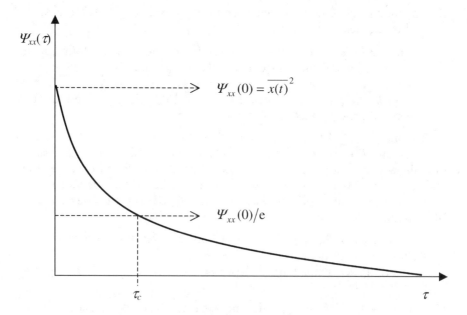

Fig. 2.8. Autocorrelation function (ACF) of a stochastic process. τ_c is the correlation time (time constant)

pling because measured values $x(t)$ and $x(t + \tau)$ at intervals $\tau > \tau_c$ can be regarded as independent from each other.

Similar models have been applied in geological exploration and environmental studies. There *semivariogram analysis* (AKIN and SIEMENS [1988], EINAX et al. [1997]) plays a comparable role than autocorrelation analysis for the characterization of stochastic processes.

Representative sampling is the first and most important prerequisite to produce accurate and precise analytical results. The fact that sampling may be a crucial step with regard to reliable chemical measurements has been well-known for years and has promoted the development of advanced practical methods as well as the forming of theoretical fundamentals of sampling. Excellent reviews are given by KRAFT [1980], KRATOCHVIL et al. [1984], GY [1992], KATEMAN [1990], and KATEMAN and BUYDENS [1993]. Nomenclature for sampling in analytical chemistry is given in KRATOCHVIL et al. [1984] and HORWITZ [1990].

2.2
Sample Preparation

Only in a few cases are test samples measurable without any treatment. As a rule, test samples have to be transformed into a measurable form that optimally corresponds to the demands of the measuring technique. Therefore, sample preparation is a procedure that converts a test sample into a *measuring sample*. Whereas test samples represent the material in its original form, measuring samples embodies a form that is able to interact with the measuring system in an optimum way. In this sense, measuring samples can be solutions, extracts, pellets, and melt-down samples, but also definite surface layers and volumes in case of micro- and nanoprobe techniques.

As an adaptation procedure, sample preparation has to fulfill at least one of the following aims:

(1) Making the test sample physically or chemically measurable by the analytical technique
(2) Elimination of interferences
(3) Improving the relation between the amounts of analyte and matrix

Measuring samples are considered as being composed of the *analyte* (PRICHARD et al. [2001]) and the *matrix*. The term "matrix" summarizes all the sample constituents apart from the analyte. The relationships between the amounts of sample, analyte, and matrix are given by:

- The sample mass

$$m_{sample} = m_{analyte} + m_{matrix} \qquad\qquad (2.12a)$$

- The absolute mass of the analyte

$$m_{analyte} = m_{sample} \times x_{analyte} \qquad\qquad (2.12b)$$

- The relative mass of the analyte (content, concentration)

$$x_{analyte} = \frac{m_{analyte}}{m_{sample}} = \frac{m_{analyte}}{m_{analyte} + m_{matrix}} \tag{2.12c}$$

Sample preparation is directed to the conversion of test samples in a physically and chemically measurable form. The measuring sample can require a definite state (gaseous, liquid, or solid) or form (aqueous or organic solution, melt-down tests, and pellets). In other cases, measuring samples have to become diluted or enriched to get an optimum concentration range. It may also be necessary to remove interfering matrix constituents which disturb the determination of the analyte.

Physical sample pretreatment. Almost all test samples, including those which are measured by so-called "direct techniques", need physical pretreatment in any form. The most applied physical techniques are:

- Change in the physical state: freezing (freeze-drying), crystallizing, condensation, melting, pressure-pelletizing, vaporization etc.
- Improvement of the form: grinding and homogenizing of solid and coarse materials
- Surface state: polishing, sputtering or etching of rough surfaces of specimen
- Preparation of electrically conductive mixtures and coatings on surfaces, respectively

Chemical sample preparation. The most important chemical and physicochemical procedures of sample pretreatment are:

- Dissolving the test sample by water, other solvents, acids, bases, melts, and gases
- Extraction of a group of analytes from complex matrices by liquid extraction (in recent time increasingly by SPME)
- Separation of interfering constituents to make the analyte determination possible or more specific (in some cases masking serves the same purpose)
- Enrichment of the analyte to improve the signal-to noise-ratio and the detection limit of the analytical method, respectively
- Transforming the analyte in a chemical measurable form by means of chemical reaction (e.g. redox reaction, complex forming)

In some cases, sample preparation techniques are linked directly with analysing techniques using so-called *multi-stage combined systems* (TöLG [1979]) which frequently work in a closed system. The principle is shown in Fig. 2.9 in contrast to a direct analytical system.

Separation technique is a particularly wide field in sample pretreatment. Some essential methods of analytical determination are based on separa-

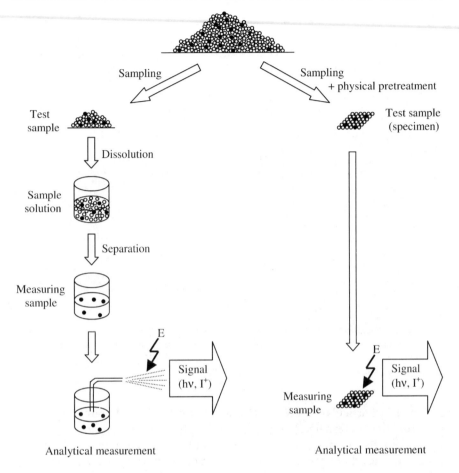

Fig. 2.9. Comparison of analytical multi-stage techniques (*left*) and direct instrumental techniques (*right*)

tion principles, particularly mass spectrometry and chromatographic techniques. In recent times these methods have been coupled amongst themselves and with other methods to achieve separations of high performance. Also other techniques of sample preparation are directly coupled with detection units. Today, such combined methods are called *hyphenated techniques* (HIRSCHFELD [1980]) and play an important role in modern analytical chemistry.

In contrast to combined systems, hyphenated techniques consist of two or more analytical systems each of which is independently applicable as an analytical technique. Usually, the connection is realized by means of an interface and the system is controlled by a computer. With regard to integrated sample treatment, separation and transfer, hyphenated methods like GC-MS, HPLC-MS, GC-IR, GC-IR-MS, GC-AAS, GC-ICP-MS, MS-MS, and

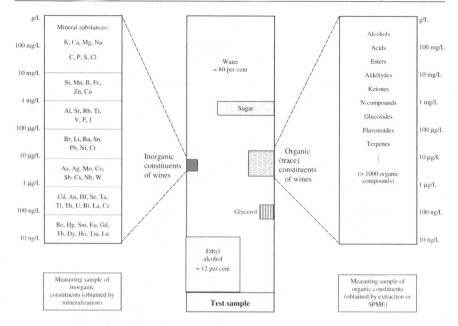

Fig. 2.10. Composition of a *test sample* of wine (schematic representation in the middle; the size of each area symbolizes the relative amount of the respective component or sort of constituents) and of the *measuring samples* of inorganic and organic components, respectively (*left* and *right* diagrams)

MS^n are of particular interest. MS^n stands for the n-fold coupling of mass spectrometers, alternatively serving as separation and detection instrument. By hyphenated techniques the dimensionality of analytical information (see Sect. 3.4) and, therefore, also the information amount (see Sect. 9.3) is significantly increased (ECKSCHLAGER and DANZER [1994]).

Sometimes it is necessary to apply two (or more) variants of sample preparation to get different measuring samples from only one test sample. This is the case if various problems have to be solved, e.g. determination of major- and ultra trace constituents, comparison of depth- and surface profiles, or analysis of inorganic and organic trace components in the same test sample. An example is shown in Fig. 2.10 where in a test sample of wine both inorganic and organic trace constituents have to be determined and, therefore, different measuring samples must be prepared.

2.3
Principles of Analytical Measurement

Measurement always involves comparison of an object with a suitable standard. Whereas physical elementary quantities like length and mass can be compared directly, chemical quantities are mostly compared indirectly. It

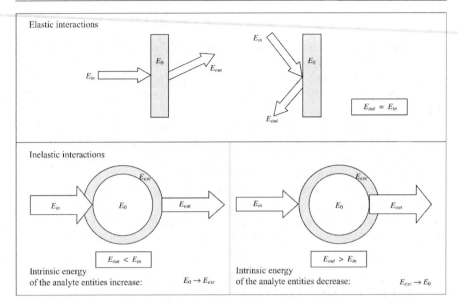

Fig. 2.11. Elastic and inelastic interactions between energy and measuring sample

is the goal of analytical measurements to make available the information on the composition and structure, respectively, which is contained in the sample in a latent way. This hidden (static) information must be released from the measuring sample by chemical reactions or energetic interactions by which real (dynamic) signals are generated.

The interactions to which the measuring samples are exposed can be of elastic or inelastic type (in the physical sense). Figure 2.11 shows schematically the difference between elastic and inelastic interactions.

Elastic interactions are characterized by the condition that no energy is exchanged between the measurement system and the measured sample:

$$\Delta E = E_{in} - E_{out} = 0 \tag{2.13}$$

In other words, $E_{out} = E_{in}$. The energy (electromagnetic radiation or particle beam) is neither absorbed nor otherwise energetically changed while coming into contact with or passing through the measuring sample. What happens instead of this is a change in the spatial structure of the radiation by refraction, reflection, diffraction or scattering. From such structural changes of the energy system E_{out} information can be obtained on the structural arrangement of the constituents in the measuring sample.

The majority of analytical methods are based on inelastic interactions. In contrast to elastic interactions, in case of inelastic interactions both the energy of the measuring system, E_{in}, and the intrinsic energy of the constituents of the measuring sample, E_0, will be changed in such a way that the following condition is fulfilled:

$$\sum \Delta E_{kin} = \sum (E_{in} - E_{out}) = \sum \Delta U = \sum (E_{exc} - E_0) \qquad (2.14)$$

where each change in the kinetic energies ΔE_{kin} definitely corresponds to changes in the intrinsic energy ΔU of the sample constituents. The amounts of kinetic and intrinsic energy, $|\Delta E_{kin}|$ and $|\Delta U|$, characterize the sort of sample constituents whereas the sum of all the energy quanta characterizes the amount of these species.

As a result of the interchanges, signals are produced which contain both information on the type of constituents and their amounts. ΔE_{kin} and ΔU determine the signal position, z, and the intensities of each signal, y, correspond to the amount of the belonging species.

The latent information of the measuring sample is transferred via an energetic carrier into analytical information which is manifested by signals. Their parameters correspond to the *coding* (*encoding*) process in information systems. For the formal representation of the analytical coding the following analytical quantities are introduced:

(a)   The categorical quantities (discrete variables) $Q = A, B, C, \ldots, N$ stand for the chemical entities being investigated (elements, isotopes, ions, modifications, complexes, functional groups, compounds, viz. simple, macromolecular or biomolecular, and occasionally sum parameters). These entities are called *analyte* (PRICHARD et al. [2001]). All the components of the sample other than the relevant are put together under the term *matrix*.

(b)   The quantity x_Q, which is a continuous variable as a rule, characterizes the amount of the analyte as an absolute or relative measure (e.g. mass, content or concentration). The analyte amount x_Q is the original goal of quantitative analysis but is frequently determined indirectly. As the quantity subjected to measurement, x_Q is called *measurand* (FLEMING et al. [1997], PRICHARD et al. [2001]).

By the operations *coding* (*analytical measurement*) and *decoding* (*analytical evaluation*) information will be transformed from the *sample domain* into the *signal domain* and vice versa as shown in Fig. 2.12. Therefore, quantities which correspond to that of the sample domain (Q and x_Q) must also exist in the signal domain. The characteristics in the signal domain are:

(c)   The variables z_i which characterize the values of the signal positions. Respective z_i-positions correspond to given Qs according to assignment rules and tables and – in rare cases – natural and empirical laws and equations, respectively. Such relationships between z_i and Q represent a special type of calibration ("qualitative" or "identity-" calibration).

(d)   On the other hand, the variables y_{z_i} characterize the signal intensity at given positions z_i. The relationship between the intensity of a definite signal and the amount of the belonging constituent Q is rarely known in form of theoretical connections. In most cases the relation between y_{z_i} and x_Q is modelled by experimental calibration.

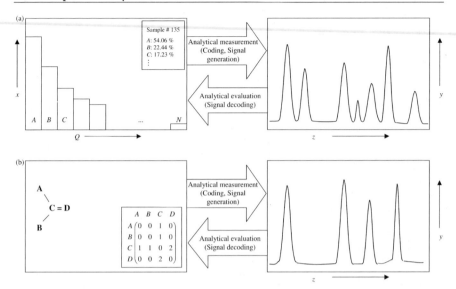

Fig. 2.12. Relationship between sample domain and signal domain in element analysis (**a**) and structure analysis (**b**). The representation in the sample domains is shown in different forms, as a block diagram and a list in case of (**a**) and as constitution formula and structure matrix, respectively, in case (**b**)

A classification of the basis of these analytical quantities with regard to their mutual relationships and dependencies from physical quantities like space co-ordinates and time will be given in Sect. 3.4 according to the dimensionality of analytical information.

Coding and decoding are reverse operations. Coding can be imagined to consist of two steps:

– Coding of the (latent) sample information on Q into energy signals $|\Delta U|_Q$
– Coding of these energy quantities $|\Delta U|_Q$ into measurable signals z_i (measuring quantities)

Both steps are characterized by the corresponding natural laws as well as empirical relations caused by measuring instruments and conditions. All such laws and rules are put together by the *measuring function (calibration function)*.

Taking the forms of intrinsic energy as the basis of the measuring principle, analytical methods can be classified according to the scheme given in Fig. 2.13.

Another scheme arranged according to the forms of energy exchanged between energy system and measuring sample is shown in Fig. 2.14 in form of the so-called "Benninghoven matrix".

Both chemical reactions and physical measurements can leave some "traces" in the measured signals. Chemical contaminations can come from

Form of intrinsic energy	Material entity					
	Elementary particle	Atoms, Ions	Molecules	Aggregated systems	Multiphase systems	Reaction systems
TRANSLATION: – Mass – Charge	Detection methods	MS IC, EPH				
NUCLEAR ENERGY: – Nuclear moments – Nucleon levels – Nucleon binding energy		AA	NMR MÖS			
ELECTRON ENERGY: – Electron spin – Electron levels – Ionization		OES	ESR XFA, PES, AES UV-VIS			
MOLECULE ENERGY: – Rotation – Vibration – Dissociation			MWS IRS, RS			
AGGREGATION ENERGY: – van der Waals energy – Lattice binding energy				TA		
PHASE ENERGY: – Adsorption energy – Partition energy					GC LC	
REACTION ENERGY: – Chemical potential – Electrochemical potential						CA ECA

Fig. 2.13. Analytical methods classified according to the changes of the intrinsic energy of analyte atoms, ions, molecules etc.

Probe (Excitation) \ Detected energy	Identical with the energy of the probe	hν	e⁻	I⁺	e. f.
Radiation hν	RF, LM, IRM, XD	LMA, XFA, MÖS, RS	PES, AES	LEMS, SNMS	
Electrons e⁻	EM (SEM), ED	EPMA	AES, EELS	MS, SNMS	
Ions I⁺	IM, (ISS), (ID)	GDOS		GDMS, SIMS, RBS	
Neutrons	ND				
Thermal energy		AAS, OES, IRS	EEM		
Electric fields	DCM			FIM, FIAP	STM, AFM
	Elastic interactions	Inelastic interactions			

Fig. 2.14. Updated "BENNINGHOVEN matrix" of analytical methods in the context of interactions between different forms of energy

the reaction system, e.g. in the form of residues of a reagent. On the other hand, physical signal functions can contain disturbing primary radiation or "contamination" coming from insufficiencies of the measuring system, e.g. inhomogeneous distribution of radiation or particle beams, finite slit width of optical systems, and noise of electronic components.

Therefore, signal functions $y(z)$ always represent a convolution of the true signal function $y(z)_{true}$ and the characteristic function of the analytical instrument $h(z)$ which characterizes all the insufficiencies of the measuring system:

$$y(z) = y(z)_{true} * h(z) \tag{2.15}$$

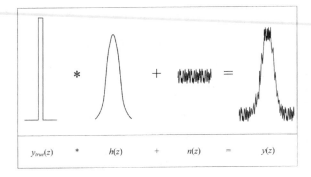

$y_{true}(z)$ $*$ $h(z)$ $+$ $n(z)$ $=$ $y(z)$

Fig. 2.15. Changing of the "true" sample signal by the process of analytical measurement

The mathematical symbol $*$ is used for the convolution function which means in detail

$$y(z) = y(z)_{true} * h(z) = \int_{-\infty}^{+\infty} y_{true}(z - \zeta)h(\zeta)d\zeta = \int_{-\infty}^{+\infty} y_{true}(\zeta)h(z - \zeta)d\zeta$$

where ζ is the integration variable. Additionally, noise $n(z)$ is added to the signal function:

$$y(z) + n(z) = y(z)_{true} * h(z) + n(z) \tag{2.16}$$

The situation is illustrated in Fig. 2.15 where a signal is shown which has been obtained in the analytical reality, distorted and disfigured by noise and broadening. All of these effects can be returned to a certain degree by techniques of signal treatment like deconvolution, signal accumulation and smoothing, etc.

2.4
Analytical Evaluation

Evaluation of signals is carried out effectively in a 'twofold manner', namely according to:

(i)   The analytical *evaluation functions* $x = f^{-1}(y)$ and $Q = f^{-1}(z)$ which are the respective reciprocal functions of the measuring functions (calibration functions) $y = f(x)$ and $z = f(Q)$, respectively (see Fig. 2.12)

(ii) The uncertainty of the analytical measurement and therefore, the uncertainty of the analytical result

Regardless of whether quantitative results of element or structure analysis are the matter of evaluation, analytical results have to be reported always in the form

$$result(x) = mean(x) \pm unc(x) \tag{2.17}$$

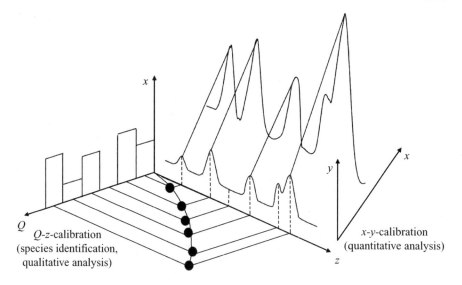

Fig. 2.16. Quasi-four-dimensional representation of the connection between qualitative and quantitative evaluation according to DANZER [1995a], DANZER and CURRIE [1998]

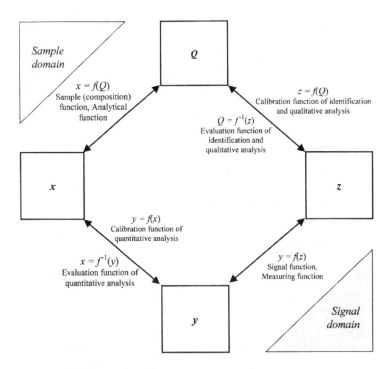

Fig. 2.17. The four analytical quantities in sample and signal domain and the six fundamental functions between them; the functions $Q = f^{-1}(x)$ and $z = f^{-1}(y)$ do not make much sense in analytical practice

see Sect. 8.1, where *mean(x)* is represented by the arithmetic mean $\bar{x}$, as a rule, but can also be given by the median, *med(x)* see Eq. (4.22) and the geometric mean, see Eq. (4.18).

Starting with the relation between the analytical function $x = f(Q)$ and the signal function $y = f(z)$ represented in Fig. 2.12, four quantities x, Q, y, and z have to be related to one another. The situation can be characterized in Fig. 2.16 by a quasi-four-dimensional representation (DANZER [1995a], DANZER and CURRIE [1998]).

The foreground of the representation depicts the relationship between the species Q and their characteristic signals z, while behind that, the relationship between signals z, their intensities y, and the species amounts x is established. Taken together, these relationships establish the composition of the sample (the sample domain) and the signal domain, too.

Another illustration of the relationships between the four quantities x, Q, y, and z together with the type and terms of the related functions between them is shown in Fig. 2.17.

Whereas the x-y-relationships are represented by continuous functions which are linear in most cases, the Q-z-relationships correspond mainly to discrete functions as schematically shown in Fig. 2.18.

In analytical practice, Q-z-evaluation refers to the position of signals on energy scales or scales that are related to energy, such as frequency-,

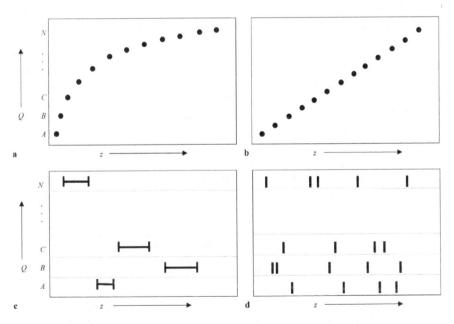

Fig. 2.18. Schematic representations for discrete functions of Q-z-relationships according to natural or empirical laws (**a** MOSELEY type, **b** KOVATS type) and empirical connections (**c** COLTHUP type, **d** atomic spectra type)

wavelength- or mass/charge coordinates of spectrometers or time coordinates of chromatograms, respectively.

The evaluation function $Q = f^{-1}(z)$ that transfers measured z-values into information on Q may be the reciprocal function of one of the following calibration functions (DANZER [1995a]):

(a) *Deterministic function* on the basis of natural laws:

$$z = f_{det}(Q) \tag{2.18a}$$

e.g. MOSELEY's law of the dependence of X-ray frequencies on the atomic number (MOSELEY [1913])

(b) *Empirical function*

$$z = f_{emp}(Q) \tag{2.18b}$$

like KOVATS' indices of homologous compounds in their dependence on retention data in gas chromatography (KOVATS [1958])

(c) *Empirical connection*

$$z = emp(Q) \tag{2.18c}$$

represented by tables and atlases, e.g. by tables of characteristic vibrations (COLTHUP et al. [1975]) and iron-atlases in atomic spectroscopy (HARRISON [1939], DE GREGORIO and SAVASTANO [1972]).

While the relations $z = f_{det}(Q)$ can be derived on the basis of natural laws, the estimation of an empirical function $z = f_{emp}(Q)$ for the purpose of identification and qualitative analysis is mostly carried out by (linear) least squares to fit the observed z-values for a set of pure component standards or a multicomponent standard. On the other hand, empirical relationships $z = emp(Q)$ in the form of tables, atlases and graphs are developed by collection and classification of experimental results.

In general, the *evaluation function* for quantitative analysis is the inverse of the calibration function $y = f(x)$ (CURRIE [1995, 1999], IUPAC ORANGE BOOK [1987, 1988]):

$$x = f^{-1}(y) \tag{2.19}$$

provided that the relationship between the measured value y and the analyte amount x has been estimated which is mainly the case in analytical chemistry. However, there are also other types of evaluation procedures, e.g. on the basis of natural laws, depending on the nature of the analytical method.

The determination of the amounts of analytes can be based on absolute, relative or reference measurements (HULANICKI [1995]) which are based on equations of the general type

$$y = S \cdot x \tag{2.20}$$

where S is the sensitivity; see Sect. 7.2. For the three mentioned types of analytical measurements the sensitivity is given by mathematically well-defined relations, namely in the case of:

(1) *Absolute measurements* by fundamental quantities like FARADAY constant and quotients of atomic and molar masses, respectively (coulometry, electrogravimetry, gravimetry, gas volumetry)

(2) *Definitive measurements* by fundamental quantities complemented by an empirical factor, e.g. titre (titrimetry), as well as by well-known empirical (transferable) constants like molar absorption coefficient (spectrophotometry), NERNST factor (potentiometry, ISE), and conductivity at definite dilution (conductometry)

(3) *Direct reference measurements* by the relation of measured value to content (concentration) of a reference material (RM)

$$S = \frac{y_{RM}}{x_{RM}} \tag{2.21}$$

On the other hand, *indirect reference measurements* are based on empirical calibration functions obtained experimentally and frequently based on linear models

$$y = a + bx + e_y \tag{2.22}$$

where the intercept a corresponds to the experimental *blank* and the slope b to the *sensitivity*.

In analytical practice, some methods using definitive measurements, in principle, are also calibrated in an experimental way (e.g. spectrophotometry, polarography) to provide reliable estimates of S.

Calibration functions corresponding to Eq. (2.22) are not generally transferable over long times and not from one laboratory to another. However in the case of blank-free or blank-corrected relations

$$y = bx + e_y \tag{2.23}$$

methods can be robustly calibrated under fixed experimental conditions. Then the experimental sensitivity coefficients (sensitivity factors) are transferable over time and between laboratories under standard operating conditions. Because of this transferability, such methods are occasionally called "*standard-free*". Such standard-free methods (not to confuse with "*calibration-free*" which characterizes absolute measurements) have been developed, e.g., in the field of OES (HARVEY [1947], DANZER [1976]), SS-MS (RAMENDIK [1990]), and XFA (SHERMAN [1955], FEI HE and VAN ESPEN [1991]) for semi-quantitative multielement analysis.

For evaluation of multisignal measurements, e.g. in OES and MS, more than one signal per analyte can be evaluated simultaneously by means of multiple and multivariate calibration. The fundamentals of experimental calibration and the relating models are given in Chap. 6.

In principle, all measurements are subject to random scattering. Additionally measurements can be affected by systematic deviations. Therefore, the uncertainty of each measurement and measured result has to be evaluated with regard to the aim of the analytical investigation. The uncertainty of a final analytical result is composed of the uncertainties of all the steps of the analytical process and is expressed either in the way of classical statistics by the addition of variances

$$\sigma_{total}^2 = \sigma_{sampling}^2 + \sigma_{preparation}^2 + \sigma_{measurement}^2 + \sigma_{evaluation}^2 \qquad (2.24)$$

or, increasingly in recent time, by means of the empirically calculated combined uncertainty. The estimation of the uncertainty of analytical results is given in Sects. 4.2 and 7.1.

Today, analytical evaluation is done on a large scale in a computerized way by means of data bases and expert systems (Sect. 8.3.6). In particular, a library search is a useful tool to identify pure compounds, confirm them and characterize constituents in mixtures. Additionally, unknown new substances may be classified by similarity analysis (ZUPAN [1986], HIPPE [1991], WARR [1993], HOBERT [1995]). The library search has its main application in such fields where a large number of components has to be related with large sets of data such as environmental and toxicological analysis (SCOTT [1995], PELLIZARRI et al. [1985]).

In comparing spectra (and other signal records) with reference data, the uncertainty of measurement of the signal position has to be considered. The comparison of test spectra and reference spectra in form of data files may be regarded from the standpoint of *set theory*. In *conventional set theory* (CANTOR [1895]), individual data (elements, here intensity values at given signal positions) follow a membership function $m(x)$ that takes only the values 1 and 0, characterizing an element belonging or not belonging to a given set. Comparing such "hard" data sets, slight shifts of test data

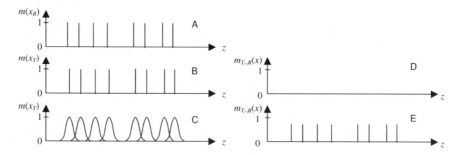

Fig. 2.19. Various sets of analytical data: (A) Hard reference data set, $m_R(x)$. (B) Hard test data set, $m_T(x)$, which is slightly shifted compared with (A). (C) Fuzzy set of test data, $m_T(x) = \exp\left(-(x-a)^2/b^2\right)$. (D) Intersection $m_{T \cap R}(x)$ of test data and reference data which is empty in this case. (E) Intersection of fuzzed test data and reference data with a membership value of about 0.8 in this case

(which may be caused by the sample or measurement) result in an empty intersection $T \cap R$ of the test and reference data as shown in Fig. 2.19D.

ZADEH [1975] extended the classical set theory to the so-called *fuzzy set theory*, introducing membership functions that can take on any value between 0 and 1. As illustrated by the intersection of the (hard) reference data set (A) and the fuzzed test data set (C), the intersection (E) shows an agreement of about 80%. Details on application of fuzzy set theory in analytical chemistry can be found in BLAFFERT [1984], OTTO and BANDEMER [1986a,b] and OTTO et al. [1992].

References

Akin H, Siemens H (1988) *Praktische Geostatistik.* Springer, Berlin Heidelberg New York

Barnard TE (1995) *Environmental Sampling*, in: *The handbook of environmental chemistry* (Ed. O. Hutzinger), Vol 2, Part G *Chemometrics in environmental chemistry – statistical methods* (Vol. Ed. J. Einax), Springer, Berlin Heidelberg New York, 1-47

Blaffert T (1984) *Computer-assisted multicomponent spectral analysis with fuzzy data sets.* Anal Chim Acta 161:135

Bohacek P (1977a) *Chemical inhomogeneity of materials and its determination. Compact materials.* Coll Czechoslov Chem Commun 4:2982

Bohacek P (1977b) *Chemical inhomogeneity of materials and its determination. Granular materials.* Coll Czechoslov Chem Commun 4:3003

Box GEP, Jenkins GM (1976) *Time series analysis.* Holden-Day, Oakland, CA

Cantor G (1895) *Beiträge zur Begründung der transfiniten Mengenlehre.* Math Ann 46: 481

Chatfield C (1989) *The analysis of time series. An introduction.* Chapman & Hill, London, 4th edn

Cochran WG (1977) *Sampling techniques.* Wiley, New York

Colthup NB, Daly LH, Wiberly SE (1975) *Introduction to infrared and Raman spectroscopy.* Academic Press, New York, London

Currie LA (1985) *The limitations of models and measurements as revealed through chemometric intercomparisons.* J Res NBS 90:409

Currie LA (1995) IUPAC, Analytical Chemistry Division, Commission on General Aspects of Analytical Chemistry: *Evaluation of analytical methods including detection and quantification capabilities.* Appl Chem 67:1699

Currie LA (1999) *Nomenclature in evaluation of analytical methods including detection and quantification capabilities (IUPAC Recommendations 1995).* Anal Chim Acta 391:105

Danzer K (1976) *Spectrographic semiquantitative determination of element traces in thin layers without calibration (Russ.).* Zh Analit Khim 31:2327

Danzer K (1995a) *Calibration – a multidimensional approach.* Fresenius J Anal Chem 351:3

Danzer K (1995b) *Sampling of inhomogeneous materials. Statistical models and their practical relevance.* Chem Anal [Warsaw] 40:429

Danzer K (2004) *A closer look at analytical signals.* Anal Bioanal Chem 380:376

Danzer K, Currie LA (1998) IUPAC, Analytical Chemistry Division, Commission on General Aspects of Analytical Chemistry: *Guidelines for calibration in analytical chemistry part 1. Fundamentals and single component calibration* (Recommendations 1998), Pure Appl Chem 70:993

Danzer K, Ehrlich G (1984) *Homogenitätscharakterisierung von Festkörpern mit Hilfe mathematischer Modelle.* Tagungsber Techn Hochsch Karl-Marx-Stadt, 12: 4. Tagung Festkörperanalytik, vol 2, p 547

Danzer K, Küchler L (1977) *Homogenitätskriterien für festkörperanalytische Untersuchungen.* Talanta 24:561

Danzer K, Marx G (1979) *Application of two-dimensional variance analysis for the investigation of homogeneity of solids.* Anal Chim Acta 110:145

Danzer K, Singer R (1985) *Application of pattern recognition methods for the investigation of chemical homogeneity of solids.* Mikrochim Acta [Wien] I:219

Danzer K, Than E, Molch D (1976) *Analytik – Systematischer Überblick.* Akademische Verlagsgesellschaft Geest & Portig, Leipzig

Danzer K, Doerffel K, Ehrhardt H, Geissler M, Ehrlich G, Gadow P (1979) *Investigations of the chemical homogeneity of solids.* Anal Chim Acta 105:1

De Gregorio P, Savastano G (1972) *Spektrum des Eisens von 2206 bis 4656 Å mit Analysenlinien.* Specola Vaticana

Doerffel K, Wundrack A (1986) *Korrelationsfunktionen in der Analytik,* in: Analytiker-Taschenbuch, Bd. 6 (Hrsg.: W. Fresenius, H. Günzler, W. Huber, I. Lüderwald, G. Tölg, H. Wisser), Springer, Berlin Heidelberg New York

Doerffel K, Küchler L, Meyer N (1990) *Treatment of noisy data from distribution analysis using models from time-series analysis.* Fresenius J Anal Chem 337:802

Eckschlager K, Danzer K (1994) *Information theory in analytical chemistry.* Wiley, New York, Chichester, Brisbane

Ehrlich G, Kluge W (1989) *Characterization of the chemical homogeneity of solids affected with periodical concentration fluctuations using the statistics of stochastic processes.* Mikrochim Acta [Wien] I:145

Einax JW, Zwanziger HW, Geiß S (1997) *Chemometrics in environmental analysis.* VCH, Weinheim

Fei He, Van Espen PJ (1991) *General aspects for quantitative energy dispersive X-ray fluorescence analysis based on fundamental parameters.* Anal Chem 63:2237

Fleming J, Albus H, Neidhart B, Wegscheider W (1997) *Glossary of analytical terms (VIII).* Accr Qual Assur 2:160

Gilbert RO (1987) *Statistical methods for environmental pollution monitoring.* Van Nostrand Reinhold, New York

Gy PM (1982) *Sampling of particulate materials – theory and practice.* Elsevier, Amsterdam

Gy PM (1992) *Sampling of heterogeneous and dynamic material systems – theories of heterogeneity, sampling and homogenizing.* Elsevier, Amsterdam

Harrison GR (1939) *M. I. T. wave-length tables of 100,000 spectrum lines.* New York

Hartung J, Elpelt B, Klösener K-H (1991) *Statistik. Lehr- und Handbuch der angewandten Statistik.* R. Oldenbourg Verlag, München, Wien, Kap. XII

Harvey CE (1947) *A method of semi-quantitative spectrographic analysis.* A.R.L., Glendale, CA

Hippe Z (1991) *Artificial intelligence in chemistry. Structure elucidation and simulation of organic reactions.* Elsevier, Amsterdam

Hirschfeld T (1980) *The hy-phen-ated methods.* Anal Chem 52:297A

Hobert H (1995) *Library search – principles and applications.* In: *Chemometrics in environmental chemistry – applications.* Vol 2, part H (Vol ed J Einax), Springer, Berlin Heidelberg New York, p 1

Horwitz W (1990) IUPAC, Analytical Chemistry Division, Commission on Analytical Nomenclature: *Nomenclature for sampling in analytical chemistry* (Recommendations 1990), Pure Appl Chem 62:1193

Hulanicki A (1995) IUPAC, Analytical Chemistry Division, Commission on General Aspects of Analytical Chemistry: *Absolute methods in analytical chemistry.* Pure Appl Chem 67:1905

Inczedy J (1982) *Homogeneity of solids: a proposal for quantitative definition.* Talanta 29:643

Ingamells CO (1974) *New approaches to geochemical analysis and sampling.* Talanta 21:141

Ingamells CO (1976) *Derivation of the sampling constant equation.* Talanta 23:263

Ingamells CO, Switzer P (1973) *A proposed sampling constant for use in geochemical analysis.* Talanta 20:547

ISO (1993) International Organization for Standardization, *Guide to the expression of uncertainty in measurement.* Geneva

IUPAC Orange Book (1987) *Compendium of analytical nomenclature* (eds H. Freiser, G.H. Nancollas), Blackwell, Oxford 2nd edn (1st edn 1978)

IUPAC Orange Book (1988) *Compendium of analytical nomenclature* (eds J. Inczédy, J.T. Lengyiel, A.M. Ure, A. Geleneser, A. Hulanicki), Blackwell, Oxford 3rd edn

Kateman, G (1990) *Sampling Strategies.* Wiss Z TH Leuna-Merseburg 32: 54

Kateman G, Buydens L (1993) *Quality control in analytical chemistry,* 2nd edn. Wiley, New York

Keith LH (ed) (1988) *Principles of environmental sampling.* ACS, American Chemical Society, Washington DC

Kellner R, Mermet J-M, Otto M, Widmer HM (eds) (1998) *Analytical chemistry. The approved text to the FECS Curriculum Analytical Chemistry.* Wiley-VCH, Weinheim, New York, Chichester

Kovats E (1958) *Gas-chromatographische Charakterisierung organischer Verbindungen. Teil 1: Retentionsindizes aliphatischer Halogenide, Alkohole, Aldehyde und Ketone.* Helvet Chim Acta 41:1915

Kraft G (1980) *Probennahme an festen Stoffen,* in: Analytiker-Taschenbuch, Bd. 1 (Hrsg.: H. Kienitz, R. Bock, W. Fresenius, W. Huber, G. Tölg), Springer, Berlin Heidelberg New York

Kratochvil B, Wallace D, Taylor JK (1984) *Sampling for chemical analysis*. Anal Chem 56:114R

Kurfürst U (1991) *Die direkte Analyse von Feststoffen mit der Graphitrohr-AAS*, in: Analytiker-Taschenbuch, Bd. 10 (Hrsg.: H. Günzler, R. Borsdorf, W. Fresenius, W. Huber, H. Kelker, I. Lüderwald, G. Tölg, H. Wisser), Springer, Berlin Heidelberg New York

Lucchesi CA (1980) *The analytical chemist as problem solver.* Internat Lab 11/12:67

Malissa H (1973) *Stereometrische Analyse mit Hilfe der Elektronenstrahlmikroanalyse.* Fresenius Z Anal Chem 273:449

Markert B (ed) (1994) *Environmental sampling for trace analysis.* VCH, Weinheim, New York Basel

Metzler P, Nickel B (1986) *Zeitreihen- und Verlaufsanalyse.* Hirzel, Leipzig

Moseley HGA (1913) *The high-frequency spectra of the elements.* Philos Mag 26:1024

Nothbaum N, Scholz RW, May TW (1994) *Probenplanung und Datenanalyse bei kontaminierten Böden.* Erich Schmidt, Berlin

Otto M, Bandemer H (1986a) *Pattern recognition based on fuzzy observations for spectroscopic quality control and chromatographic fingerprinting.* Anal Chim Acta 184:21

Otto M, Bandemer H (1986b) *Calibration with imprecise signals and concentrations based on fuzzy theory.* Chemom Intell Lab Syst 1:71

Otto M, Stingeder G, Piplits K, Grasserbauer M, Heinrich M (1992) *Comparison of depth profiles in SIMS by a fuzzy method.* Mikrochim Acta 106:163

Parczewski A (1981) *The use of empirical mathematical models in the examination of homogeneity of solids.* Anal Chim Acta 130:221

Parczewski A, Danzer K (1993) *Some limitations of global characterization of inhomogeneity of solids.* Polish J Chem 67:961

Parczewski A, Danzer K, Singer R (1986) *Investigation of the homogeneity of solids with a linear regression model.* Anal Chim Acta 191:461

Pellarrini ED, Hartwell T, Crowder JA (1985) *A comparative evaluation of GC/MS data analysis processing.* US EPA Project Report PB-85-125664

Prichard E, Green J, Houlgate P, Miller J, Newman E, Phillips G, Rowley A (2001) *Analytical measurement terminology - handbook of terms used in quality assurance of analytical measurement.* LGC, Teddington, Royal Society of Chemistry, Cambridge

Ramendik GI (1990) *Elemental analysis without standard reference samples: The general aspect and the realization in SSMS and LMS.* Fresenius J Anal Chim 337:772

Scott DR (1995) *Empirical pattern recognition/expert system approach for classification and identification of toxic organic compounds from low resolution mass spectra.* In: Chemometrics in environmental chemistry - applications. Vol 2, part H (Vol ed J Einax), Springer, Berlin Heidelberg New York, p 25

Sherman J (1955) *The theoretical derivation of fluorescent X-ray intensities from mixtures.* Spectrochim Acta 7:283

Singer R, Danzer K (1984) *Homogenitätsuntersuchungen von Festkörpern mit Hilfe linearer Regressionsmodelle.* Z Chem 24:339

Taggart AF (1945) *Handbook of mineral dressing, ores and industrial minerals.* Wiley, New York

Tölg G (1979) *Neue Wege zur analytischen Charakterisierung von Reinststoffen.* Fresenius Z Anal Chem 294:1

Tyson JF (1989) *Modern analytical chemistry.* Anal Proc 26:251

Warr WA (1993) *Computer-assisted structure elucidation.* Anal Chem 65:1045A; 1087A

Wegscheider W (1996) *Proper sampling: a precondition for accurate analyses.* In: *Accreditation and quality assurance in analytical chemistry* (Ed. by H. Günzler) Springer, Berlin Heidelberg New York

Zadeh LA (1975) *Fuzzy sets and their applications to cognitive and decision processes.* Academic Press, New York

Zupan J (1986) *Computer-supported spectroscopic databases.* Ellis Horwood, Chichester

3 Signals in Analytical Chemistry

The analytical process is a procedure of gaining information. At first, samples contain only latent information on the composition and structure, namely by their intrinsic properties (MALISSA [1984]; ECKSCHLAGER and DANZER [1994]). By interactions between the sample and the measuring system this information is transformed step by step into signals, measured results and useful chemical information.

3.1
Signals and Information

Information is always connected with signals. In general, signals are definite states or processes of material systems (ECKSCHLAGER and DANZER [1994]; DANZER [2004]). They can, therefore, be differentiated into static and dynamic signals. Examples of static signals are script, colors, images, figures, and buildings. On the other hand, dynamic signals result from electrical, thermal, optical, acoustic, or chemical interactions. Nowadays, these signals are converted in each case into electric signals, in which form they may be treated and transmitted. Finally, the essential signal characteristics are recorded in a suitable form. The process of signal generation and evaluation is given by the chemical measurement process (CMP, CURRIE [1985, 1995]) which is shown in its analogy to the information process in Fig. 3.1.

Both the CMP and information process in Fig. 3.1 have been simplified in comparison with the representations in Figs. 2.1 and 2.2 because sample preparation has not been considered here and the measurement begins with the measuring sample as information source from which the signals are obtained.

From the signal-theoretical point of view, signals must fulfil the following three functions to gain analytical information:

- A *syntactic function* which describes the relationship between equivalent signals, the formation of signal sequences, and the transformation of signals (knowledge of the genesis of signals). The syntactic function characterizes the *structure of signals and signal sequences.*

- A *semantic function* which describes the meaning and significance of signals and thus their unambiguous connection with the object to be

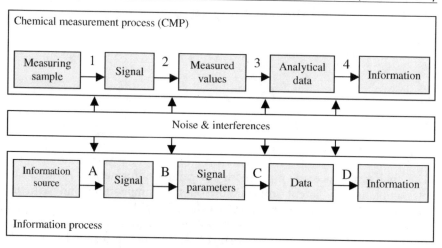

Fig. 3.1. Chemical measurement process (*above*) and information process (*below*) 1 Measurement, 2 Signal validation, 3 Evaluation/calibration, 4 Data evaluation, A Coding, B Selection, C Decoding, data evaluation

characterized (knowledge of coding and decoding, assignment rules, and calibration). The semantic function characterizes the *meaning of signals*.

- A *pragmatic function* which determines the relationship between signals and persons who receive them. The pragmatic function, therefore, characterizes the *importance and benefit of signals*.

All three of these functions should be harmonized to obtain the expected information to the full extent and to avoid misinterpretation of signals.

3.2
Analytical Signals

The term "*analytical signal*" has a different meaning in signal theory and analytical chemistry. In signal theory, analytical signals are of the type

$$z(t) = x(t) + j\hat{x}(t) \tag{3.1}$$

where $x(t)$ is the signal function and $\hat{x}(t) = H\{x(t)\}$ is the Hilbert transform as imaginary part (WOLF [1999]).

Here, the term *analytical signal* is used for all the signals which are produced by analytical methods and used (treated, evaluated and interpreted) in any form in analytical chemistry. Analytical signals can result from test samples, reference samples, or data banks (reference spectra and other recordings).

Analytical signals are generated by interactions between species of the analyte, to be precise between certain forms of intrinsic energy of them (see

Molecules	Chemical reactions	
Atoms		
Ions (solvatized)	....	
Electrons	Electrochemical processes	
Electron beams	Particle–matter interactions	(a) Inelastic interactions (energy transfer with sample species): *SPECTROSCOPY*
Ion beams		(b) Elastic interactions (energy transfer): *DIFFRACTION, MICROSCOPY*
Electromagnetic radiation	Radiation–matter interactions	
Heat	Thermal interactions	
Directed movement energy	Interaction between phases (partition, adsorption)	

Fig. 3.2. Overview of various forms of interactions taking place between measuring samples and different forms of matter and energy to produce analytical signals

Fig. 2.13), and an external system of matter and energy, respectively. These systems and the resulting interactions are classified in Fig. 3.2.

The type of interaction is characterized by the analytical technique applied. In detail, analytical signals can appear in form of:

- The product of chemical reactions, namely *precipitations* (their appearance, color, and crystal shape), *colors* in solution, *gases* (their appearance, bubbles, and color), and *sublimates*
- Changes of *colors in flames* and other light sources, *emission of radiation*
- Differences between physical quantities like temperature, potential, voltage, and absorbance etc.

Analytical signals can be classified according to different criteria. According to their *genesis*, analytical signals are frequently differentiated into chemical and physical signals, notwithstanding that chemical signals are quantified ultimately by physical quantities (weight, volume, absorbance etc).

Analytical signals can also be differentiated according to their *time characteristics* into dynamic, transient and static signals, and their *complexity* into single signals, periodic signals, signal sequences and signal functions.

Most of the signals with which analysts have to do are *manifest signals*, i.e. real observable signals from which real measured values can be taken. But it is also possible to evaluate *latent signals*, i.e. non-real signals which can be obtained mathematically from real signals, in the simplest cases by calculating differences, ratios, and sums of manifest signals. Latent signals are realized in the form of latent variables, e.g. differences of mass numbers in mass spectra as well as isotope pattern of intensities of molecular peaks. A special type of latent variables is represented by principle components, eigenvalues, factors, and other chemometrically obtainable quantities.

Hidden signals can be given in the form of non-resolved fine structures of spectral bands or signals covered by background and noise. Hidden signals can be detected by improving the experimental technique or by applying chemometric procedures to improve the signal resolution and/or signal-to-noise ratio.

In some cases it makes sense to record a signal not only in its original form but also in its integrated form or in the form of its first, second or higher derivation. In this way, the quantification of signal intensity can be improved and the position of the signal can be measured more precisely, respectively.

The generation of analytical signals is a complex process that takes place in several steps. Methods of instrumental analysis often need five steps, namely (1) the *genesis*, (2) the *appearance*, (3) the *detection* and *conversion*, (4) the *registration*, and (5) the *presentation* of signals; see Fig. 3.3.

The *genesis* of signals is directly connected with the interaction between the entities of the sample (see Fig. 2.13) and the form of matter and energy represented in Fig. 3.2. This interaction produces the signal as a result of a chemical reaction, an electrochemical, physicochemical or physical process, e.g. by a neutralization or precipitation reaction, an electrolytical process, or by interactions between radiation and particles on the one hand and the sample species on the other.

The *appearance* of signals takes place in form of a substance of a given state (gaseous, liquid, precipitate), change of colors by absorption or reflection of light, or emitted and absorbed radiation. Spectroscopic signals

Method	Genesis	Appearance	Detection	Conversion	Registration	Presentation	Storage
Emission spectrometry	Interaction thermal- & radiation energy – species	Emitted radiation quanta, $h \cdot v$	Photo multiplier tubes, photoplate			Spectrum	
Absorption spectrometry	Interaction radiation energy – species	Absorbed radiation quanta, $h \cdot v$	Thermal conductivity bridge	Current	Pen recorder, display $i = f(t)$	Spectral signal, spectrum	
Mass spectrometry	Ionization	Accelerated ions	Ionsensitive multiplier tubes, dynodes			Spectrum	Digitalized data
Nuclear magnetic resonance spectroscopy	Interaction magnetic fields – nuclei	Resonance of radiation quanta, $h \cdot v$	Radiofrequency pulses			Spectrum in time or frequency (FT) domain	
Chromatography	Interaction between phases	Retention at given time or volume	Thermal conductivity, flame-ionization, MS			Chromatogram	
Gravimetry	Chemical reaction	Precipitation	Weighing form – Balance	–	–	Result	
Titrimetry	Chemical reaction	Equivalence point	Volume	Voltage	Pen recorder, display	Titration curve	
Potentiometry	Electrochemical reaction	Exchange of charge	Electrode potentials		$E = f(t)$	Response curve	
Polarography	Electrochemical reaction	Exchange of charge	Diffusion current	Current, voltage	$i = f(E)$	Polarogram	

Fig. 3.3. Illustration of the steps of signal generation for different analytical principles

appear in form of energy quanta $h \cdot \nu$ which are characterized qualitatively and quantitatively by wavelengths, frequencies, and intensities. In this form a signal might be visible or not but it cannot be stored, treated or evaluated. Therefore, signals must be brought in an applicable form by a suitable detection procedure.

By the *detection* step the signal is transformed into a form that is electrically measurable. By means of photocells photons and ions can be generated into electrons the number of which is greatly increased in multiplier tubes. Several types of detection are shown in Fig. 3.3. It can be seen that, independently of the genesis (chemical or physical), the manifested signal is always converted into the form of a physical quantity like mass, volume, mostly, however, into an electric quantity (current, voltage). Via the *conversion* step the signals come into a form where current or voltage can be registered as a continuous function of time or a time-proportional quantity.

For the analytical representation the signals have to be transformed from time functions into conventional measuring functions. These are characterized by analytical quantities on abscissa and ordinate axes where the values of them may be relativized in some cases (e.g. MS). Such a transformation of quantities is mostly carried out on the basis of instrument-internal adjustment and calibration.

The *presentation* of signals takes place in the form of records of single signals and of signal functions (displays and prints of spectra, chromatograms, or images). The presentation occurs in the signal domain as schematically shown in Fig. 2.12 on the right side. Because the signal function cannot be reversibly stored in this analogous form, an analogue-to-digital conversion may follow where a sequence of numbers is obtained that is discretely arranged on a time axis. Signals in digital form are resistant to disturbances and can be reversibly stored. In particular, there is no limitation in the treatment of digital data. Therefore, some additional procedures can be followed where the signals are chemometrically evaluated and improved, respectively.

3.3
Types and Properties of Analytical Signals

Signals used in analytical chemistry have a definite origin from particular species or given structural relationships between constituents of samples. The relation of the sample domain and the signals domain, i.e. the coding and decoding process as represented in Fig. 2.12, must be as unambiguous as possible.

The reliability of signals depends on the information amount (see Chap. 9), particularly on the signal resolving power of the analytical method and, therefore, on the fact how narrow the signals are. The smaller the signal half-width (width of the signal at half the maximum height) is, the more unambiguous are the connections to the belonging species.

This has led to such cases in the history of chemistry that spectroscopic signals have been unidentified till newly discovered elements was found (e.g. rubidium, caesium, indium, helium, rhenium) or new species (highly ionized atoms, e.g. in northern lights [aura borealis], luminous phenomena in cosmic space and sun aura, such as "nebulium", "coronium", "geocoronium", "asterium", which was characterized at first to be new elements; see BOWEN [1927]; GROTRIAN [1928]; RABINOWITSCH [1928]).

There are a lot of forms in which signals are represented in general, and particularly in analytical chemistry. The common form of a signal is bell-shaped (of GAUSS-, LORENTZ- or VOIGT type; see, e.g. KELLY and HORLICK [1974]). But there are also various other types of signal forms, e.g. differentiated (ESR, AES), integrated (polarography, NMR), bar diagrams (MS), and dynamic signal functions the form of which may be changing (chromatography). By means of modern instrumental methods, multispecies analysis is carried out where, therefore, signal functions are obtained that contain a certain number of single signals.

Signal functions may have a very different character even in case of one and the same analytical method as Fig. 3.4 shows.

Figure 3.4 shows (i) a line spectrum (one-dimensional dispersive spectrographic record), (ii) a spectrometric record, (iii) an interferogram obtained by a FOURIER transform spectrometer, and (iv, v) two- and three-dimensional double dispersive spectra recorded e.g. by Echelle spectrometers. In principle, all forms may be obtained by OES.

In general, signal functions will be obtained in analogous form and consist of a large number of arranged measured points which form a data vector. There are three types of signal functions, which contain:

- Exactly one signal for each species or phenomenon provided it is present in the sample (e.g. chromatography)
- A number of signals per species (e.g. XRF, MS), sometimes up to several thousand per element (OES; see, e.g. HARRISON [1939]; DE GREGORIO and SAVASTANO [1972])
- Not only original signals (one or several) but additional combined signals (overtones, coupled oscillations, e.g. NIR) and latent signals in form of relations between original signals (differences, e.g. MS)

Occasionally, typical pattern can be observed which can be formed according to special rules like multiplets in ESR-, NMR-, and OES spectroscopy or isotopic ratios in MS (molecular peak pattern). There can also be randomly formed pattern within such spectra, being rich in signals like OES (e.g. the known sodium doublet (Na-D) 589.6 and 589.0 nm, and the magnesium "quintet" 277.67, 277.83, 277.98, 278.14, and 278.30 nm). The identification of species is always made easier when pattern – whatever type – can be compared instead of a number of signals that are irregularly arranged.

Only such signals are used in analytical chemistry, as a rule, which can reliably be related to the species or phenomenon under investigation. To

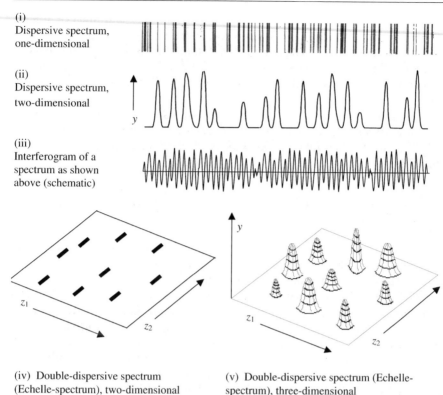

(i)
Dispersive spectrum,
one-dimensional

(ii)
Dispersive spectrum,
two-dimensional

(iii)
Interferogram of a
spectrum as shown
above (schematic)

(iv) Double-dispersive spectrum
(Echelle-spectrum), two-dimensional

(v) Double-dispersive spectrum (Echelle-spectrum), three-dimensional

Fig. 3.4. Different types of signal recordings (schematic) obtained by various types of instruments and registering in OES

guarantee the correct recognition and identification of signals, several procedures are applicable. As mentioned in Sect. 2.4, a reliable evaluation function $Q = f^{-1}(z)$ is the basis of an unambiguous recognition of the analyte. If, nevertheless, uncertainties remain, there are some theoretical and experimental possibilities to guarantee reliably the identity. Such procedures are, e.g. addition of the analyte in a definite form, isotope dilution, and isotope substitution (e.g. deuterization of characteristic groups).

Signal functions are inspected in several ways:

- By "XX" (expert's experience), or objectively
- By comparison with reference lists or diagrams in form of tables or atlases, respectively, and increasingly computerized (mainly in the case of spectra)
- By comparison with reference spectra (RS), mainly by automated spectra search (data banks)

For library searches, test spectra (TS) have to be available in digitalized form. The demands on the quality of reference spectra are high. They have to be free from systematic deviations and impairing noise and must be unambiguously coded. Both reference and test spectra must be pretreated in the same way concerning background correction, standardization, and encoding. Sometimes it may be advantageous to use test spectra which are fuzzed to a certain degree to compare them successfully with reference spectra (see Sect. 2.4, Fig. 2.19, BLAFFERT [1984]).

Efficient techniques which can be applied also for spectra comparisons are *subtraction* (TS–RS) and *cross-correlation* of TS and RS (DANZER et al. [1991]). The *cross-correlation function* (CCF) is calculated according to the relationship

$$\psi_{y_{TS}y_{RS}}(\tau) = \overline{y_{TS}(t)\, y_{RS}(t+\tau)} = \lim_{T\to\infty} \frac{1}{2T} \int\limits_{-T}^{+T} y_{TS}(t)\, y_{RS}(t+\tau)\, \mathrm{d}t \quad (3.2)$$

The CCF has a sharp maximum at $\tau=0$ in case of total agreement of the TS and RS. The correlation coefficient then becomes unity. Values less than 1 characterize a certain degree of agreement.

Within the signal domain the signal functions can be represented in two sub-domains, namely the *original (signal) domain* and the *image domain* which are commonly called *time domain* and *frequency domain*. The records of spectra in the time and frequency domain may look like (ii) and (iii) in Fig. 3.4. The transition between the both domains is carried out mathematically by means of the FOURIER transformation[1]. The relations between signal functions in original (time) and Fourier domains (frequency $\omega = 1/t$) are represented in Fig. 3.5.

Fourier pairs not only exist in time-/frequency domain but also in any other domain combined by a quantity q and the belonging dimension-inverted quantity $1/q$.

Signals are characterized by typical parameters. In Fig. 3.6 the fundamental signal parameters of a common bell-shaped signal are given.

Signals are determined by the following parameters:

- *Position* z_A: the quantity which corresponds to the energy or motion, respectively, of species of the analyte A. As detailed in Sect. 3.5, the signal position is not a totally fixed parameter but is variable to a certain degree.
- *Intensity* y_A: the quantity which corresponds to the amount of the analyte A (species or phenomenon). Both *maximum intensity*, $y_{A_{max}}$, and *integral intensity*, $y_{A_{int}}$, are used depending on the analytical procedure. In the case of maximum intensity, frequently the *net intensity* $y_{A_{net}}$

$$y_{A_{net}} = y_{A_{max}} - y_0 \tag{3.3}$$

---

[1]   There are other transformations, too, e.g. LAPLACE transformation which is frequently used in technical systems.

Signal function in the time domain

Signal function in the frequency domain

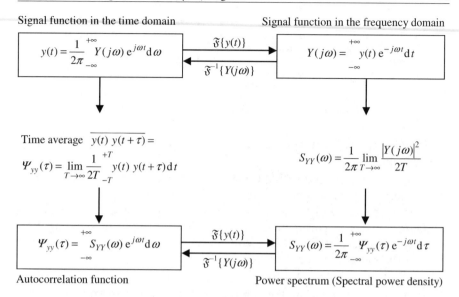

$$y(t) = \frac{1}{2\pi} \int_{-\infty}^{+\infty} Y(j\omega)\, e^{j\omega t} d\omega$$

$$\mathfrak{F}\{y(t)\}$$

$$\mathfrak{F}^{-1}\{Y(j\omega)\}$$

$$Y(j\omega) = \int_{-\infty}^{+\infty} y(t)\, e^{-j\omega t} dt$$

Time average $\overline{y(t)\, y(t+\tau)} =$

$$\Psi_{yy}(\tau) = \lim_{T\to\infty} \frac{1}{2T} \int_{-T}^{+T} y(t)\, y(t+\tau) dt$$

$$S_{YY}(\omega) = \frac{1}{2\pi} \lim_{T\to\infty} \frac{|Y(j\omega)|^2}{2T}$$

$$\Psi_{yy}(\tau) = \int_{-\infty}^{+\infty} S_{YY}(\omega)\, e^{j\omega t} d\omega$$

$$\mathfrak{F}\{y(t)\}$$

$$\mathfrak{F}^{-1}\{Y(j\omega)\}$$

$$S_{YY}(\omega) = \frac{1}{2\pi} \int_{-\infty}^{+\infty} \Psi_{yy}(\tau)\, e^{-j\omega t} d\tau$$

Autocorrelation function

Power spectrum (Spectral power density)

Fig. 3.5. Connection between characteristic signal functions in time and frequency domain

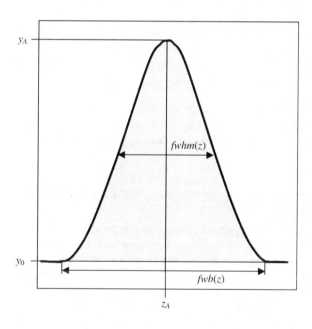

Fig. 3.6. Essential signal parameters: z_A signal position, y_A signal (gross) intensity, y_0 signal background, $fwhm(z)$ full width at half maximum (signal half width), $fwb(z)$ full width at background

which characterizes the number of analyte species in the measuring sample, is applied. The integral intensity corresponds to the area under the signal and is estimated by

$$y_{A_{int}} = \int\limits^{fwb(z)} y(z) \, dz \tag{3.4}$$

In some cases, the integral intensity is recorded additionally to the original signal function.

- *Shape, symmetry*, and *width*: these parameters can be influenced both by properties of the analyte species and their interactions, e.g. with neighboring species, and by disturbing effects of the measuring system (e.g. tailing effects in chromatography and peak widening in general). Peak shape can be considered as a convolution of the "true" signal function and the measuring function (see Fig. 2.15).

- *Fine structure* and *splitting*: under given conditions of the measuring technique (e.g. state of the sample, high-resolution instruments, magnetic fields) signals may produce a fine structure or split up to multiplet pattern. Splitting follows given rules.

- *Background* and *noise*: sometimes signals are affected by a background (zero signal) y_0 which can be corrected experimentally or by calculation. On the other hand, noise is an inherent component of signals, and may appear weak or strong. Noise has effects in both z- and y-directions and causes random variations of the analytical quantities z and y. Noise in the z-direction is mostly and rightly disregarded but can become significant in special cases, e.g. in remote sensing where laser pulses may be blurred in both z- and y-direction (KOZLOV [2004]).

A signal function $y(z)$, which is mostly treated as a time-dependent function, $y(t)$, can be regarded as consisting of the original "true" signal function $y_{true}(t)$ which is superposed by a noise function $n(t)$; see Fig. 2.15:

$$y(t) = y_{true}(t) + n(t) \tag{3.5}$$

Noise is generally added to the signal function; the types of noise are schematically shown in Fig. 3.7.

In analytical instruments, all types of noise can be produced simultaneously – white and flicker noise as well as interference. The following procedures can be applied to improve the signal-to-noise ratio S/N: signal averaging (accumulation), digital filtering (WILLIAMS [1986]), autocorrelation, and smoothing of the signal function.

The applicability of these techniques depends on the type and form of the power spectra of both signal and noise. The only method that can be universally applied, is signal averaging. If the signal function is measured n times, the S/N ratio increases by $\sqrt{n}$.

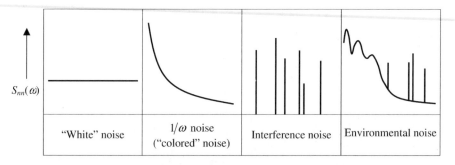

Fig. 3.7. Power spectra of the most important types of noise (schematic representation in the frequency domain)

3.4
Dimensionality of Analytical Signals and Information

The dimensionality of a functional relationship will be defined here (axiomatically) by the number of dependent and independent variables in such a function. Therefore, functions of the type $a = f(b)$ are *two-dimensional*, of the type $a = f(b_1, b_2)$ *three-dimensional*, and of the type $a = f(b_1, b_2, \ldots b_n)$ are $(n + 1)$-*dimensional*. The representation of various realizations of only one variable (either y or z) is *one-dimensional* (DANZER et al. [2002]).

Dimensionality in analytical chemistry comprises different types of dimensions displayed by different sorts of variables, which can be discrete or continuous:

- *Chemical dimensions (dimensions in the sample domain)*: type (Q), number (n) and amount (x) of analytes, i.e. distinct chemical species; see Fig. 2.12 (left)
- *Measuring dimensions (signal-theoretical dimensions in the signal domain)*: signal position z (given by the signal's energy- or time characteristics) and signal intensity y; see Fig. 2.12 (right)
- *Physical dimensions*: time (t) and space, characterized by the variables latitude (l_x), longitude (l_y), and altitude (l_z)
- *Statistical dimensions*: number of variables (manifest or latent) taken into account in evaluation. Statistical dimensions define the type of data handling and evaluation, e.g. univariate, bivariate, multivariate

As a result of analytical measurements, signals are obtained and, in the case of instrumental measurements, signals functions, $y = f(z)$. The record of the signal intensity as a function of the signal position, Fig. 3.8, represents a *two-dimensional signal function* which can be back-transformed into *two-dimensional analytical information*, $x = f(Q)$.

The abscissa in Fig. 3.8 may represent an energy-related scale, e.g., wavelength-, frequency-, or mass/charge coordinates of spectrometers or re-

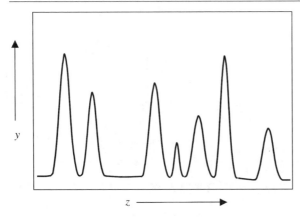

Fig. 3.8. Two-dimensional analytical information in form of a signal function

tention time coordinates of chromatograms (DANZER and CURRIE [1998]). The evaluation of such signal functions is carried out in both dimensions as a rule.

An evaluation of only one of the two quantities concerns:

(a) The *intensity* y_A of a given signal in position z_A which is known to be characteristic for a certain species A; see Fig. 3.9 (*left-hand side*)

(b) The appearance of certain signals z_i that indicate the presence of the corresponding species in the sample (Fig. 3.10)

In this case, the z-axis is inspected as to whether a signal (and therefore the belonging species) is found $(+)$ or not $(-)$. Therefore, one-dimensional analytical evaluation of signals corresponds to the quantitative analysis of one given analyte by typical single species methods like AAS, as shown in Fig. 3.9 (left side) on the one hand, and qualitative multielement analysis, frequently carried out in form of *inspection analysis*, by typical multielement methods like OES. The basis of testing the presence of a certain species Q involves the exceeding of the critical value, $y_Q > y_{Q_{crit}}$ (CURRIE [1995]).

Quasi-multidimensional information as shown in Fig. 3.9 (right side) can be obtained in different ways, mostly by *sequential* measurements, for example:

• By changing detecting channels, thus, measuring one analyte after another (e.g., spectrophotometry)

• By changing excitation source (e.g., hollow cathode lamps in AAS)

• By measurement the same element, which is temporally separated in different species (e.g. GC-AAS)

The most frequent case in analytical chemistry is the evaluation of two-dimensional signal functions in the form of spectra, chromatograms, thermograms, current-voltage curves, etc. (Fig. 3.8).

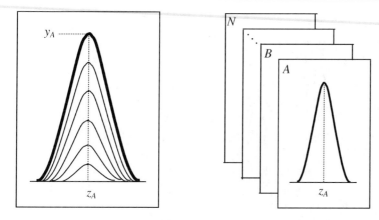

Fig. 3.9. One-dimensional information (in y-direction) of a single signal at a fixed z-value (*left-hand side*) and quasi-multidimensional information on several analytes $A, B, \ldots, N$ as a sequence of one-dimensional information (*right-hand side*)

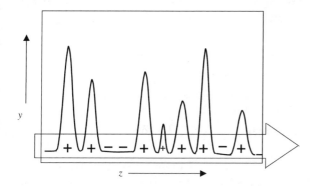

Fig. 3.10. One-dimensional information in z-direction (signal inspection): qualitative evaluation, signal identification

Additional dimensions in signal functions can be given by:

- *Successive separation steps*, e.g. in two-dimensional chromatography (two-dimensional thin-layer chromatography) that result in three-dimensional signal functions $y = f(z_1, z_2)$, as schematically shown in Fig. 3.4(v).

- *Two-dimensional excitation experiments* (two wavelengths excitation in fluorescence spectroscopy or two frequency experiments in 2D-NMR) also generate three-dimensional signal functions.

- *Time-dependent analytical measurements*, which give three-dimensional information of the type $y = f(z, t)$ as shown schematically in Fig. 3.11a. The same characteristic holds for distribution analysis in one spatial direction, i.e., line scans, $y = f(z, l_x)$. Such signal functions are frequently represented in form of multiple diagrams as shown in Fig. 3.11b.

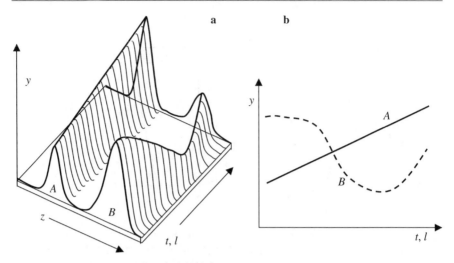

Fig. 3.11. Temporally (t) or spatially (l) dependent analytical measurements, e.g., a line scan of electron microprobe analysis represented in a three-dimensional (**a**) and quasi-three-dimensional way (**b**)

- *Distribution-analytical investigations* in which spatial directions of the sample represent additional variables such as:
    - *Line scans* of microprobes $y = f(z, l_x)$ for several analytes giving images like that represented in Fig. 3.11
    - *Surface scans* of microprobes $y = f(z, l_x, l_y)$ for a given analyte, see Fig. 3.12a,b
    - *Component images* $z = f(l_x, l_y)$ for one or two analytes by black and white representations like Fig. 3.12c, for more components by color images
    - So-called "3D" *images* of SIMS, which in fact represent four- or five-dimensional analytical information depending on the actual function $y = f(l_x, l_y)$, $y = f(l_x, l_y, l_z)$, or $y = f(z, l_x, l_y, l_z)$, respectively
- *Coupled (hyphenated) analytical techniques* of a separation method and a method of analytical determination, e.g. GC-MS, see Fig. 3.13, by which higher-dimensional signal functions are obtained. Strictly speaking, the dimension is $2(n + 1)$ in the case that n mass spectra are recorded.

Alternative characterization of the dimensionality. The SIMS example demonstrates that the dimensionality of analytical information and of signal functions occasionally follow other principles than those given above, where the dimensionality of a functional relationship is determined by the number of dependent and independent variables in such a function.

(a)   (b)

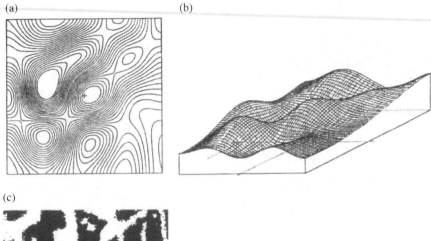

(c)

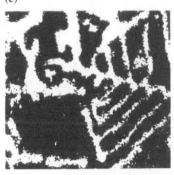

Fig. 3.12. Different types of surface scans: **a** two-dimensional isoline representation of the Mn distribution on a steel surface measured by M-OES; **b** the same Mn distribution in a three-dimensional representation (DANZER [1995]); **c** EPMA scan of Si in a binary eutectic Al-Si alloy

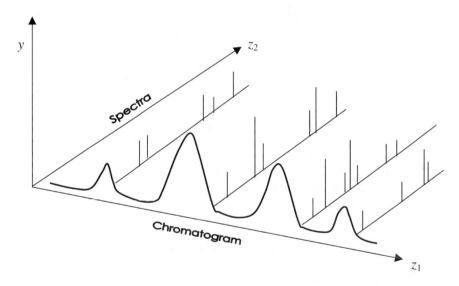

Fig. 3.13. Higher-dimensional system of signal functions in case of coupled techniques, e.g., GC-MS

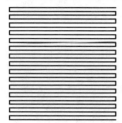

Fig. 3.14. Line with a fractal dimension of about 1.8

Table 3.1. Selection of signal functions in analytical chemistry and their dimensionality

Signal function	Information	Dimension
$z_{y,l_x,l_y,l_z,t}$	Qualitative analysis (y, l_x, l_y, l_z, $t{=}const$) for several analytes, e.g., by a sequence of spot tests	1
$y_{z,l_x,l_y,l_z,t}$	Quantitative average analysis (z, l_x, l_y, l_z, $t{=}const$) for one given analyte, e.g., by gravimetry or AAS	1
$y = f(z)_{l_x,l_y,l_z,t}$	Quantitative average analysis (l_x, l_y, l_z, $t{=}const$) for several analytes, e.g., by ICP-OES	2
$y = f(l_x)_{z,l_y,l_z,t}$	Line analysis: determination of the variation (distribution) of the amount of one analyte (given z) along a line l_x across the sample, e.g., by EPMA	2
$y = f(t)_{z,\,l_x,l_y,l_z}$	Quantitative process analysis: recording the amount of one given analyte z in dependence of time, e.g., by flow techniques	2
$y = f(z,l_x)_{l_y,l_z,t}$	Line analysis of several analytes: determination of the variation (distribution) of the amounts along l_x, see Fig. 3.11, e.g., by EPMA	3
$z = f(l_x,\,l_y)_{y,l_z,t}$	Surface analysis: qualitative representation of the presence (distribution) of several analytes across a surface area, see Fig. 3.12c obtained by EPMA	3
$y = f(l_x,\,l_y)_{z,l_z,t}$	Surface analysis: quantitative representation of the variation (distribution) of the amount of a given analyte z across a surface area, see Fig. 3.12b obtained by micro spark OES	3
$y = f(l_x,\,l_y,\,l_z)_{z,t}$	Volume distribution of one given analyte: SIMS data (representation mostly by categorial y, e.g., color-coded)	4
$\vdots$		$\vdots$
$y = f(z,l_x,l_y,l_z,t)$	Quantitative determination of several analytes in dependence of space and time: spatial-resolved dynamic quantitative multicomponent analysis, e.g., by a series of 3D-SIMS images for several elements and at different times	6

In contrast with this, in signal theory the dimensionality of signal functions is frequently determined only by the number of independent variables. The dependent quantities are ignored. Therefore, a signal curve representing the dependence of the signal intensity from a given independent variable (mostly time) is characterized as being one-dimensional. Also, in analytical practice such a characterization of analytical functions is sometimes used, especially in relation to analytical images. So, spectra are sometimes considered as one-dimensional images and analytical element distribution like those given in Fig. 3.12b,c as two-dimensional element images.

A different point of view is also adopted by BOOKSH and KOWALSKI [1994], who start from an *order of instruments* or methods in such a way that a classification is given according to the type of data which is generated. An instrument (or method) is called a zero-order instrument if it generates a single datum per sample. This terminology comes from mathematics where a single number is considered a zero-order tensor. Examples for zero-order instruments are ion sensitive electrodes (ISE) and single-wavelength photometers. First-order instruments are represented by common spectrometers, chromatographs and similar equipment. They usually produce data vectors (e.g., spectral intensities at multiple wavelengths). Second-order instruments yield a data matrix per sample as is the case in coupled (hyphenated) techniques like GC-MS, MS-MS, HPLC-MS, etc. Similar to the way of looking at dimensions in signal theory, the tensorial theory of instruments neglects one of the dimensions. But whereas in signal theory the ordinate values (e.g., the intensity y) are ignored, in tensorial theory the abscissa values z (e.g., wavelengths or retention times) are disregarded. Notwithstanding, the classification according to the order of instruments and methods, respectively, represents a suitable basis for theoretical fundamentals of analytical chemistry.

> From the most general point of view, the theory of fractals (MANDELBROT [1977]), one-, two-, three-, m-dimensional figures are only borderline cases. Only a straight line is strictly one-dimensional, an even area strictly two-dimensional, and so on. Curves such as in Fig. 3.11 may have a fractal dimension of about 1.1 to 1.3 according to the principles of fractals; areas such as in Fig. 3.12b may have a fractal dimension of about 2.2 to 2.4 and the figure given in Fig. 3.14 drawn by one line may have a dimension of about 1.9 (MANDELBROT [1977]). Fractal dimensions in analytical chemistry may be of importance in materials characterization and problems of sample homogeneity (DANZER and KÜCHLER [1977]).

Analytical functions of different dimensionality have been listed in some detail by DANZER et al. [1987]. The most important signal functions of practical relevance are given in Table 3.1.

Dimensionality of analytical data. Analytical data are present either in the form of measured values y_i or analytical results x_i. Multivariate data, i.e., results of m variables (e.g., analyte concentrations) measured at n different samples, are mostly represented in the form of data sets and data matrices:

$$X = \begin{pmatrix} x_{11} & x_{12} & \cdots & x_{1m} \\ x_{21} & x_{22} & \cdots & x_{2m} \\ \vdots & \vdots & & \vdots \\ x_{n1} & x_{n2} & \cdots & x_{nm} \end{pmatrix} \tag{3.6}$$

where m is the number of components (distinct analytes) determined in n objects (samples). The matrix at Eq. (3.6) has the dimension $n \times m$, and the related data set is called *m-dimensional*. It is the aim of multivariate data analysis to reduce the dimensions of data sets (matrices) by exclusion of redundant variables (components which are correlated) up to such a number that can be displayed two- or three-dimensionally. Details of how to achieve reduction of dimensionality in data sets will be given in Sect. 8.3.1 and are available in the chemometric literature (e.g., SHARAF et al. [1986]; MASSART et al. [1988]; FRANK and TODESCHINI [1994]; DANZER et al. [2001]).

3.5
Mathematical Model of Signal Generation

In an ideal case, the signal $y_A = f(z_A)$, as shown in Fig. 3.6, is determined only by the analyte A (or the phenomenon of interest), namely both the position, $z_A = f(A)$, and intensity, $y_A = f(x_A)$. But in real samples, matrix constituents are present which can principally interfere with the analyte signal. In structure analysis the same holds for the neighboring relationships (the "environment" of the species A of interest). Therefore, signal parameters are additionally influenced by the matrix (or the "neighborhood", respectively), namely the species $B, C, \ldots, N$, and follow then the complex relationships $z_A = f(A; B, C, \ldots, N)$, $y_A = f(x_A; x_B, x_C, \ldots, x_N)$. Additionally, influencing factors $a, b, \ldots, m$, background, y_0, and noise (random deviations e_A) may become relevant and have to be considered.

The *signal position* is given by a characteristic value z_{A0} which is determining for the species A according to evaluation rules for identification and qualitative analysis; see Sect. 2.4, Eq. (2.18a–c). Additional changes in position such as:

- *Chemical shifts* Δz
- *Fine structures* such as rotation structure of vibration bands
- *Multiplet* splitting

can arise from neighboring effects, interactions, and couplings, in accordance with natural laws and rules (multiplet structure). Influencing factors such as temperature, pressure, solvent, etc., can also alter the signal position which holds in general:

$$z_A = f(A; B, C, \ldots, N; a, b, \ldots, m) \tag{3.7}$$

The *real signal position* and *structure* then is given by

$$z_A = \mathcal{M} \left\{ z_{A0} + \sum_{i=B}^{N} \delta z_{Ai} + \sum_{j=a}^{m} \Delta z_{Aj} \right\} \tag{3.8}$$

where $\mathcal{M}$ is the multiplet structure according to relevant rules. In the case of even-numbered multiplets, z_{A0} can become latent and is then represented by the mean of the multiplet. The sum of δz_{AQ} comprises the resulting chemical shift, that of Δz_{Aj}, the additional shift by influencing factors.

The signal intensity is influenced by the entire matrix ($Q = A, B, C, \ldots, N$) with preference of A. In ideal case, y_A would be only caused by x_A:

$$y_A = f(x_A) = S_{AA} x_A + e_A \tag{3.9}$$

where the factor S_{AA} represents the *sensitivity* $S_{AA} = \partial y_A / \partial x_A$ of the determination of A; see Sect. 7.2. But in reality, the accompanying species $i = B, C, \ldots, N$ can contribute to the signal intensity to variable degree:

$$y_A = f(x_A; x_B, x_C, \ldots, x_N) = S_{AA} x_A + \sum_{i=B}^{N} S_{Ai} x_i + e_A \tag{3.10}$$

The factors S_{Ai} represent the *cross sensitivities (partial sensitivities* according to KAISER [1972]) characterizing how the signal $y(z_A)$ is influenced by the species i:

$$S_{Ai} = \frac{\partial y_A}{\partial x_i} \tag{3.11}$$

The deviation e_A represents the unavoidable experimental error of the determination of A.

Only in exceptional cases is it possible to estimate all or part of the influences of interfering species because their amount is mostly unknown. In such cases, the sum of $S_{Ai} x_i$ is considered an additional term of deviation e_i caused by the interferents:

$$y_A = f(x_A; x_B, x_C, \ldots, x_N) = S_{AA} x_A + e_i + e_A = S_{AA} x_A + e_{Ai} \tag{3.12}$$

where $e_{Ai} = e_i + e_A$ is the summarized deviation term of the deviations of interferences e_i and the experimental deviations.

In addition, the signal intensity is influenced by effects of various factors $a, b, \ldots, m$ of the operating conditions, e.g., temperature, pressure, pH value, and instrumental adjustments. Taking these factors into account the model becomes

$$y_A = f(x_A; x_B, x_C, \ldots, x_N; a, b, \ldots, m)$$
$$= S_{AA} x_A + \sum_{i=B}^{N} S_{Ai} x_i + \sum_{j=a}^{m} I_{Aj} x_j + e_A \tag{3.13}$$

The quantity I_{Aj} represent the specific strength of the influence factor j on the signal y_A, $I_{Aj} = \partial y_A / \partial x_j$. Their products with the given influence values x_j yield the actual factor effects. Analogous to the interferences i, the influences of the factors j are also hard to quantify. It can be done by means of a multifactorial design on the basis of a model like Eq. (3.13); see Sect. 5.1. But usually the sum of influences is also considered as a further additive term of deviations e_j:

$$y_A = f(x_A;\ x_B, x_C, \ldots, x_N;\ a, b, \ldots, m)$$
$$= S_{AA}\, x_A + e_i + e_j + e_A = S_{AA}\, x_A + e_{Aij} \tag{3.14}$$

The *significance of influences*, namely both the interferences and several factors can be studied in two ways:

(i)   By *hypothesis testing* where the null hypotheses H_0: $e_i = 0$, and H_0: $e_j = 0$, differently formulated H_0: $e_{Ai} = e_A$, H_0: $e_{Aj} = e_A$, and H_0: $e_{Aij} = e_A$ are verified by FISHER's F-test (see Sect. 4.3.3). In this way, all the factors can only be summarily tested.

(ii) By *experimental design* each influence variable can be examined separately but in interaction to the other variables. This can be done by using a multifactorial design according to a linear model that contains all the factors of interest (see Sect. 5.1). Mostly models of the type

$$y_A^* = f(x_B, x_C, \ldots, x_N;\ a, b, \ldots, m) \tag{3.15}$$

are applied, i.e., the study is carried out at a constant analyte concentration. The values of the influence factors then are varied between sensible levels. The differences of the levels should be, on the one hand, large enough to recognize the effects and, on the other hand, not too large to avoid concealing of (mostly nonlinear) effects.

Taking into account that an experimental blank y_{A0} has to be considered, Eqs. (3.12)–(3.14) can be summarized as follows:

Gross signal		Blank		Analyte signal	Signal contributions by other species	Signal contributions by influence factors	Random deviations	

$$y_A = y_{A0} + S_{AA}x_A + \sum_{i=B}^{N} S_{Ai}x_i + \sum_{j=a}^{m} I_{Ai}x_j + e_{A'} \tag{3.16a}$$

$$y_A = y_{A0} + S_{AA}x_A \underbrace{+ e_i}\quad \underbrace{+ e_j}\quad + e_{A'} \tag{3.16b}$$

$$y_A = y_{A0} + S_{AA}x_A \underbrace{\qquad\qquad + e_{Aij}\qquad\qquad} \tag{3.16c}$$

Taking into account that Eqs. (3.16) may also contain some unknown influences u_k which reveal themselves in interlaboratory comparisons but may normally be hidden in the term $e_{A'}$, the fundamental signal model is given by

$$y_A = y_{A0} + S_{AA}\, x_A + \sum_{i=B}^{N} S_{Ai}x_i + \sum_{j=a}^{m} I_{Aj}x_j + \sum_{k} u_k + e_A \qquad (3.17)$$

The terms of the signal model at Eq. (3.17) are of particular interest for both theoretical and practical aspects of analytical chemistry, viz. in detail:

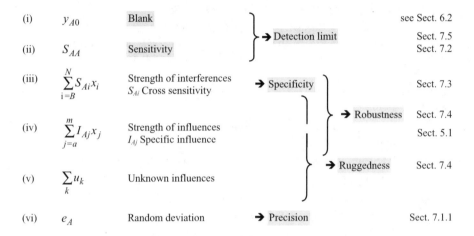

(i)	y_{A0}	Blank	→ Detection limit	see Sect. 6.2 / Sect. 7.5
(ii)	S_{AA}	Sensitivity		Sect. 7.2
(iii)	$\sum_{i=B}^{N} S_{Ai}x_i$	Strength of interferences S_{Ai} Cross sensitivity	→ Specificity	Sect. 7.3
			→ Robustness	Sect. 7.4
(iv)	$\sum_{j=a}^{m} I_{Aj}x_j$	Strength of influences I_{Aj} Specific influence		Sect. 5.1
			→ Ruggedness	Sect. 7.4
(v)	$\sum_{k} u_k$	Unknown influences		
(vi)	e_A	Random deviation	→ Precision	Sect. 7.1.1

In the following chapters, these analytical quantities and parameters will be considered in more detail in the given relationships.

References

Blaffert T (1984) *Computer-assisted multicomponent spectral analysis with fuzzy data sets.* Anal Chim Acta 161:135

Booksh KS, Kowalski BR (1994) *Theory of analytical chemistry.* Anal Chem 66:782A

Bowen IS (1927) *The origin of the nebulium spectrum.* Nature 1927:473

Currie LA (1985) *The limitations of models and measurements as revealed through chemometric intercomparisons.* J Res NBS 90:409

Currie LA (1995), IUPAC, Analytical Chemistry Division, Commission on General Aspects of Analytical Chemistry. *Nomenclature in evaluation of analytical methods including detection and quantification capabilities.* Pure Appl Chem 67:1699

Danzer K (1995) *Sampling of inhomogeneous materials. Statistical models and their practical relevance.* Chem Anal [Warsaw] 40:429

Danzer K (2004) *A closer look at analytical signals.* Anal Bioanal Chem 380:376

Danzer K, Currie LA (1998), IUPAC, Analytical Chemistry Division, Commission on General Aspects of Analytical Chemistry. *Guidelines for calibration in analytical chemistry. Part 1. Fundamentals and single component calibration (IUPAC Recommendations 1998)* Pure Appl Chem 70:993

Danzer K, Küchler L (1977) *Homogenitätskriterien für festkörperanalytische Untersuchungen*.Talanta 24:561

Danzer K, Eckschlager K, Wienke D (1987) *Informationstheorie in der Analytik. I. Grundlagen und Anwendung auf Einkomponentenanalysen*. Fresenius Z Anal Chem 327:312

Danzer K, Wienke D, Wagner M (1991) *Application of chemometric principles for evaluation of emission spectrographic data*. Acta Chim Hungar 128:623

Danzer K, Hobert H, Fischbacher C, Jagemann K-U (2001) *Chemometrik – Grundlagen und Anwendungen*. Springer, Berlin, Heidelberg, New York

Danzer K, van Staden JF, Burns DT (2002), IUPAC, Analytical Chemistry Division, Commission on General Aspects of Analytical Chemistry. *Concepts and applications of the term "dimensionality" in analytical chemistry (IUPAC Technical Report)* Pure Appl Chem 74:1449

De Gregorio P, Savastano G (1972) *Spektrum des Eisens von 2206 bis 4656Å mit Analysenlinien*. Specola Vaticana

Eckschlager K, Danzer K (1994) *Information theory in analytical chemistry*. Wiley, New York, Chichester, Bribane

Frank IE, Todeschini R (1994) *The data analysis handbook*. Elsevier, Amsterdam

Grotrian W (1928) *Über den Ursprung der Nebellinien*. Naturwiss 16:177

Harrison GR (1939) *M. I. T. wave-length tables with intensities in arc, spark or discharge tube of more than 100,000 spectrum lines*. Wiley, New York

Kaiser H (1972) *Zur Definition von Selektivität, Spezifität und Empfindlichkeit von Analysenverfahren*. Fresenius Z Anal Chem 260:252

Kelly PC, Horlick G (1974) *Bayesian approach to resolution with comparisons to conventional resolution techniques*. Anal Chem 46:2130

Kozlov VV (2004) *Quantum regulations for traffic on intercontinental Telecom highways*. Humboldt Kosmos 84:31

Malissa H (1984) *Analytical chemistry today and tomorrow*. Fresenius Z Anal Chem 319:357

Mandelbrot BB (1977) *Fractals – form, chance, and determination*. Freeman, San Francisco, CA

Massart DL, Vandeginste BGM, Deming SN, Michotte Y, Kaufman L (1988) *Chemometrics – a textbook*. Elsevier, Amsterdam

Rabinowitsch E (1928) *Physikalische Methoden im chemischen Laboratorium. IV. Bedeutung der Spektroskopie für die die chemische Forschung*. Z Angew Chem 16:555

Sharaf MA, Illman DL, Kowalski BR (1986) *Chemometrics*. Wiley, New York

Williams CS (1986) *Designing digital filters*. Prentice Hall, New York

Wolf D (1999) *Signaltheorie*. Springer, Berlin Heidelberg New York

4 Statistical Evaluation of Analytical Results

Both qualitative observations and quantitative measurements cannot be reproduced with absolute reliability. By reason of inevitable deviations, measured results vary within certain intervals and observations, mostly in form of decision tests, may fail. The reliability of analytical tests depends on the sample or the process to be controlled and the amount of the analyte, as well as on the analytical method applied and on the economical expenditure available.

4.1
Reliability of Analytical Observations and Measurements

Analytical measurements are fundamentally subject of uncertainty where various types of deviations (errors) can appear and these may be influenced to varying degree. Even when instrument readings are sufficiently accurate, repeated measurements of a sample lead, in general, to measured results which deviate by varying amounts from each other and from the true value of the sample.

Before the different types of deviations are characterized in detail, some essential terms have to be considered.

In analytical chemistry, the term *error* (used in the sense of deviation) is defined by the *difference between the test result* (x_{test}) *and the true value* $(x_{true},$ i.e., the accepted reference value, see ISO 3534-1 [1993]; FLEMING et al. [1997]). The term may be related both to measured value y and analytical value x which correspond to each other according to the *sensitivity factor b* of an analytical procedure.

According to their character and magnitude, the following types of deviations can be distinguished.

Random deviations (errors) of repeated measurements manifest themselves as a distribution of the results around the mean of the sample where the variation is randomly distributed to higher and lower values. The expected mean of all the deviations within a measuring series is zero. Random deviations characterize the reliability of measurements and therefore their *precision*. They are estimated from the results of replicates. If relevant, it is distinguished in *repeatability* and *reproducibility* (see Sect. 7.1)

Systematic deviations (errors) displace the individual results of measurement *one-sided* to higher or lower values, thus leading to incorrect results. In contrast to random deviations, it is possible to avoid or eliminate systematic errors if their causes become known. The existence and magnitude of systematic deviations are characterized by the *bias*. The *bias* of a measured result is defined as a consistent difference between the measured value y_{test} and the true value y_{true}:

$$bias(y) = y_{test} - y_{true} \qquad (4.1a)$$

The absence of a bias characterizes the *accuracy* of a measured result. The same holds for the analytical value:

$$bias(x) = x_{test} - x_{true} \qquad (4.1b)$$

Outliers are random errors in principle. However, they have to be eliminated because of their disproportionate deviation, so that the mean will not be misrepresented.

Gross errors are generated by human mistakes or by instrumental or computational error sources. Depending on whether they are short- or long-term effects, they may have systematic or random character. Frequently, it is easy to perceive and to correct for them. They will not play any role in the following discussion.

The relation between systematic and random deviations as well as the character of outliers is shown in Fig. 4.1. The scattering of the measured values is manifested by the range of random deviations (*confidence interval* or *uncertainty interval*, respectively). Measurement errors outside this range are described as outliers. Systematic deviations are characterized by the relation of the true value μ and the mean $\overline{y}$ of the measurements, and, in general, can only be recognized if they are situated beyond the range of random variables on one side.

Conventionally, a measured result is said to be *correct* if the true value is situated within the confidence interval of the observed mean (μ_1, case 1 in Fig. 4.1). If the true value is located outside of the range of random deviations (μ_2, case 2 in Fig. 4.1) then the result is *incorrect*.

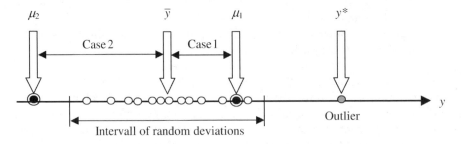

Fig. 4.1. Schematic representation of systematic and random deviations

It is not always possible to tell strictly the difference between random and systematic deviations, especially as the latter are defined by random errors. The total deviation of an analytical measurement, frequently called the "total analytical error", is, according to the law of error propagation, composed of deviations resulting from the measurement as well as from other steps of the analytical process (see Chap. 2). These uncertainties include both random and systematic deviations, as a rule.

4.1.1
Systematic Deviations

Systematic deviations, in addition to random deviations, may be produced at all steps of the analytical process, to be precise, for example during

- *Sampling* by incorrect (non-representative) treatment of individual sample fractions.
- *Sample preparation* by incomplete dissolution, separation or enrichment operations.
- *Measurement* caused by concurrent reactions or incomplete reaction processes in the case of chemical principles, and by instrumental deviations and wrong adjustment in the case of physical methods. A frequently encountered reason for the occurrence of systematic deviations is erroneous calibration due to unsuitable calibration standards, matrix effects, or insufficient methodical or theoretical foundation.
- Even *data evaluation*, often thought free of errors to a large extent, can generate systematic deviations by reason of incorrect or incomplete algorithms.

According to their influence on the measurand, one can distinguish (see Fig. 4.2):

(i) *Additive deviations* altering the measured values by a constant value α. Instead of the true value x_{true}, the falsified value

$$x_{test} = x_{true} + \alpha \tag{4.2}$$

is measured. A reason may be, for example, an unrecognized blank.

(ii) *Multiplicative deviations*, proportional to the measured value and changing the slope of the calibration curve and thus sensitivity, because, instead of the true value y the misrepresented value

$$x_{test} = \beta \cdot x_{true} \tag{4.3}$$

is measured. They are often generated by erroneous calibration factors.

(iii) *Nonlinear, response-depending errors* causing an incorrect value

$$x_{test} = (x_{true})^{\gamma} \tag{4.4}$$

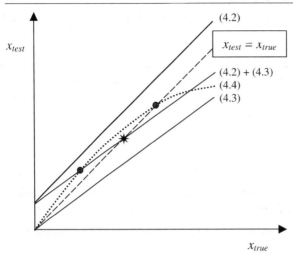

Fig. 4.2. Effects of systematic errors on test values according to Eqs. (4.2)–(4.4); the line of 45° (*broken*) corresponds to ideal recovery (line of true measurements, $x_{test} = x_{true}$)

to be measured instead of the true value. As a result, the calibration linear relationship between measured value and analytical value is lost. In OES, for example, self-absorption of resonance signals generates such an effect.

Several of the types of systematic deviations mentioned can frequently occur in combination with each other:

$$x_{test} = \alpha + \beta\,(x_{true})^{\gamma} \tag{4.5}$$

In analytical practice, they are best recognized by the determination of x_{test} as a function of the true value x_{true}, and thus, by analysis of *certified reference materials* (CRMs). If such standards are not available the use of an independent analytical method or a balancing study may provide information on systematic errors (DOERFFEL et al. [1994]; KAISER [1971]). In simple cases, it may be possible, to estimate the parameters α, β, and γ, in Eq. (4.5) by eliminating the unknown true value through appropriate variation of the weight of the test portions or standard additions to the test sample. But in the framework of quality assurance, the use of reference materials is indispensable for validation of analytical methods.

Commonly the absence of systematic errors is tested by *recovery studies*. The estimation of recovery by analyzing only one standard, as frequently be done, can give misleading results as can be seen from Fig. 4.2. In case of combined and nonlinear systematic deviations there can occur points of accidental congruence that can be hit. Therefore, if at all possible, there should be estimated the entire *recovery function*. However, at least two different standards should be analyzed.

4.1.2
Random Variations

Repeated measurements of the same measurand on a series of identical measuring samples result in random variations (random errors), even under carefully controlled constant experimental conditions. These should include the same operator, same apparatus, same laboratory, and short interval of the time between measurements. Conditions such as these are called repeatability conditions (Prichard et al. [2001]). The random variations are caused by measurement-related technical facts (e.g., noise of radiation and voltage sources), sample properties (e.g., inhomogeneities), as well as chemical or physical procedure-specific effects.

By careful proceeding of measurements random variations can be minimized, but fundamentally not eliminated. The appearance of random errors follow a natural law (often called the "GAUSS law"). Therefore, random variations may be characterized by mathematical statistics, namely, by the laws of probability and error propagation.

Classifying varying measured values by their magnitude does not, as a rule, result in a uniform distribution over the whole variation range, but gives rise to a frequency distribution around the mean value, as shown, e.g., by the bar graph in Fig. 4.3a.

Increasing the number of repeated measurements to infinity, while decreasing more and more the width of classes (bars), normally leads to a symmetrical bell-shaped distribution of the measured values, which is called *Gaussian* or *normal distribution*.

The frequency density of the measured values $p(y)$ shown in Fig. 4.3b is given by the relation

$$p(y) = \frac{1}{\sigma_y \sqrt{2\pi}} \exp\left(-\frac{(y - \mu_y)^2}{2\sigma_y^2}\right) \qquad (4.6a)$$

with the parameters μ_y being the maximum (*mean value*) and μ_y being half the distance of the inflection points of the curve (*standard deviation*). The same characteristics hold for the analytical values

$$p(x) = \frac{1}{\sigma_x \sqrt{2\pi}} \exp\left(-\frac{(x - \mu_x)^2}{2\sigma_x^2}\right) \qquad (4.6b)$$

where their parameters μ_x and σ_x are related to that of y according to the linear calibration function (see Chap. 6).

Using standardized values with $z = (x - \mu_x)/\sigma_x$ or $z = (y - \mu_y)/\sigma_y$ leads to

$$p(z) = \frac{1}{\sqrt{2\pi}} \exp\left(-\frac{1}{2} z^2\right) \qquad (4.6c)$$

with a mean of zero and a standard deviation of 1, see Fig. 4.3d.

In analytical practice, not the *population* but *random samples* are studied.

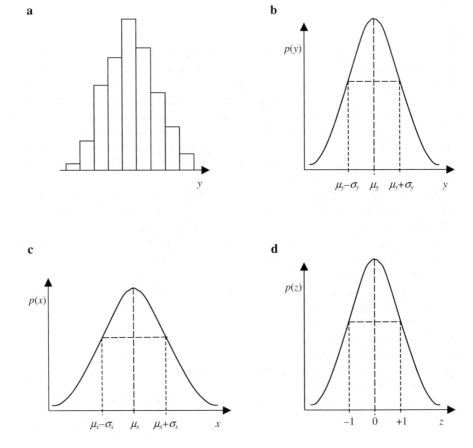

Fig. 4.3. Schematic frequency distribution of measured values y (**a**), GAUSSian normal distributions of measured values y (**b**) as well as of analytical values x(**c**), and standard normal distribution (**d**)

The term "*sample*" is used here in a statistical way and not in chemical sense as mentioned above (Sect. 2.1). A sample is a part of the population taken from it by random sampling which has to be representative. Examples from the daily life are predictions by TED (TEle Dialogue) for election results or popularity of persons or groups of people. In a technical process the population may be given by the total daily production. For the reason of quality assurance several a day test samples are taken from which conclusions on the total quality are derived. To distinguish between population and samples from the statistical point of view, different symbols are used. Greek symbols (μ, σ) characterize the population and normal letters (e.g., $\bar{x}$, s) the sample. The distributions of samples differ from the normal distribution and are characterized instead of this by the so-called t-distribution (STUDENT [1908]). Figure 4.4 shows the relation between the normal distribution of a given population and some possible sample distributions.

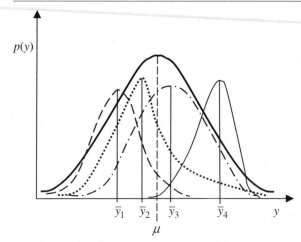

Fig. 4.4. Distribution of the population (parents distribution) and the distributions of four samples taken from this population

The parameters of the normal distribution (the *population parameters*) are calculated as follows, namely the mean

$$\mu_y = \frac{1}{N} \sum_{i=1}^{N} y_i \tag{4.7}$$

and the variance

$$\sigma_y^2 = \sum_{i=1}^{N} \frac{(y_i - \mu_y)^2}{N} \tag{4.8}$$

where N is the total number of objects of the population. On the other hand, the *sample parameters* which represent the estimates of μ and σ^2, respectively, are calculated by

$$\bar{y} = \frac{1}{n} \sum_{i=1}^{n} y_i \tag{4.9}$$

and

$$s_y^2 = \sum_{i=1}^{n} \frac{(y_i - \bar{y})^2}{n - 1} \tag{4.10}$$

where n is the number of objects (sub-samples) of the respective sampling procedure. The variances σ^2 and s^2, respectively, are additive quantities. Therefore, the variance of a mean $\bar{y}$ is given by

$$s_{\bar{y}}^2 = \frac{s_y^2}{n} \tag{4.11}$$

From the variance some measures of dispersion are derived. The most commonly used are

- *Standard deviation* as the square root of the variance

$$s_y = \sqrt{\frac{\sum_{i=1}^{n}(y_i - \overline{y})^2}{n - 1}} \tag{4.12}$$

The standard deviation can also be obtained from repetition measurements of different samples with various contents:

$$s_y = \sqrt{\frac{\sum_{i=1}^{n_A}\sum_{j=1}^{m}(y_{ij} - \overline{y}_i)^2}{n - m}} \tag{4.13}$$

where m is the number of samples (series), n_A the number of repetitions for each of the m samples and with it $n = m \cdot n_A$ the total number of determinations (y_{ij} single measured values, $\overline{y}_i$ sample means (means of series), the number of degrees of freedom is $df = n - m$ in this case). For this proceeding, the standard deviation of the samples should not differ significantly within the actual range of content. In the case of double-measurements of m samples it holds that

$$s_y = \sqrt{\frac{\sum_{j=1}^{m}(y_j' - y_j'')^2}{2m}} \tag{4.14}$$

where y_j' and y_j'' are connected values of double-measurements. The number of degrees of freedom is $df = m$.

- *Relative standard deviation* $s_{y_{rel}}$, s_{rel}, or $rsd(y)$

$$s_{y_{rel}} = \frac{s_y}{\overline{y}} \tag{4.15}$$

as a measure that is independent of the magnitude of the observation (the term "*coefficient of variation*" instead of relative standard deviation is *not* recommended by IUPAC, see IUPAC ORANGE BOOK [1997, 2000])

- *Confidence intervals* – measured values, following a GAUSSian distribution, see Eq. (4.6a), may occur principally in the whole range of definition, $-\infty < y < +\infty$, even though great positive and negative deviations of μ show only very small probability $p(y)$, see Fig. 4.3b. Therefore, it is useful to define dispersion ranges including a certain number of measured values with a given high level of significance P (and therefore a slight risk of error, $\alpha = 1 - P$). The statistical reliability is determined by the integration limits $\pm u(P) \cdot \sigma$. The integration intervals given in Fig. 4.5 correspond to the levels of significance listed in Table 4.1.

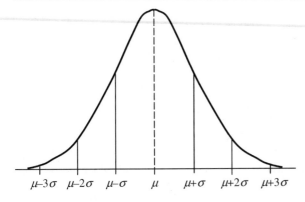

Fig. 4.5. Integration intervals of the Gaussian distribution

Table 4.1. Corresponding integration intervals and levels of significance

$\mu \pm \sigma$:	$P = 0.683$	$\mu \pm 1.96\sigma$:	$P = 0.95$
$\mu \pm 2\sigma$:	$P = 0.955$	$\mu \pm 2.58\sigma$:	$P = 0.99$
$\mu \pm 3\sigma$:	$P = 0.907$	$\mu \pm 3.29\sigma$:	$P = 0.999$

In the case of finite sample size in analytical practice, the quantiles of STU-DENT's t-distribution are used as realistic limits.

The *distance* between a mean $\overline{y}$ and its *confidence limits* is calculated according to

$$\Delta\overline{y} = s_{\overline{y}} \, t_{1-\alpha,\nu} = \frac{s_y \, t_{1-\alpha,\nu}}{\sqrt{n}} \tag{4.16}$$

with s_y being the standard deviation of a measurement series of n individual values from which $\overline{y}$ has been determined; ν is the number of degrees of freedom. The level of significance $P = 1 - \alpha$ has explicitly to be fixed.

The *total (two-sided) confidence interval* includes the mean and is given in the form

$$cnf(\overline{y}) = \overline{y} \pm \Delta\overline{y} \tag{4.17}$$

In general, this interval includes $P \cdot 100\%$ of all measured values, i.e., in the case of $n = 20$ individual measurements, one value outside the confidence interval corresponds to the statistical expectation for $P = 1 - \alpha = 0.95$.

One-sided confidence intervals $cnf(\overline{y}) = \overline{y} + \Delta\overline{y}$ and $cnf(\overline{y}) = \overline{y} - \Delta\overline{y}$, respectively, are of importance for the control of limiting values and for test statistics.

Of interest for analytical chemistry are at least two further distributions, the *logarithmic normal distribution* for analytical results at the trace- and ultra-trace level, and the *Poisson distribution* for discrete results (e.g., counts of impulse summator in XRF).

In the first case it is not the measured values themselves that are normally distributed but their logarithms. Consequently, the parameters of logarithmic-normal-distributed results are estimated as *geometric mean*

$$\overline{y}_{geom} = \sqrt[n]{\prod_{i=1}^{n} y_i} = \text{antilg}\left(\frac{1}{n}\sum_{i=1}^{n} \lg y_i\right) \tag{4.18}$$

and the *variance of the logarithms*

$$s_{\lg y}^2 = \frac{\sum_{i=1}^{n}(\lg y_i - \lg \overline{y})^2}{n-1} = \frac{\sum_{i=1}^{n}(\lg y_i)^2 - \dfrac{\left(\sum_{i=1}^{n}\lg y_i\right)^2}{n}}{n-1} \tag{4.19}$$

Because this is an unwieldy quantity, for the characterization of the unsymmetrical scattering of the results the *dispersion factor (scattering factor)* v is used:

$$v = \text{antilg } s_{\lg y} = 10^{s_{\lg y}} \tag{4.20}$$

In case of unsymmetric distributed measured values, the *dispersion factor* v can be used to estimate a relative dispersion measure that has the character of $rsd(y)$:

$$s_{(\lg y)_{rel}} = v - 1 \tag{4.21}$$

Sometimes there exists doubt about normal and log-normal distribution and the actual character of the distribution is unknown. Then so-called *"distribution-free"* (robust) parameters can be applied. The mostly used of them is the (common) *median*[1]

$$med\{y_i\} = \begin{cases} y_{(n+1)/2} & \text{if } n \text{ is odd} \\ (y_{n/2} + y_{n/2+1})/2 & \text{if } n \text{ is even} \end{cases} \tag{4.22}$$

For this type of estimation the results of the measurements have to be arranged in numerical order. Together with the median also robust measures of dispersion, such as *median absolute deviation*, $mad\{y_i\} = med\{|y_i - med\{y_i\}|\}$, and interquantile ranges (SACHS [1992]) should be used.

In case of unsymmetric distributions both geometric mean and median are smaller than the arithmetic mean. In the same way as the distribution converges towards a normal one, geometric mean and median turn into the arithmetic mean.

The discrete POISSON distribution is only characterized by one parameter, the mean Y. The standard deviation is given by $s_Y = \sqrt{Y}$ and the relative standard deviation by $s_{Y_{rel}} = 1/\sqrt{Y}$.

In general, the total variation of a measured result y is composed of several variation components $y_1, y_2, \ldots, y_m$. In the course of the analytical process, all the steps of the analytical procedure (e.g., sampling, sample preparation, separation, and measurement) and of single operations (e.g.,

---

[1]   There exist a large number of "medians" on which an overwiew can be found, e.g., in DANZER [1989]; HUBER [1981]; HAMPEL et al. [1986]; ROUSSEEUW and LEROY [1987].

difference and comparative measurements) contribute to the total error. The combination of the individual errors is determined partly by statistics and partly by functional relationships of the form $y = f(y_1, y_2, \ldots, y_m)$. In the case of independent variables $y_1, y_2, \ldots, y_m$, the total error can be estimated according to the GAUSSIAN *law of error propagation*

$$\sigma_y^2 = \left(\frac{\partial f}{\partial y_1}\right)^2 \sigma_{y_1}^2 + \left(\frac{\partial f}{\partial y_2}\right)^2 \sigma_{y_2}^2 + \cdots + \left(\frac{\partial f}{\partial y_m}\right)^2 \sigma_{y_m}^2 \qquad (4.23)$$

whereby it follows for the simplest and most often used relations:

$$\left.\begin{array}{l} y = y_1 + y_2 \\ y = y_1 - y_2 \end{array}\right\} \quad \sigma_y^2 = \sigma_{y_1}^2 + \sigma_{y_2}^2 \qquad (4.24a)$$

$$\left.\begin{array}{l} y = y_1 \cdot y_2 \\ y = \dfrac{y_1}{y_2} \end{array}\right\} \quad \left(\frac{\sigma_y}{y}\right)^2 = \left(\frac{\sigma_{y_1}}{y_1}\right)^2 + \left(\frac{\sigma_{y_2}}{y_2}\right)^2 \qquad (4.24b)$$

as well as Eq. (4.11). In the case of difference measurements of values of equal order , e.g. for blank correction, the total error is mostly given by $\sigma_y = \sqrt{2}\sigma_{y_1}$ because of $\sigma_{y_1} \approx \sigma_{y_2}$.

In case of correlated parameters, the corresponding covariances have to be considered. For example, correlated quantities occur in regression and calibration (for the difference between them see Chap. 6), where the coefficients of the linear model $y = a + b \cdot x$ show a negative mutual dependence.

4.2
Uncertainty Concept

Traditionally, analytical chemists and physicists have treated uncertainties of measurements in slightly different ways. Whereas chemists have oriented towards classical error theory and used their statistics (KAISER [1936]; KAISER and SPECKER [1956]), physicists commonly use empirical uncertainties (from knowledge and experience) which are consequently added according to the law of error propagation. Both ways are combined in the modern uncertainty concept. *Uncertainty of measurement* is defined as "*Parameter, associated with the result of a measurement that characterizes the dispersion of the values that could reasonably be attributed to the measurand*" (ISO 3534-1 [1993]; EURACHEM [1995]).

Such a parameter may be, e.g., *standard deviation*, or a given multiple of it, or a one-sided confidence interval attributed to a fixed level of confidence. In general, uncertainty of measurement comprises many components. These uncertainty components are subdivided into

(i) Such that may be evaluated from the *statistical distribution* of the results of series of measurements and can be characterized by experimental standard deviations.

(ii) Such that can be evaluated from *assumed probability distributions* based on experience or other information (these components can also be characterized by standard deviations or corresponding parameters).

Occasionally, both these uncertainty components are denoted (i) as *"type A"*- and (ii) as *"type B"* uncertainties.

It is an important fact, that *"it is understood that the result of the measurement is the best estimate of the value of the measurand, and that all components of uncertainty, including those arising from systematic effects, such as components associated with corrections and reference standards, contribute to the dispersion"* (ISO 3534-1 [1993]). Therefore, uncertainty marks the limits within which a result is accurate, i.e. precise and true (FLEMING et al. [1996]).

Uncertainty is estimated in various steps which can be summarized as shown in Fig. 4.6.

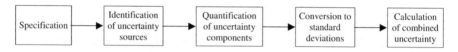

Fig. 4.6. The uncertainty estimation process, according to EURACHEM [1995]; FLEMING et al. [1996]

At first, a clear statement should be made of the measured value y and which relationship exists between y and the parameters $p_1, p_2, \ldots, p_m$ on which it depends. If possible, that should be done in form of a mathematical equation, $y = f(p_1, p_2, \ldots, p_m)$. From this the sources of uncertainty for each part of the process should be derived and listed. Some of the parameters on their part depend from other variables p_{ij}. Also these dependencies have to be considered in form of equations or schemes, where pictograms, spreadsheets, and cause-effect diagrams (as schematically shown in Fig. 4.7) may be applied as useful tools.

The quantification should start with a rough estimation of the order of magnitude of each uncertainty contribution p_i and p_{ij}. Insignificant one can be neglected because the uncertainty components are added according to a squared model. The significant values should be refined in subsequent stages and converted to parameters $u(p_i)$ which correspond to standard deviations.

In the case that the parameters p_i are independent from each other, the combined uncertainty is given by

$$u\left(y(p_1, p_2, \ldots)\right) =$$
$$\sqrt{\left(\frac{\partial y}{\partial p_1}\right)^2 \cdot \left(u(p_1)\right)^2 + \left(\frac{\partial y}{\partial p_2}\right)^2 \cdot \left(u(p_2)\right)^2 + \cdots} \qquad (4.25)$$

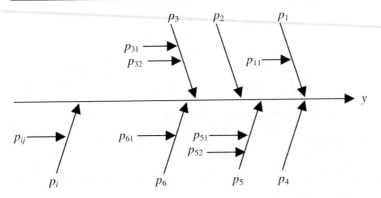

Fig. 4.7. Diagram of primary and secondary effects of parameters p_i and p_{ij} on the measured value y

If parameters are not independent but correlated with each other, then the relationship is more complex because the covariance has to be considered:

$$u\left(y(p_1, p_2, \cdots, p_m)\right) =$$

$$\sqrt{\left(\frac{\partial y}{\partial p_1}\right)^2 \cdot (u(p_1))^2 + \left(\frac{\partial y}{\partial p_2}\right)^2 \cdot (u(p_2))^2 + \cdots \sum_{i,j=1}^{m} \left(\frac{\partial y}{\partial p_i}\right) \cdot \left(\frac{\partial y}{\partial p_j}\right) \cdot s(p_i, p_j)} \tag{4.26}$$

The terms $s(p_i, p_j)$ are the covariances statistically defined by

$$s(p_i, p_j) = s(p_i) \cdot s(p_j) \cdot r(p_i, p_j) \tag{4.27a}$$

and in connection with the combined uncertainty by

$$s(p_i, p_j) = u(p_i, p_j) = u(p_i) \cdot u(p_j) \cdot r(p_i, p_j) \tag{4.27b}$$

According to Eq.(4.27a,b), the covariance is composed of the product of the standard deviations or uncertainties, respectively, of the parameters p_i and p_j and the *correlation coefficient* between them:

$$r(p_i, p_j) = \frac{\sum\limits_{k=1}^{n} (p_{i_k} - \overline{P}_i)(p_{j_k} - \overline{P}_j)}{\sqrt{\sum\limits_{k=1}^{n} (p_{i_k} - \overline{P}_i)^2 \sum\limits_{k=1}^{n} (p_{j_k} - \overline{P}_j)^2}} \tag{4.28}$$

which is defined in the range $-1 \leq r(p_i, p_j) \leq +1$. If values of $r(p_i, p_j)$ approximate to $+1$ or -1, then a strong correlation (positive or negative, respectively) is indicated. On the other hand, an approximation to 0 indicates missing correlation.

From the combined uncertainty the extended uncertainty is calculated. The *extended combined uncertainty* $U(y)$ represents an interval that contains a (sufficiently) large part of the values of a series of measurements that could reasonably be attributed to the measurand

$$U(y) = k \cdot u(y) \tag{4.29}$$

in which $u(y) = u\left(y(p_1, p_2, \cdots, p_m)\right)$ as calculated according to Eqs. (4.25) or (4.26), respectively. The *coverage factor* usually is chosen $k = 2 \ldots 3$. Table 4.1 shows that in case of normally distributed measured values, $k = 2$ covers an interval in which 95.5% of the values are found ($k = 3$ correspondingly 99.7%). If the type of distribution is unknown, $k = 2$ does not guarantee a sufficient level of confidence.

The *uncertainty interval* includes the estimated mean $\overline{y}$ and is given by

$$unc(\overline{y}) = \overline{y} \pm U(\overline{y}) \tag{4.30}$$

In general, the uncertainties of measured values $u(y)$ are converted into that of analytical values $u(x)$ by means of the sensitivity S and its uncertainty $u(S)$:

$$u(x) = \frac{u(y, S)}{S} \quad . \tag{4.31}$$

Therefore, the *uncertainty interval* of $\overline{x}$ are given by

$$unc(\overline{x}) = \overline{x} \pm U(\overline{x}) = \frac{unc(\overline{y}, S)}{S} = \frac{\overline{y} \pm U(\overline{y}, S)}{S} \tag{4.32}$$

The significance of the uncertainty concept in analytical chemistry has increased in the last century, notwithstanding that at first some conformity was missed. But inconsistencies have been dispelled (see THOMPSON [1995]; AMC [1995]) and operational approaches have been presented by HUND et al. [2001]. Numerous examples of application have been given in EURACHEM [1995].

The uncertainty concept is composed of both chemists' and physicists' approaches of handling of random deviations and substitutes so classical error theories in an advantageous way.

4.3
Statistical Tests

Statistical tests make it possible, objectively to compare and interpret experimental data. They base on a test statistic to verify a statistical hypothesis about a:

(i)   Parameter
(ii) Distribution
(iii) Goodness of fit

Generalizations which go beyond the given data are normally not possible. Statistical tests can be carried out with measured values, y, and analytical values, x, respectively, if there exists a linear relationship between them. If this not apply, it depends on the problem, which of the measures has to be tested[2], y or x.

4.3.1
Null Hypotheses

Statistical tests are based on hypotheses, so-called *null hypotheses* H_0, the statements of which are verified by statistical tests. In this context, the following rules apply:

(i)   Null hypotheses must always be formulated in an *affirmative* way, i.e., $H_0 : \mu_1 = \mu_2$ (two subsets with the means $\overline{x}_1$ and $\overline{x}_2$ belong to the same population $\mu_1 = \mu_2$), or $H_0 : \sigma_1^2 = \sigma_2^2$ (the variances of two subsets are identical).

(ii) Every null hypothesis H_0 has an *alternative hypothesis* H_A which is confirmed if the null hypothesis is rejected, i.e., if the test leads to a negative result, e.g., $H_A : \mu_1 \neq \mu_2$ (the compared means $\overline{x}_1$ and $\overline{x}_2$ differ significantly, and, thus, belong to different populations).

(iii) A non-rejection of a null hypothesis does *not* mean *its acceptance*. If a test does not result in a significant difference of two compared parameters, this merely means, that for reasons of the existing data, the difference are not conclusive. Evidence for correspondence is not provided. Such evidence can only be confirmed indirectly, see e.g., DOERFFEL [1990]; DOERFFEL et al. [1990].

(iv) Each test result is only valid for a *certain* (freely chosen) level of significance P, underlying the test procedure. Thus, a test result carries a *risk* of $\alpha = 1 - P$. In general, a level of significance $P = 0.95$ is chosen (corresponding to $\alpha = 0.05$). In cases where great significance or consequence is attached to the test result, a higher level of significance ($P = 0.99$) has to be chosen. The following decisions have to be made:

 – H_0 for $P = 0.95$ not rejected: the difference is considered to be non conclusive.

 – H_0 for $P = 0.95$ rejected: the difference is regarded as guaranteed in the normal case.

 – H_0 for $P = 0.99$ rejected: the difference is highly significant.

(v)   Every statistical test may possibly result in two different kinds of error, as shown in Table 4.2, i.e.,

 –   To reject the null hypothesis erroneously although it is true (*error of first kind, false-negative, risk α*).

 –   Not to reject the null hypothesis by erroneously though the alternative hypothesis is true (*error of second kind, false-positive, risk β*).

---

[2]   Therefore, both quantities will be used in the following to formulate the tests

Table 4.2. Types of errors for statistical tests of null hypotheses

H_0 is by the test \ H_0 is	True	False
Not rejected	Test result OK	Error of second kind "consumer risk" "false alarm"
Rejected	Error of first kind "producer risk" "false alarm"	Test result OK

Means and standard deviations calculated according to Eqs. (4.9) and (4.12) or the following, respectively, are only characteristic for sample subsets if certain pre-requisites are satisfied, the fulfilment of these can be verified by tests. The observed values of measurement series have to

(i)   Be *normally distributed*
(ii) *Vary randomly* and show *no systematic trend*
(iii) Be *free of outliers*

This, on the one hand, is in order to determine correct estimates of the mean and the variation of the measurement series, and, on the other hand, to allow the statistical comparison of these parameters to those of other measurement series.

4.3.2
Test for Measurement Series

(1) *Rapid test for normal distribution: Range test of* DAVID *et al.* [1954]

$$\hat{q}_R = \frac{R}{s} \tag{4.33}$$

with $R = y_{\max} - y_{\min}$ being the range and s being the standard deviation. If the calculated test value $\hat{q}_R$ is not within the limits of the tabulated values (which can be found, e.g., by SACHS [1992]; GRAF et al. [1987]), a normal distribution of the measured values may not be assumed. Then, it can be tested if a normal distribution is obtained after transformation of the measured values, e.g., a logarithmic one. If there is still no success, the measurement series should be evaluated by methods of *robust statistics* (see e.g. DANZER [1989]).

The existence of a normal distribution can only be confirmed by a goodness-of-fit test (e.g., χ^2, according to KOLMOGOROFF [1941] and SMIRNOFF [1948]).

(2) *Test for a trend*

For a measurement series $y_1, y_2, \ldots, y_n$ (in the sequence of the measurement, not classified) the test parameter

$$\Delta^2 = \sum_{i=2}^{n} \frac{(y_{i-1} - y_i)^2}{n - 1} \tag{4.34}$$

has to be formed (von Neumann et al. [1941]). The measured values can be regarded as independent and, therefore, as varying at random for $\Delta^2 \approx 2s^2$. A trend has to be considered if $\Delta^2 < 2s^2$. For an exact test Δ^2/s^2 has to be calculated and compared to the limits given e.g. in Sachs [1992].

(3) *Test for outliers*

Among numerous tests for outliers presented in the literature, in analytical practice the following have turned out to be especially useful:

a) For measurement series of limited size ($n \leq 25$): Test for outliers according to Dean and Dixon [1951]; Dixon [1953]:

   The measured values are sorted in ascending or descending order, depending on whether the suspected outlier value y_1^* deviates to higher or lower values. The test statistic is formed depending on the data set

$$\hat{Q} = \frac{|y_1^* - y_b|}{|y_1^* - y_k|} \tag{4.35}$$

   where indices b and k may have the following values depending on the measurement series (n is the number of measured values):

   $b = 2$ for $3 \leq n \leq 10$, $b = 3$ for $11 \leq n \leq 25$
   $k = n$ for $3 \leq n \leq 7$, $k = n - 1$ for $8 \leq n \leq 13$,
   $k = n - 2$ for $14 \leq n \leq 25$

   The comparison is done with the significance limits $Q(n, \alpha)$, given, e.g. in Sachs [1992]; Graf et al. [1987].

b) For nearly any measurement series ($3 \leq n \leq 150$): Test for outliers according to Grubbs [1969]:

$$\hat{G} = \frac{|y^* - \bar{y}|}{s} \tag{4.36}$$

   The test statistic is compared with the significance limits $G(n, \alpha)$, given e.g. in Graf et al. [1987].

4.3.3
Comparison of Standard Deviations

Two standard deviations, s_1 and s_2, with corresponding degrees of freedom, ν_1 and ν_2, are compared by means of the F-test. The test statistic

$$\hat{F} = \frac{s_1^2}{s_2^2} \tag{4.37}$$

normally with $s_1 > s_2$, is compared to the corresponding quantile of the F-distribution; see DOERFFEL [1990] SACHS [1992]; GRAF et al. [1987]. For $\hat{F} > F_{1-\alpha,\nu_1,\nu_2}$ it is proven that s_1 is significantly larger than s_2.

Comparison of Several Standard Deviations:

(1) For measurement series of equal size ($n_1 = n_2 = n_3 = \ldots$): *Hartley test* (F_{max}-*test*, HARTLEY [1950]):

$$\hat{F}_{max} = \frac{s_{max}^2}{s_{min}^2} \tag{4.37'}$$

Tables with the significance limits are given in SACHS [1992]; GRAF et al. [1987].

(2) COCHRAN *test* (COCHRAN [1941]):

$$\hat{G}_{max} = \frac{s_{max}^2}{s_1^2 + s_2^2 + \cdots + s_k^2} \tag{4.38}$$

preferentially applied to compare k samples if one of the variances is essentially larger than all others. Tables are given in SACHS [1992]; GRAF et al. [1987].

(3) BARTLETT *test* (BARTLETT [1937a, b]):

$$\hat{\chi}^2 = \frac{2.3026}{c} \left(\nu \lg s^2 - \sum_{i=1}^{k} \nu_i \lg s_i^2 \right) \tag{4.39}$$

with $\nu = n - k = \sum \nu_i$ being the total number of degrees of freedom (k number of groups), $s^2 = \sum \nu_i s_i^2 / \nu$ being the weighted variance, s_i^2 being the variances of the i-th group with degrees of freedom ν_i and

$$c = 1 + \frac{\sum\limits_{i=1}^{k} \frac{1}{\nu_i} - \frac{1}{\nu}}{3(k - 1)} \tag{4.40}$$

which has the character of a correction factor; if the number of degrees of freedom ν_i is not too small, c approximates 1. The test statistic χ^2 will be compared to the corresponding quantile of the χ^2-distribution $\chi_{\alpha,\nu}^2$; see DOERFFEL [1990]; SACHS [1992]; GRAF et al. [1987].

4.3.4
Comparison of Measured Values

(1) Comparison of two means $\overline{y}_1$ and $\overline{y}_2$ of different measurement series with n_1 resp. n_2 measurements having the standard deviations s_1 resp. s_2 is done by the *t-test*:

$$\hat{t} = \frac{|\bar{y}_1 - \bar{y}_2|}{s_{av}} \cdot \sqrt{\frac{n_1 \, n_2}{n_1 + n_2}} \tag{4.41}$$

with the (weighted) average standard deviation

$$s_{av} = \sqrt{\frac{(n_1 - 1)s_1^2 + (n_2 - 1)s_2^2}{n_1 + n_2 - 2}} \tag{4.42}$$

The t-test in this form can only be applied under the condition that the variances of the two sample subsets, s_1^2 and s_2^2, do not differ significantly. This has to be checked by the F-test beforehand. The test statistic $\hat{t}$ has to be compared to the related quantile of the t-distribution $t_{1-\alpha,v}$ where $v = n_1 + n_2 - 2$.

If the variances s_1^2 and s_2^2, differ significantly, a generalized t-test (t_W−test) according to WELCH [1937] can be applied:

$$t_W = \frac{|\bar{y}_1 - \bar{y}_2|}{\sqrt{\frac{s_1^2}{n_1} + \frac{s_2^2}{n_2}}} \tag{4.43}$$

The comparison is done again with $t_{1-\alpha,v}$, but in this case it is

$$v = \frac{\left(\frac{s_1^2}{n_1} + \frac{s_2^2}{n_2} \right)^2}{\frac{\left(\frac{s_1^2}{n_1} \right)^2}{n_1 - 1} + \frac{\left(\frac{s_2^2}{n_2} \right)^2}{n_2 - 1}} \tag{4.44}$$

If more than two means have to be compared, the t-test cannot be applied in a multiple way. Instead of this, an indirect comparison by analysis of variance (ANOVA) has to be used, see (3) below.

(2) Comparison of an experimental mean with a prescribed resp. true value: *simplified t-test:*

The decision, if a mean of a sample, $\bar{y}$, differs randomly or significantly from a prescribed mean, μ, leads to the question, if the confidence interval of the experimental mean, $cnf(\bar{y})$, includes the mean μ or not, i.e., is the absolute difference $|\bar{y} - \mu|$ smaller or larger than the interval: $\Delta\bar{y} = s \cdot t_{1-\alpha,v}/\sqrt{n}$:

$$\hat{t} = \frac{|\bar{y} - \mu|}{s} \sqrt{n} \tag{4.45}$$

(3)   Comparison of several means: *analysis of variance (ANOVA):*

Because no direct way exists to test simultaneously more than two means for significant differences, the comparison of several means, $\overline{y}_1, \overline{y}_2, \ldots, \overline{y}_k$, is traced back to the comparison of variances.

One-way analysis of variance is based on a linear model like the following:

$$y_i = \overline{y} + a_i + e_i \tag{4.46}$$

where three different models have to be distinguished (EISENHART [1947]):

(i)   Model I with fixed effects (*model "fixed"*) which is suitable for comparisons of means on the basis of the null hypothesis H_0: $\mu_1 = \mu_2 = \ldots = \mu_k$

(ii) Model II with random effects (*model "random"*) that bases on the null hypothesis H_0: $\sigma_a^2 = 0$ (*"proper" analysis of variance*)

(iii) Model *"mixed"*

Regarding the comparison of means, a_i and e_i can be interpreted as to correspond to the deviations between the measurement series and within them, $\sigma_{between}^2$ and σ_{within}^2, respectively. Therefore, the comparison is carried out according to

$$\hat{F} = \frac{\sigma_{between}^2}{\sigma_{within}^2} = \frac{\frac{1}{k-1} \sum_{i=1}^{k} n_i \left(\overline{y}_i - \overline{\overline{y}}\right)^2}{\frac{1}{n-k} \sum_{i=1}^{k} s_i^2 (n-1)} \tag{4.47}$$

with k number of means to be compared, $\overline{y}_i$ mean of the i-th measurement series with n_i individual values, s_i standard deviation of the i-th measurement series, $n = \sum n_i$ total number of all individual measurements, $\overline{\overline{y}} = \frac{1}{n} \sum n_i \overline{y}_i$ weighted total mean. The degrees of freedom are ν_1 and ν_2.

$\hat{F}$ exceeds the corresponding quantile of the F-distribution $F_{1-\alpha,\nu_1,\nu_2}$ if at least one of the means differs significantly from the others. This global statement of variance analysis may be specified in the way to detect which of the mean(s) differ(s) from the others. This can be done by pairwise multiple comparisons (TUKEY [1949]: GAMES and HOWELL [1976]: see SACHS [1992]).

In the simplest case, if all sample sub groups have the same size, $n_1 = n_2 = \ldots = n_k$, DIXON's test for outliers can be used (see Sect. 4.3.2). Then, in Eq. (4.35), instead of the individual values, the means are entered.

4.4
Reliability of Qualitative Analytical Tests

In principle, there is no fundamental contrast between qualitative, (semi-quantitative), and quantitative analyses. The analytical signal is generated in the same way, only the detection and evaluation is done on the basis of a more rough scale, in qualitative analysis only in form of a *yes/no* decision.

Today an increasing importance of qualitative analysis can be stated in certain fields. This is due to an increasing number of materials under study, especially active agents of interest on the one hand, and the many and diverse ways of synthesis (e.g., by combinatorial chemistry) on the other hand as well as the increasing demands on quality. Because analytical laboratories in research and routine control would be overtaxed in their capacity if full quantitative analyses were done generally, screening methods become more and more significant.

Screening techniques can be understood as to be *filtering procedures* of samples. The principle of screening consists in giving an overview on constituents in certain samples, namely:

(i) Whether a threshold value (specification limit) is exceeded or not

(ii) Whether constituents are present in addition to such that are well-known, in general, to specify the type of the sample (e.g., identification of the type of steel by recognizing typical alloying constituents additionally to that always present like C, S, P, Si, and Mn)

Although the term "screening methods" is often used as a synonym for qualitative analysis, it is mostly related to (i) the scheme of which is given in Fig. 4.8; see also TRULLOLS et al. [2004].

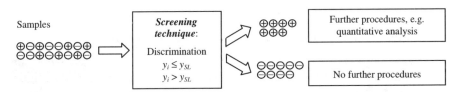

Fig. 4.8. Principle of screening techniques. $\oplus$ samples for which $y_i > y_{SL}$, $\ominus$ samples for which $y_i \leq y_{SL}$; y_{SL} is the specification limit

Screening methods may be classified according to the detection of signals into:

(a) *Sensorial detection* (mostly visibility) traced back to changes in absorption or transmission (e.g., color of solution or test strips, turbidity).

(b) *Instrumental detection*, where in principle any instrumental technique may be used. Screening by instrumental analysis can be done by means of reduced calibration expense where only a comparison with a reference sample containing the analyte at the specification level, x_{SL}, giving so a response y_{SL}, is carried out.

According to the practical equipment there are useful tools, so-called *test kits*, which are units that contain all the reagents and a simple instrumentation in form of plates, tubes and wells. The test kits work rapidly, are easily to handle and field-portable. Frequently, biochemical principles are applied, especially *immunoassay techniques* which use body-antibody reactions.

The discrimination limit in screening may have different character. It may be a detection limit or a pre-set value set by a responsible authority, e.g., an official agency, medical institutions, or clients.

Although screening tests are evaluated qualitatively, as a rule, quantitative aspects of test statistics and probability theory have to be considered. In this respect, validation of qualitative analytical procedures has been included in international programs and concepts, see TRULLOLS et al. [2004].

The application of screening methods requires that proper reference samples are available. These must contain the analyte exactly at or close to the specification level. Such reference samples have to be verified by quantitative analysis using reference methods.

The reliability of screening methods is usually expressed in terms of probability theory. In this regard, the *conditional probability*, $P(B|A)$, characterizing the probability of an event B given that another event A occurs, plays an important role.

(1) *True positive rate*[3]: probability that the test result is positive in case that the analyte A is present (above the specification limit) in the test sample

$$TPR = P(T^+ | A) = \frac{tp}{tp + fn} \tag{4.48}$$

where tp are the true positives and fn the false negatives.

(2) *True negative rate*: probability that the test result is negative when the analyte is not present, $\overline{A}$ (i.e. equal or below the specification limit, respectively)

$$TNR = P(T^- | \overline{A}) = \frac{tn}{tn + fp} \tag{4.49}$$

where tn are the true negatives and fp the false positives.

---

[3]   In clinical chemistry and medical diagnostics the true positive rate is called "*sensitivity rate*" and the true negative rate "*specificity rate*" (O'RANGERS and CONDON [2000]) without any relation to the general definition of the terms *sensitivity* and *specificity* and their use in analytical chemistry (see Sects. 7.2 and 7.3).

(3) *False positive rate*: probability that the test result is positive when the
     analyte is not present, $\overline{A}$

$$FPR = P(T^+\,|\overline{A}) = \frac{fp}{tn + fp} \tag{4.50}$$

(4) *False negative rate*: probability that the test result is negative when the
     analyte is present

$$FNR = P(T^-\,|A) = \frac{fn}{tp + fn} \tag{4.51}$$

Some other important quantities in test theory are:

(5) *Prevalency*: probability that A is present in n samples:

$$PV = P(A) = \frac{tp + fn}{n} \tag{4.52}$$

where $n = tp + tn + fp + fn$ is the total number of samples (tests).

(6) *Concordancy rate*: proportion of correct test results:

$$CR = \frac{tp + tn}{n} \tag{4.53a}$$

where $CR \leq 1$.

The correctness of test is also expressed by

(7) Test efficiency:

$$TE = P(A\,|T^+) + P(\overline{A}\,|T^-) \leq 2 \tag{4.53b}$$

The problem of test efficiency and concordance may be illustrated by a
scheme (Table 4.3) which corresponds to that given in Table 4.2.

Table 4.3. Scheme of test results for screening procedures; *tp* true positive, *fp* false positive, *tn* true negative, *fn* false negative, *n* total number of tests

Test result / Situation	Positive (T^+)	Negative (T^-)	Sum of situations
A is present (A)	tp	fn / Error of second kind	tp + fn
A is not present ($\overline{A}$)	fp / Error of first kind	tn	tn + fp
Sum of test results	tp + fp	tn + fn	n = tp + tn + fp + fn

From the decision variants and probabilities given above, the *prediction values* as applied in Eq. (4.53b) play an important role in decision theory, namely:

- *Positive prediction rate*: proportion of samples which are positively tested when the analyte A is present in the sample

$$PPR = P(A\,|T^+) = \frac{P(A) \cdot P(T^+\,|A)}{P(A) \cdot P(T^+\,|A) + [1 - P(A)] \cdot [1 - P(T^-\,|\overline{A})]}$$
(4.54)

$$= \frac{P(A) \cdot P(T^+\,|A)}{P(A) \cdot P(T^+\,|A) + P(\overline{A}) \cdot P(T^+\,|\overline{A})}$$

and

- *Negative prediction rate*: proportion of samples which are negatively testes if the analyte is not present $(\overline{A})$ in the sample

$$NPR = P(\overline{A}\,|T^-) = \frac{P(\overline{A}) \cdot P(T^-\,|\overline{A})}{P(\overline{A}) \cdot P(T^-\,|\overline{A}) + P(A) \cdot [1 - P(T^+\,|A)]}$$
(4.55)

$$= \frac{P(\overline{A}) \cdot P(T^-\,|\overline{A})}{P(\overline{A}) \cdot P(T^-\,|\overline{A}) + P(A) \cdot P(T^-\,|A)}$$

False positive and false negative decisions result from unreliabilities that can be attributed to uncertainties of quantitative tests. According to Fig. 4.9 the belonging of test values to the distributions $p(y_{LSP})$ or $p(y_{SCR})$, respectively, may be affected by the risks of error α and β (see Sect. 4.3.1) which corresponds to false positive (α) and false negative (β) test results.

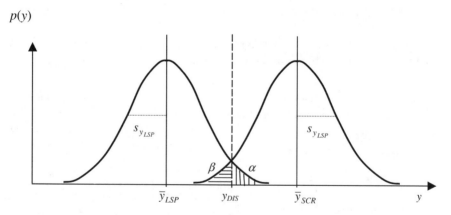

Fig. 4.9. Distributions of measured values which belong to the content at the limit of specification x_{LSP} and at the screening limit x_{SCR}, y_{DIS} is the limit of discrimination

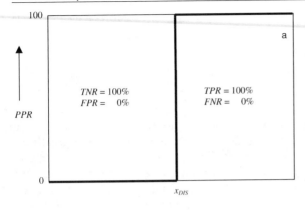

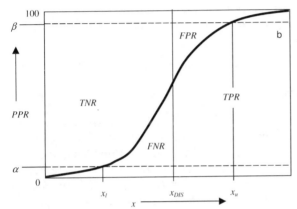

Fig. 4.10. Performance characteristic curves of screening tests of binary classifications:
a ideal curve shape,
b experimentally obtained curve
following SIMONET et al. [2004];
PPR is the probability of obtaining positive responses, TPR true positive rate (Eq. 4.48), TNR true negative rate (Eq. 4.49), FPR false positive rate (Eq. 4.50), FNR false negative rate (Eq. 4.51), x_l and x_u are the lower and upper limits of the unreliability region

Another way to characterize the performance of screening methods for binary classification is the construction of performance curves. A *performance characteristic curve (PCC)* represents the probability of positive test results vs the concentration of the analyte as shown in Fig. 4.10.

The performance curve presents graphically the relationship between the probability of obtaining positive results PPR, i.e. $x_i \geq x_{LSP}$ on the one hand and the content x within a region around the limit of discrimination x_{DIS} on the other. For its construction there must be carried out a larger number of tests ($n \geq 30$) with samples of well-known content (as a rule realized by doped blank samples). As a result, curves such as shown in Fig. 4.10 will be obtained, where Fig. 4.10a shows the ideal shape that can only be imagined theoretically if "infinitely exact" decisions, corresponding to measured values characterized by an "infinitely small" confidence interval, exist.

The curve illustrates the sharpness of tests depending on the discrimination limit. In this way, TPR and TNR may be recognized and the unreliability region around the limit of specification can be estimated. Beyond the limits of the unreliability interval, it is possible to classify samples correctly apart from a defined risk of error (PULIDO et al. [2002]; SIMONET et al. [2004]; TRULLOLS et al. [2004]).

4.5
Statistical Quality Control

The principles of quality assurance are commonly related to product and process control in manufacturing. Today the field of application greatly expanded to include environmental protection and quality control within analytical chemistry itself, i.e., the quality assurance of analytical measurements. In any field, features of quality cannot be reproduced with any absolute degree of precision but only within certain limits of tolerance. These depend on the uncertainties of both the process under control and the test procedure and additionally from the expense of testing and controlling that may be economically justifiable.

4.5.1
Quality Criteria for Analytical Results

Analytical methods, particularly those used by accredited laboratories, have to be validated according to official rules and regulations to characterize objectively their reliability in any special field of application (WEGSCHEIDER [1996]; EURACHEM/WELAC [1993]). Validation has to control the performance characteristics of analytical procedures (see Chap. 7) such as *accuracy, precision, sensitivity, selectivity, specificity, robustness, ruggedness*, and *limit values* (e.g., *limit of detection, limit of quantitation*).

Within the scope of quality agreements in production, environment, or laboratory, quality is often stipulated to a *standard value* x_0 *(target value)*. For demands regarding quality, this standard value may be an upper limit (e.g. in case of pollution and contamination) or a lower limit (e.g. for active reagents). The statistical situation is the same when a quality criterion has to exceed or fall below a standard value. The problem is illustrated here by the practical situation of manufacturer and customer as shown in Fig. 4.11.

If an analytical test results in a lower value $x_i \leq x_0$, then the customer may reject the product as to be defective. Due to the variation in the results of analyses and their evaluation by means of statistical tests, however, a product of good quality may be rejected or a defective product may be approved according to the facts shown in Table 4.2 (see Sect. 4.3.1). Therefore, manufacturer and customer have to agree upon statistical limits (critical values) which minimize *false-negative* decisions (errors of the first kind which characterize the manufacturer risk) and *false-positive* decisions (errors of the second kind which represent the customer risk) as well as test expenditure. In principle, analytical precision and statistical security can be increased almost to an unlimited extent but this would be reflected by high costs for both manufacturers and customers.

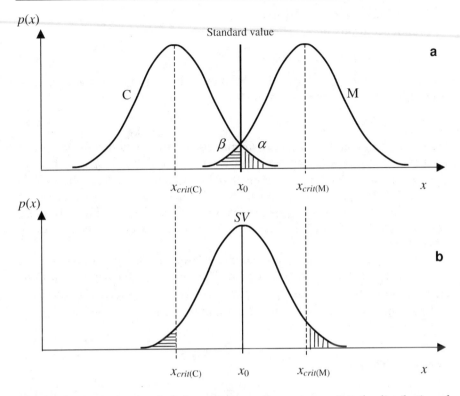

Fig. 4.11. Distributions of analytical results for quality assurance; C is the distribution of analytical results at the critical customer limit, M at the critical manufacturer limit, *SV* is the distribution of the standard value

With the aid of consequences resulting from wrong decisions, *risks R* can be given for the manufacturer M as well as for the customer C:

$$R_M = w \cdot \overline{\alpha} \cdot F_M \tag{4.56a}$$

$$R_C = (1 - w)\,\overline{\beta} \cdot F_C \tag{4.56b}$$

with $w = P(x = x_{crit(M)})$ representing the a priori probability that a determined quality value x is equal to the critical manufacturer limit and $(1 - w) = P(x = x_{crit(C)})$ that is equal to the critical customer limit. The parameters F_M and F_C characterize consequences (e.g. costs) resulting from wrong decision for the manufacturer and customer, respectively; $\overline{\alpha}$ is the (one-sided) manufacturer risk (error of the first kind) and $\overline{\beta}$ the (one-sided) customer risk (error of the second kind).

The relationship between the target value x_0 and the distribution of the critical values are illustrated in Fig. 4.11 for manufacturer (M) and customer (C). For $x > x_0$ the quality is better than agreed while $x < x_0$ indicates poorer quality.

The critical values $x_{crit(C)}$ and $x_{crit(M)}$ being *tolerance limits* are derived as confidence limits of x_0. They are given by

$$x_{crit(M)} = x_0 + s\, t_{1-\overline{\alpha},v}/\sqrt{n} \qquad (4.57a)$$

for the upper tolerance limit with the standard deviation s of the analytical procedure and n parallel determinations, as well as

$$x_{crit(C)} = x_0 - s\, t_{1-\overline{\beta},v}/\sqrt{n} \qquad (4.57b)$$

for the lower tolerance limit. If manufacturer and customer have agreed upon x_0 not to be a target value but a *guarantee value*, $x_{0,guar}$, with fixed significance level $\overline{P} = 1 - \overline{\alpha}$, then that has to supply at least the quality $x_{crit(M)}$. The statement whether an analytical value $\overline{x} = x_{0,guar}$ is still accepted ($\overline{x} \geq x_{0,guar}$) or not ($\overline{x} > x_{0,guar}$) is then also part of the agreement. In these cases, the customer's risk is that in $\overline{\alpha} \cdot 100\%$ of all cases the quality can be worse than x_0.

Quality limits are often stipulated by laws, regulations, or standards; however, they are frequently determined by mutual agreements, too. Considering the particular risks R_M resp. R_C, especially the values $\overline{\alpha}$ and $\overline{\beta}$ have to be reconciled, whereby, in practice, $\overline{\alpha} = \overline{\beta}$ is often taken as the basis.

4.5.2
Attribute Testing

In contrast to *variable testing* (comparison of measured values or analytical values), *attribute testing* means testing of product or process quality (non-conformity test, good-bad test) by samples. Important parameters are the sample size n (the number of units within the random sample) as well as the acceptance criterion n_{accept}, both of which are determined according to the lot size, N, and the proportion of defective items, p, within the lot, namely by the related distribution function or by operational characteristics.

According to the number of defective items n_{defect}, which have been determined in the sample results are required for,:

$$n_{defect} \leq n_{accept} \quad \text{acceptance of the test items}$$
$$n_{defect} > n_{accept} \quad \text{rejection} \ .$$

Attribute testing can also be effected on the basis of two or more (m) samples. The disadvantage for the higher expenditure to determine the sample sizes $n_1, n_2, \ldots, n_m$, the acceptance and rejection rates $n_{accept,1}, n_{accept,2}, \ldots, n_{accept,m}$ resp. $n_{rejet,1}, n_{reject,2}, \ldots, n_{reject,m}$ must be set against the advantage of the possibility of coming to unambiguous decisions in clearly defined situations with only a first small random sample $i = 1$.

In contrast to classical statistical tests, three test outcomes exist:

(i) $n_{defect,i} \leq n_{accept,i}$ \qquad acceptance

(ii) $n_{defect,i} > n_{accept,i}$ \qquad rejection

(iii) $n_{accept,i} < n_{defect,i} < n_{reject,i}$ inspection of another sample

This model leads to sequential tests, which generally use three outcomes and which represent the most effective variant of quality control.

4.5.3
Sequential Analysis

The principle of sequential analysis consists of the fact that, when comparing two different populations A and B with pre-set probabilities of risks of error, α and β, just as many items (individual samples) are examined as necessary for decision making. Thus the sample size n itself becomes a random variable.

Sequential investigations may be used for both attribute testing and quantitative measurements (variable testing). The fact that it is enough to perform only as many tests or measurements as are absolutely necessary, is of great advantage in such cases if individual samples are either difficult to obtain or expensive, or if the same is true for the measurement.

Based on the actual result of every individual measurement, it has to be checked whether a decision may now be taken or whether studies have to be continued. According to the final aim of sequential analysis, one can differentiate between *closed sequential test plans* that always come to a decision A > B or A < B (possibly with a large test expenditure) and *open sequential test plans*, which also allow – after a certain test expenditure – the statement A = B.

Data evaluation can be done by calculation or graphically. An example of graphical sequential analysis is given in Fig. 4.12 for attribute testing.

In the case of attribute testing, limit curves for acceptance and rejection are normally straight lines. In case of variable testing, they are mostly nonlinear functions (GRAF et al. [1987]).

For variable testing, the analytical values x are represented on the ordinate axis, and a decision is taken when the acceptance or rejection curve is exceeded.

Decisions to be taken after every individual sample are as follows:

(a) For *attribute testing*:

$n_{defect,n} \leq n_{accept,n} = g_{accept}(n)$ acceptance

$n_{defect,n} \leq n_{reject,n} = g_{reject}(n)$ rejection

$g_{accept}(n) < n_{defect,n} < g_{reject}(n)$ continue testing: inspect another item

with $n_{defect,n}$ being the number of defective items within n tested, $n_{accept,n}$ resp. $n_{reject,n}$ being the actual acceptance resp. rejection criterion for the current sample size n; $g_{accept}(n)$ and $g_{reject}(n)$ being the acceptance resp. rejection functions.

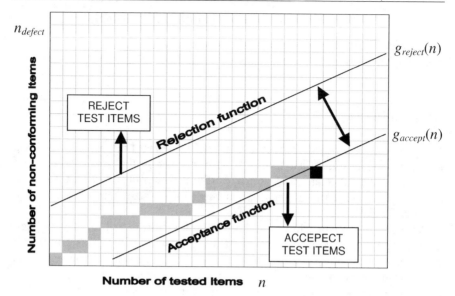

Fig. 4.12. Graphical evaluation of sequential analysis; $g_{accept}(n)$ acceptance function, $g_{reject}(n)$ rejection function

(b)   For *variable testing* (quantitative measurements):

$x_n \leq g_{accept}(n, s, \overline{\alpha})$ acceptance

$x_n \geq g_{reject}(n, s, \overline{\beta})$ rejection

$g_{accept}(n, s, \overline{\alpha}) < x_n < g_{reject}(n, s, \overline{\beta})$ continue testing:
                                                 inspect another item

with x_n being the actual analytical value after n measurements, $g_{accept}(n, s, \overline{\alpha})$ and $g_{reject}(n, s, \overline{\beta})$ being the acceptance and rejection functions (s standard deviation of the analytical procedure, and $\overline{\alpha}$ and $\overline{\beta}$ manufacturer and customer risk, respectively).

Sequential test plans are also suitable for the direct comparison of two products or procedures based on subjective parameters as well as on measured results. This can be done without any computation, but only according to the following three criteria:

– A is better than B
– B is better than A
– There is no difference between A and B

A graphical evaluation scheme of such comparisons is given in Fig. 4.13.

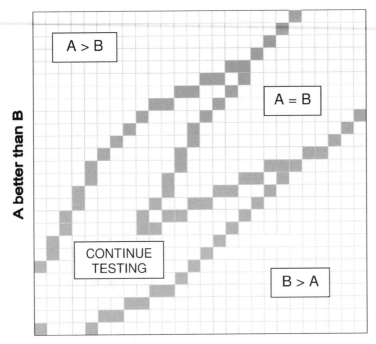

B better than A

Fig. 4.13. Sequential test plan for direct characteristic comparison

4.5.4
Statistical Quality Control

Continuous quality control is based on principles that firstly were used in the system of *"quality control charts"* (QCC, SHEWHART [1931]). Today, admittedly the monitoring of the characteristics of a process or product in order to detect deviations from the target value is not tied to charts but is mostly done by computer, although it is frequently still called a "control chart" system.

On the one hand, statistical quality control is an important tool for *quality assurance within analytical chemistry* itself (monitoring of test methods), and on the other for quality control of processes and products *by means of analytical methods*.

In each case, the aim of quality control charts[4] (QCC) is the representation of quality target values x_0 (standard values) and their statistical limits. A control chart usually consists of five lines as shown in Fig. 4.14.

QCC contain – as quality target values Q – standard or reference values, x_0, resp. optimum values as well as their limits. The inner pair of limits are called *warning limits* and the outer pair *control limits (action limits)*. When

---

[4]   may it now represented by a paper card or a computer

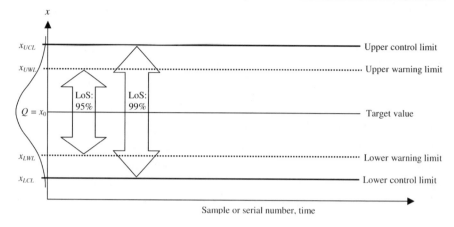

Fig. 4.14. General scheme of a quality control chart; LoS: level of significance

the warning limits are exceeded, attention must be increased to recognize a possible tangency or exceeding of the control limit. The latter requires corrective action.

The results of sample tests are marked in the charts as a series of points. Their sequence allows discerning quickly typical situations as some are shown in Fig. 4.15.

Case A represents a normal inconspicuous process, while the other cases show out-of-control situations (OCS), namely for various reasons. Case B show the typical OCS: a value is situated outside the upper control limit and demands instantaneous action. An OCS is also indicated if a given number of successive points are above resp. below the target value (namely 7 out of 7, see case C, 10 of 11, 12 of 14, 16 out of 20). The same holds true for steadily increasing or decreasing tendencies (case D). Whereas a single crossing of the warning limit only demands closer attention, 2 out of 3 successive values exceeding the warning limit, as shown by example E, indicate OCS.

The cases F, G and H illustrate exceptional situations, demanding instant attention as the process is threatening to get out of control. F shows periodic alterations, G a long-term trend, while in case H a particular great number of values are close to the control limit.

According to the character of the target value Q several types of control charts can be distinguished:

- Single value charts
- Mean value charts (charts on $\bar{x}$, median charts, blank charts)
- Variation charts (standard deviation charts, s resp. s_{rel}, range charts, R)
- Recovery rate charts
- Cusum charts
- Combination charts (e.g., charts on $\bar{x} - s$ and on $\bar{x} - R$)
- Correlation charts (e.g., charts on $\bar{x}_A$ and $\bar{x}_B$).

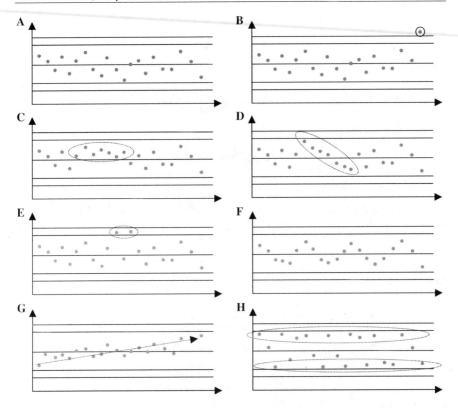

Fig. 4.15. Conspicuous plots of control charts

Single value charts are only used for special purposes, e.g. as *original value chard* for the determination of warning and control limits or, for data analysis of *time series* (SHUMWAY [1988]; MONTGOMERY et al. [1990]). All the other types of charts are used relatively often and have their special advantages (BESTERFIELD [1979]; MONTGOMERY [1985]; WHEELER and CHAMBERS [1990]).

For the limits of means ($\bar{x} = x_0 = Q$), the following bounds are chosen: $UCL = \bar{x} + t_{P=0.99,v}\, s/\sqrt{n}$, $UWL = \bar{x} + t_{P=0.95,v}\, s/\sqrt{n}$, $LWL = \bar{x} - t_{P=0.95,v}\, s/\sqrt{n}$, and $LCL = \bar{x} - t_{P=0.99,v}\, s/\sqrt{n}$. The limits $WL = \bar{x} \pm 2s/\sqrt{n}$ and $CL = \bar{x} \pm 3s/\sqrt{n}$ are also often used.

Parameters for the most important of the above mentioned quality control charts are given in Table 4.4.

Appropriately designed, Cusum control charts give sensitive and instructive impressions on process changes. Cumulative sums $S_n = \sum_{i=1}^{n} (x_i - x_0)$ also contain information on actual as well as on previously obtained values. Therefore their display enables one to perceive earlier changes leading to OCS than by means of the chart of original values (see WOODWARD and GOLDSMITH [1964]; MARSHALL [1977]; DOERFFEL [1990]).

Table 4.4. Control parameters for QCCs

Target value	Upper limits	Lower limits
Mean value $\bar{x}$	$L_U = \bar{x} + t_{\alpha,\nu}\, s/\sqrt{n}$	$L_L = \bar{x} - t_{\alpha,\nu}\, s/\sqrt{n}$
Standard deviation s	$L_U = s\,\sqrt{\dfrac{\chi^2_{n-1,1-\alpha/2}}{n-1}}$	$L_L = s\,\sqrt{\dfrac{\chi^2_{n-1,\alpha/2}}{n-1}}$
Range $\bar{R}$	$L_U = D_U\,\bar{R}$	$L_L = D_L\,\bar{R}$
Recovery rate $\bar{RR} = \beta$ (slope of the recovery function, see Eq. (4.3))	$L_U = \bar{RR} + t_{\alpha,\nu}\, s_{RR}/\sqrt{n}$	$L_L = \bar{RR} - t_{\alpha,\nu}\, s_{RR}/\sqrt{n}$

The corresponding factors of the t- and χ^2-distribution can be found in all textbooks on statistics, e.g., DIXON and MASSEY [1969]; DAVIES and GOLDSMITH [1984]; GRAF et al. [1987]; ARNOLD [1990]; SACHS [1992]. For D_U and D_L see e.g. GRAF et al. [1987].

References

AMC (1995) The Analytical Methods Committee: *Uncertainty of measurement – implications for its use in analytical science.* Analyst 120:2303

Arnold SF (1990) *Mathematical statistics.* Prentice-Hall, Englewood Cliffs, NJ

Bartlett MS (1937a) *Properties of sufficiency and statistical tests.* Proc Roy Soc A 160:268

Bartlett MS (1937b) *Some examples of statistical methods of research in agriculture and applied biology.* J R Statist Soc Suppl 4:137

Besterfield DH (1979) *Quality control.* Prentice-Hall, London

Cochran WG (1941) *The distribution of the largest of a set of estimated variances as a fraction of their total.* Ann Eugen [London] 11:47

Danzer K (1989) *Robuste Statistik in der analytischen Chemie.* Fresenius Z Anal Chem 335:869

David HA, Hartley HO, Pearson ES (1954) *The distribution of the ratio, in a single normal sample of range to standard deviation.* Biometrika 41:482

Davies OL, Goldsmith PL (1984) *Statistical methods in research and production.* Longman, London

Dean RB, Dixon WJ (1951) *Simplified statistics for small numbers of observations.* Anal Chem 23:636

Dixon WJ (1953) *Processing data for outliers.* Biometrics 9:74

Dixon WJ, Massey, FJ (1969) *Introduction to statistical analysis.* McGraw-Hill, New York

Doerffel K (1990) *Statistik in der analytischen Chemie.* Deutscher Verlag für Grundstoffindustrie, Leipzig

Doerffel K, Eckschlager K, Henrion G (1990) *Chemometrische Strategien in der Analytik.* Deutscher Verlag für Grundstoffindustrie, Leipzig

Doerffel K, Geyer R, Müller H (eds) (1994) *Analytikum – Methoden der analytischen Chemie und ihre theoretischen Grundlagen, 9th edn.* Deutscher Verlag für Grundstoffindustrie, Leipzig, Stuttgart

Eisenhart C (1947) *The assumptions underlying the analysis of variance.* Biometrics 3:1

EURACHEM (1995) *Quantifying uncertainty in analytical measurement.* Teddington

EURACHEM/WELAC (1993) *Guide 1: Accreditation of chemical laboratories.* London, Laboratory of the Government Chemist

Fleming J, Neidhart B, Tausch C, Wegscheider W (1996) *Glossary of analytical terms (I).* Accr Qual Assur 1:41

Fleming J, Albus H, Neidhart B, Wegscheider W (1997) *Glossary of analytical terms (VIII).* Accr Qual Assur 2:160

Games PA, Howell JF (1976) *Pairwise multiple comparison procedures with unequal N's and/or variances: a Monte Carlo study.* J Educat Statist 1:113

Graf U, Henning H-U, Stange K, Wilrich P-Th (1987) *Formeln und Tabellen der angewandten mathematischen Statistik.* Springer, Berlin Heidelberg New York (3 Aufl)

Grubbs FE (1969) *Procedures for detecting outlying observations in samples.* Technometrics 11:1

Hampel FR, Roncetti EM, Rousseeuw PJ, Stahel WA (1986) *Robust statistics: the approach based on influence functions.* Wiley, New York

Hartley HO (1950) *The maximum F-ratio as a short cut test for heterogeneity of variance.* Biometrika 37:308

Huber PJ (1981) *Robust statistics.* Wiley, New York

Hund E, Massart DL, Smeyers-Verbeke J (2001) *Operational definitions of uncertainty.* Trends Anal Chem 29:394

Inczédy J, Lengyiel JT, Ure AM, Geleneser A, Hulanicki A (eds) (1997) *Compendium of analytical nomenclature, 3rd edn* (IUPAC Orange Book). Blackwell, Oxford

ISO 3534-1 (1993), International Organization for Standardization (BIPM, IEC, IFCC, ISO, IUPAC, IUPAP, OIML), *International vocabulary of basic and general terms in metrology.* Geneva

IUPAC Orange Book (1997, 2000): printed version: *Compendium of analytical nomenclature (Definitive Rules 1997)*: see Inczédy et al. (1997); web version (as from 2000): www.iupac.org/publications/analytical_compendium/

Kaiser H (1936) *Grundriß der Fehlertheorie.* Z techn Physik 7:219

Kaiser H, Specker H (1956) *Bewertung und Vergleich von Analysenverfahren.* Fresenius Z Anal Chem 149:46

Kaiser R (1971) *Systematische Fehler in der Analyse.* Fresenius Z Anal Chem 256:1

Kolmogoroff AN (1941) *Confidence limits for an unknown distribution function.* Ann Math Statist 12:461

Marshall AG (1977) *Cumulative sum charts for monitoring of radioactivity background count rates.* Anal Chem 49:2193

Montgomery DC (1985) *Introduction to statistical quality control.* Wiley, New York

Montgomery DC, Johnson LA, Gardiner JS (1990) *Forecasting and time series analysis.* McGraw-Hill, New York

O'Rangers JJ, Condon RJ (2000) in: *Current issues in regulatory chemistry.* JP Kay, JD MacNeil, JJ O'Rangers (eds). AOAC Int., Gaithersburg, MD, p 207

Prichard E, Green J, Houlgate P, Miller J, Newman E, Phillips G, Rowley A (2001) *Analytical measurement terminology – handbook of terms used in quality assurance of analytical measurement.* LGC, Teddington, Royal Society of Chemistry, Cambridge

Pulido A, Ruisánchez I, Boqué R, Rius FX (2002) *Estimating the uncertainty of binary test results to assess their compliance with regular limits.* Anal Chim Acta 455:267

Rousseeuw PJ, Leroy AM (1987) *Robust regression and outlier detection.* Wiley, New York

Sachs L (1992) *Angewandte Statistik, 7th edn.* Springer, Berlin Heidelberg New York [1383]

Shewhart WA (1931) *Economic control of the quality of manufactured products.* Van Nostrand, New York

Shumway RH (1988) *Applied statistical time series analysis.* Prentice Hall, Englewood Cliffs, NJ

Simonet BM, Riós A, Valcarcel M (2004) *Unreliability of screening methods.* Anal Chim Acta 516:67

Smirnoff NW (1948) *Tables for estimating the goodness of fit of empirical distributions.* Ann Math Statist 19:279

Student[5] (1908) *The probable error of a mean.* Biometrika 6:1

Thompson M (1995) *Uncertainty in an uncertain world.* Analyst 120:117N

Trullols E, Ruisanchez I, Rius, FX (2004) *Validation of qualitative analytical methods.* Trends Anal Chem 23:137

Tukey JW (1949) *Comparing individual means in the analysis of variance.* Biometrics 5:99

von Neumann J, Kent RH, Bellinson HB, Hart BI (1941) *The mean square successive difference.* Ann Math Statist 12:153

Wegscheider W (1996) *Validation of analytical methods.* In: H. Günzler (ed) *Accreditation and quality assurance in analytical chemistry.* Springer, Berlin Heidelberg New York

Welch BL (1937) *The significance of the difference between two means when the population variances are unequal.* Biometrika 29:350

Wheeler DJ, Chambers DS (1990) *Understanding statistical process control.* Addison-Wesley, Avon, UK

Woodward RH, Goldsmith PL (1964) *Cumulative sum techniques.* Oliver & Boyd, London

---

[5]   This is a pseudonym of WS GOSSET.

5 Studying Influences and Optimizing Analytical Procedures

According to Sect. 3.5 the influencing of an analytical signal by the environment of the experiment (influence factors), i.e. interferences and other stimuli can be estimated either in a way basing on chemical facts (Eq. 3.16a) or in a statistical way (Eq. 3.16b,c). Therefore, two ways are feasible to study the significance of influences as a requirement of a subsequent optimization.

5.1
Testing the Significance of Influencing Factors

In analytical chemistry, the optimality criterion is frequently the relative increase of that share of the analytical gross signal that is caused by the analyte itself, namely $S_{AA}x_A/y_A$. According to Eq. (3.16a):

$$y_A = y_{A0} + S_{AA}x_A + \sum_{i=1}^{N} S_{Ai}x_i + \sum_{j=1}^{m} I_{Aj}x_j + e_{A'} \tag{5.1}$$

which is tantamount to the minimization of the interferences, $\sum S_{Ai}x_i$, and influencing factors, $\sum I_{Aj}x_j$. The quantities S_{Ai} are the cross sensitivities of the interferents i and I_{Aj}, the specific strength of the influence factors j. On the basis of this relationship the significance of interferents and factors can be studied using so-called *multifactorial designs*.

On the other hand, Eqs. (3.16a) and (5.1), respectively, can also be interpreted as being composed of the analyte signal and diverse (more or less) anonymous deviations e_i, e_j or e_{Aij}, respectively, see Eq. (3.16b,c):

$$y_A = y_{A0} + S_{AA}x_A + e_i + e_j + e_{A'} \tag{5.2}$$

On this basis, an *analysis of variance* (ANOVA) can be carried out to test the significance of the variations $e_i = \sum S_{Ai}x_i$, $e_j = \sum I_{Aj}x_j$, or more in detail, $e_B = S_{AB}x_B$, $e_C = S_{AC}x_C$ etc.

5.1.1
Analysis of Variance (ANOVA)

ANOVA was developed by FISHER [1925, 1935] as a statistical procedure that investigates influences (effects) of factors on a target quantity y according to a linear model which holds in the simplest case

Table 5.1. Measurement and evaluation scheme of one-way ANOVA

		Levels i of the factor a			
		1	2	...	m
Number j	1	y_{11}	y_{21}	...	y_{m1}
of single	2	y_{12}	y_{22}	...	y_{m2}
measurement	:	:	:		:
	n	y_{1n}	y_{2n}	...	y_{mn}
Sum		$S_1 = \sum y_{1j}$	$S_2 = \sum y_{2j}$	...	$S_m = \sum y_{mj}$
Mean		$\overline{y}_1$	$\overline{y}_2$	...	$\overline{y}_m$
Overall mean			$\overline{\overline{y}}$		

Table 5.2. Variance components in one-way ANOVA

Source of variation	Sum of squares	Degrees of freedom	Variance	F-test
Between the factor levels	$SS_a = n \sum_{i=1}^{m} (\overline{y}_i - \overline{\overline{y}})^2$	$v_a = m - 1$	$s_a^2 = \dfrac{SS_a}{m - 1}$	$\hat{F} = \dfrac{s_a^2}{s_{res}^2}$
Residual (analytical random error)	$SS_{res} = \sum_{i=1}^{m} \sum_{j=1}^{n} (y_{ij} - \overline{y}_i)^2$	$v_{res} = m(n - 1)$	$s_{res}^2 = \dfrac{SS_{res}}{m(n - 1)}$	
Total	$SS_{total} = SS_a + SS_{res}$	$v_{total} = m \cdot n - 1$		

$$y_{ij} = \overline{\overline{y}} + \alpha_i + e_{ij} \tag{5.3}$$

with y_{ij} being the actual value, $\overline{\overline{y}}$ the overall mean, α_i an additive influence of the factor a at level i, and e_{ij} the residual deviation (*one-way analysis of variance*, see Sect. 4.3.4). By means of ANOVA it is possible to compare both variances and means where two different models exist:

- *Model I* (model "*fixed*") that compares means on the basis of the null hypothesis H_0: $\mu_1 = \mu_2 = \ldots = \mu_m$ and
- *Model II* (model "*random*") by which variances are compared on the basis of the null hypothesis H_0: $\sigma_1^2 = \sigma_2^2 = \ldots = \sigma_m^2$ (corresponding to H_0: $\sigma_a^2 = 0$)

and additionally a mixed one (see Sect. 4.3.4 and EISENHART [1947]).

Variance analysis should advantageously be carried out on the basis of balanced experiments where the number of observations per factor level is equal ($n_1 = n_2 = \ldots = n_m = n$).

The measurement scheme of *One-way analysis of variance* is given in Table 5.1 for $i = 1 \dots m$ levels of the factor a (in analytical practice frequently a factor is studied only on two levels to compare, e.g., two laboratories, two operators, two different techniques, etc).

The variance components are calculated according to Table 5.2.

The estimated $\hat{F}$ value has to be compared with the quantile of the F-distribution, $F_{1-\alpha,\nu}$, the tables of which can be found in textbooks of statistics (e.g., HALD [1960]; NEAVE [1981]; DIXON and MASSEY [1983]; GRAF et al. [1987]; SACHS [1992]). The influence of the factor a is significant when $\hat{F}$ exceeds $F_{1-\alpha,\nu}$. In case of unbalanced experiments the different size of measurement series and, therefore, degrees of freedom have to be considered as a result of which both the evaluation scheme and the variance decomposition become more complicated (see DIXON and MASSEY [1983]; GRAF et al. [1987]).

By means of *Two-way ANOVA* two factors can be studied simultaneously. The model

$$y_{ij} = \overline{\overline{y}} + \alpha_i + \beta_j + (\alpha\beta)_{ij} + e_{ij} \qquad (5.4)$$

considers the influences α_i of the factor a at levels i, β_j of the factor b at levels j, and additionally the interactions (correlations), $(\alpha\beta)_{ij}$, of both the factors (y_{ij} actual value, $\overline{\overline{y}}$ overall mean, e_{ij} residual deviation, i.e., experimental error). The scheme of measurement and evaluation of two-way ANOVA is given in Table 5.3 and the corresponding variance decomposition in Table 5.4.

On the basis of two-way ANOVA two null hypotheses can be tested, namely

- H_a: $\alpha_1 = \alpha_2 = \dots = \alpha_m = 0$ by means of $\hat{F}_a$ compared with $F_{1-\alpha,\nu_1=m-1,\nu_2=(m-1)(n-1)}$,
- H_b: $\beta_1 = \beta_2 = \dots = \beta_m = 0$ by means of $\hat{F}_b$ compared with $F_{1-\alpha,\nu_1=n-1,\nu_2=(m-1)(n-1)}$.

Table 5.3. Evaluation scheme of two-way ANOVA (*ab* model: single measurements in each point; the index point marks that levels over which is actually added up or averaged, respectively)

	Levels i of the factor a				Sum	Mean
	1	2	$\dots$	m		
Levels j 1	y_{11}	y_{21}	$\dots$	y_{m1}	$S_{\bullet 1} = \sum y_{i1}$	$y_{\bullet 1}$
of the 2	y_{12}	y_{22}	$\dots$	y_{m2}	$S_{\bullet 2} = \sum y_{i2}$	$y_{\bullet 2}$
factor b :	:	:		:	:	:
n	y_{1n}	y_{2n}	$\dots$	y_{mn}	$S_{\bullet m} = \sum y_{im}$	$y_{\bullet n}$
Sum	$S_{1\bullet} = \sum y_{1j}$	$S_{2\bullet} = \sum y_{2j}$	$\dots$	$S_{m\bullet} = \sum y_{mj}$	$S_{\bullet\bullet}$	–
Mean	$y_{1\bullet}$	$y_{2\bullet}$	$\dots$	$y_{m\bullet}$	–	$y_{\bullet\bullet}$

Table 5.4. Variance components in two-way ANOVA (*ab* model)

Source of variation	Sum of squares	Degrees of freedom	Variance	F-test
Between the levels of factor *a*	$SS_a = \sum_{i=1}^{m} \frac{S_{\cdot j}^2}{n} - \frac{S_{\cdot\cdot}^2}{m\cdot n}$	$\nu_a = m - 1$	$s_a^2 = \frac{SS_a}{m-1}$	$\hat{F}_a = \frac{s_a^2}{s_{res}^2}$
Between the levels of factor *b*	$SS_b = \sum_{i=1}^{n} \frac{S_{i\cdot}^2}{m} - \frac{S_{\cdot\cdot}^2}{m\cdot n}$	$\nu_b = n - 1$	$s_b^2 = \frac{SS_b}{n-1}$	$\hat{F}_b = \frac{s_b^2}{s_{res}^2}$
Total	$SS_{total} = \sum_{i=1}^{m}\sum_{j=1}^{n} y_{ij}^2 - \frac{S_{\cdot\cdot}^2}{m\cdot n}$	$\nu_{total} = m\cdot n - 1$	$s_{total}^2 = \frac{SS_{total}}{m\cdot n - 1}$	
Residual (analytical random error)	$SS_{res} = SS_{total} - SS_a - SS_b$	$\nu_{res} = (m-1)(n-1)$	$s_{res}^2 = \frac{SS_{res}}{(m-1)(n-1)}$	

As a rule, interactions $(\alpha\beta)_{ij}$ cannot be estimated if only single observations at each point of the measurement matrix are carried out. In case of single measurements, the residual error contains both experimental error and interactions, the separation of which is possible only in special cases, e.g., testing of homogeneity of solids when certain assumptions can be made. DANZER and MARX [1979] have investigated the homogeneity of steel samples by means of a destructive OES procedure. Consequently, no repeated measurements could be carried out and the residual error must be corrected by interaction terms estimated according to MANDEL [1961].

If possible, two-way ANOVA should be applied doing repetitions at each level. In case of double measurements the *2ab* model represented in Tables 5.5 and 5.6 is taken as the basis of evaluation and variance decomposition.

On the basis of this *2ab* ANOVA it is possible to test three null hypotheses, namely

- H_a: $\alpha_1 = \alpha_2 = \ldots = \alpha_m = 0$ by means of $\hat{F}_a$ compared with $F_{1-\alpha, \nu_1 = m-1, \nu_2 = (m-1)(n-1)}$,
- H_b: $\beta_1 = \beta_2 = \ldots = \beta_m = 0$ by means of $\hat{F}_b$ compared with $F_{1-\alpha, \nu_1 = n-1, \nu_2 = (m-1)(n-1)}$, and
- H_{ab}: $(\alpha\beta)_{11} = (\alpha\beta)_{12} = \ldots = (\alpha\beta)_{mm} = 0$ by means of $\hat{F}_{ab}$ compared with $F_{1-\alpha, \nu_1 = n-1, \nu_2 = m\, n}$.

In the case that interactions prove to be insignificant, it should be gone over to the *ab* model the estimations of which for the various variance components is more reliable than that of the *2ab* model. A similar scheme can be used for three-way ANOVA when the factor *c* is varied at two levels. In the general, three-way analysis bases on block-designed experiments as shown in Fig. 5.1.

Table 5.5. Evaluation scheme of two-way ANOVA (*2ab* model: double measurements in each point)

		Levels i of the factor a				Sum	Mean
		1	2	...	m		
Levels j	1	y_{111}, y_{112}	y_{211}, y_{212}	...	y_{m11}, y_{m12}	$S_{\bullet 1 \bullet}$	$y_{\bullet 1 \bullet}$
of the	2	y_{121}, y_{122}	y_{221}, y_{222}	...	y_{m21}, y_{m22}	$S_{\bullet 2 \bullet}$	$y_{\bullet 2 \bullet}$
factor b	⋮	⋮	⋮		⋮	⋮	⋮
	n	y_{1n1}, y_{1n2}	y_{2n1}, y_{2n2}	...	y_{mn1}, y_{mn2}	$S_{\bullet m \bullet}$	$y_{\bullet n \bullet}$
Sum		$S_{1 \bullet \bullet}$	$S_{2 \bullet \bullet}$	...	$S_{m \bullet \bullet}$	$S_{\bullet \bullet \bullet}$	–
Mean		$y_{1 \bullet \bullet}$	$y_{2 \bullet \bullet}$	...	$y_{m \bullet \bullet}$	–	$y_{\bullet \bullet \bullet}$

Table 5.6. Variance components in two-way ANOVA (*2ab* model)

Source of variation	Sum of squares	Degrees of freedom	Variance	F-test
Between the levels of factor a	$SS_a = \sum\limits_{i=1}^{m} \dfrac{S_{\bullet j \bullet}^2}{2n} - \dfrac{S_{\bullet \bullet \bullet}^2}{2mn}$	$\nu_a = m - 1$	$s_a^2 = \dfrac{SS_a}{m-1}$	$\hat{F}_a = \dfrac{s_a^2}{s_{res}^2}$
Between the levels of factor b	$SS_b = \sum\limits_{i=1}^{n} \dfrac{S_{i \bullet \bullet}^2}{2m} - \dfrac{S_{\bullet \bullet \bullet}^2}{2mn}$	$\nu_b = n - 1$	$s_b^2 = \dfrac{SS_b}{n-1}$	$\hat{F}_b = \dfrac{s_b^2}{s_{res}^2}$
Interaction between a and b	$SS_{ab} = SS_{total} - SS_a - SS_b - SS_{res}$	$\nu_{ab} = (m-1)(n-1)$	$s_{ab}^2 = \dfrac{SS_{ab}}{(m-1)(n-1)}$	$\hat{F}_{ab} = \dfrac{s_{ab}^2}{s_{res}^2}$
Residual (analytical random error)	$SS_{res} = \sum\limits_{i=1}^{m}\sum\limits_{j=1}^{n}\sum\limits_{k=1}^{2} y_{ijk}^2 - \sum\limits_{i=1}^{m}\sum\limits_{j=1}^{n} S_{ij \bullet}^2$	$m \cdot n$	$s_{res}^2 = \dfrac{SS_{res}}{m\,n}$	
Total	$SS_{total} = \sum\limits_{i=1}^{m}\sum\limits_{j=1}^{n} y_{ij}^2 - \dfrac{S_{\bullet \bullet}^2}{2mn}$	$\nu_{total} = 2m \cdot n - 1$	$s_{total}^2 = \dfrac{SS_{total}}{2mn - 1}$	

Following the scheme given in Fig. 5.1, the influence of three factors a, b and c can be studied on the basis of the linear model

$$y_{ijk} = \overline{\overline{y}} + \alpha_i + \beta_j + \gamma_k + (\alpha\beta)_{ij} + (\alpha\gamma)_{ik} + (\beta\gamma)_{jk} + (\alpha\beta\gamma)_{ijk} + e_{ijk} \quad (5.5)$$

The estimation of all the terms of Eq. (5.5) is possible for the balanced case and q repeated measurements in each cell of the data block represented in Fig. 5.1. Schemes for this and some reduced variants of three-way ANOVA are given in SCHEFFÉ [1961]; AHRENS [1967]; DUNN and CLARK [1974]; GRAF et al. [1987]; SACHS [1992].

By means of the three-way variance analysis according to the model at Eq. (5.5) and Table 5.7 the influence of the factors a, b and c can be tested as well as that of the interactions, i.e. the null hypotheses:

Table 5.7. Variance components in three-way ANOVA (abc model with repetitions)

Source of variation	Sum of squares	Degrees of freedom	Variance	F-test
Between the levels of factor a	$SS_a = npq \sum_{i=1}^{m} (\bar{y}_{i\bullet\bullet\bullet} - \bar{y}_{\bullet\bullet\bullet\bullet})^2$	$m-1$	$s_a^2 = \dfrac{SS_a}{m-1}$	$\hat{F}_a = \dfrac{s_a^2}{s_{res}^2}$
Between the levels of factor b	$SS_b = mpq \sum_{j=1}^{n} (\bar{y}_{\bullet j\bullet\bullet} - \bar{y}_{\bullet\bullet\bullet\bullet})^2$	$n-1$	$s_b^2 = \dfrac{SS_b}{n-1}$	$\hat{F}_b = \dfrac{s_b^2}{s_{res}^2}$
Between the levels of factor c	$SS_c = mnq \sum_{k=1}^{p} (\bar{y}_{\bullet\bullet k\bullet} - \bar{y}_{\bullet\bullet\bullet\bullet})^2$	$p-1$	$s_c^2 = \dfrac{SS_c}{p-1}$	$\hat{F}_c = \dfrac{s_c^2}{s_{res}^2}$
Interaction between a and b	$SS_{ab} = pq \sum_{i=1}^{m}\sum_{j=1}^{n} (\bar{y}_{ij\bullet\bullet} - \bar{y}_{i\bullet\bullet\bullet} - \bar{y}_{\bullet j\bullet\bullet} + \bar{y}_{\bullet\bullet\bullet\bullet})^2$	$(m-1)(n-1)$	$s_{ab}^2 = \dfrac{SS_{ab}}{(m-1)(n-1)}$	$\hat{F}_{ab} = \dfrac{s_{ab}^2}{s_{res}^2}$
Interaction between a and c	$SS_{ac} = nq \sum_{i=1}^{m}\sum_{k=1}^{p} (\bar{y}_{i\bullet k\bullet} - \bar{y}_{i\bullet\bullet\bullet} - \bar{y}_{\bullet\bullet k\bullet} + \bar{y}_{\bullet\bullet\bullet\bullet})^2$	$(m-1)(p-1)$	$s_{ac}^2 = \dfrac{SS_{ac}}{(m-1)(p-1)}$	$\hat{F}_{ac} = \dfrac{s_{ac}^2}{s_{res}^2}$
Interaction between b and c	$SS_{bc} = mq \sum_{j=1}^{n}\sum_{k=1}^{p} (\bar{y}_{\bullet jk\bullet} - \bar{y}_{\bullet j\bullet\bullet} - \bar{y}_{\bullet\bullet k\bullet} + \bar{y}_{\bullet\bullet\bullet\bullet})^2$	$(n-1)(p-1)$	$s_{bc}^2 = \dfrac{SS_{bc}}{(n-1)(p-1)}$	$\hat{F}_{bc} = \dfrac{s_{bc}^2}{s_{res}^2}$
Interaction between a, b and c	$SS_{abc} = q \sum_{i=1}^{m}\sum_{j=1}^{n}\sum_{k=1}^{p} (\bar{y}_{ijk\bullet} - \bar{y}_{ij\bullet\bullet} - \bar{y}_{i\bullet k\bullet} - \bar{y}_{\bullet jk\bullet} + \bar{y}_{i\bullet\bullet\bullet} + \bar{y}_{\bullet j\bullet\bullet} + \bar{y}_{\bullet\bullet k\bullet} - \bar{y}_{\bullet\bullet\bullet\bullet})^2$	$\nu_{abc} = (m-1)(n-1)(p-1)$	$s_{abc}^2 = \dfrac{SS_{abc}}{\nu_{abc}}$	$\hat{F}_{abc} = \dfrac{s_{abc}^2}{s_{res}^2}$
Residual error	$SS_{res} = \sum_{i=1}^{m}\sum_{j=1}^{n}\sum_{k=1}^{p}\sum_{l=1}^{q} (y_{ijkl} - \bar{y}_{ijk\bullet})^2$	$mnp(q-1)$	$s_{res}^2 = \dfrac{SS_{res}}{mnp(1-q)}$	
Total	$SS_{total} = \sum_{i=1}^{m}\sum_{j=1}^{n}\sum_{k=1}^{p}\sum_{l=1}^{q} (y_{ijkl} - \bar{y}_{\bullet\bullet\bullet\bullet})^2$	$mnpq-1$	$s_{total}^2 = \dfrac{SS_{total}}{mnpq-1}$	

$\bar{y}_{i\bullet\bullet\bullet}$ mean over the i levels of factor a, $\bar{y}_{\bullet j\bullet\bullet}$ mean over the j levels of factor b, etc.
$\bar{y}_{ij\bullet\bullet}$ mean over the i levels of factor a and the j levels of factor b; $\bar{y}_{i\bullet k\bullet}$ mean over the i levels of factor a and the k levels of factor c, etc.
$\bar{y}_{ijk\bullet}$ mean over the i levels of factor a, the j levels of factor b, and the k levels of factor c, etc.
$\bar{y}_{\bullet\bullet\bullet\bullet}$ total mean

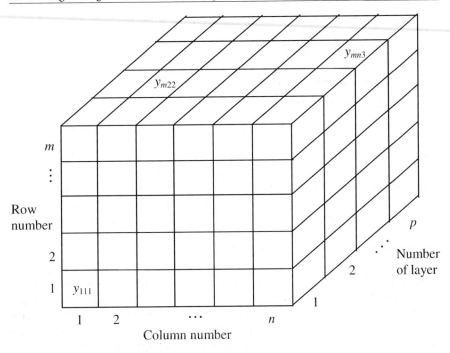

Fig. 5.1. Principle of three-way ANOVA: data are arranged in rows, columns, and layers (some examples of data pixels are given)

- H_a: $\alpha_1 = \alpha_2 = \ldots = \alpha_m = 0$ by means of $\hat{F}_a$ $\overleftrightarrow{comp}$ [1]
  $F_{1-\alpha, v_1=m-1, v_2=mnp-1}$,
- H_b: $\beta_1 = \beta_2 = \ldots = \beta_m = 0$ by means of $\hat{F}_b$ $\overleftrightarrow{comp}$ $F_{1-\alpha, v_1=n-1, v_2=mnp-1}$,
- H_c: $\gamma_1 = \gamma_2 = \ldots = \gamma_m = 0$ by means of $\hat{F}_c$ $\overleftrightarrow{comp}$ $F_{1-\alpha, v_1=p-1, v_2=mnp-1}$,
- H_{ab}: $(\alpha\beta)_{11} = (\alpha\beta)_{12} = \ldots = (\alpha\beta)_{mn} = 0$ by means of $\hat{F}_{ab}$ $\overleftrightarrow{comp}$
  $F_{1-\alpha, v_1=(m-1)(n-1), v_2=mnp-1}$,
- H_{ac}: $(\alpha\gamma)_{11} = (\alpha\gamma)_{12} = \ldots = (\alpha\gamma)_{mp} = 0$ by means of $\hat{F}_{ac}$ $\overleftrightarrow{comp}$
  $F_{1-\alpha, v_1=(m-1)(p-1), v_2=mnp-1}$,
- H_{bc}: $(\beta\gamma)_{11} = (\beta\gamma)_{12} = \ldots = (\beta\gamma)_{np} = 0$ by means of $\hat{F}_{ab}$ $\overleftrightarrow{comp}$
  $F_{1-\alpha, v_1=(n-1)(p-1), v_2=mnp-1}$, and
- H_{abc}: $(\alpha\beta\gamma)_{111} = (\alpha\beta\gamma)_{121} = \ldots = (\alpha\beta\gamma)_{mnp} = 0$ by $\hat{F}_{ab}$ $\overleftrightarrow{comp}$
  $F_{1-\alpha, v_1=(m-1)(n-1)(p-1), v_2=mnp-1}$.

---

[1] $\overleftrightarrow{comp}$ symbolizes "compared with"

In some cases interactions are improbable and information on them is not needed. Then reduced variants of three-way ANOVA can be applied by which the effects of the main factors can be estimated more reliable (see DUNN and CLARK [1974]; GRAF et al. [1987]; SACHS [1992]). Concentrating on the main effects, the design of the experiments can be aimed at a minimum number of observations.

5.1.2
Experimental Design

Methods of variance analysis are helpful tools to evaluate effects of factors on the results of experiments afterwards. On the other hand, it may be advantageous to plan experiments in a comparative way (comparative experiments).

Statistical experimental design is characterized by the three basic principles: *Replication*, *Randomization* and *Blocking* (block division, planned grouping). *Latin square design* is especially useful to separate nonrandom variations from random effects which interfere with the former. An example may be the identification of (slightly) different samples, e.g. sorts of wine, by various testers and at several days. To separate the day-to-day and/or tester-to-tester (laboratory-to-laboratory) variations from that of the wine sorts, an $m \times m$ Latin square design may be used. In case of $m = 3$ all three wine samples (a, b, c) are tested be three testers at three days, e.g. in the way represented in Table 5.8:

Table 5.8. Latin square design for $m = 3$

	Tester 1	Tester 2	Tester 3
1st day	a	b	c
2nd day	b	c	a
3rd day	c	a	b

The results of the experiments are evaluated by means of three-way ANOVA in its simplest form, $m = n = p$ and $q = 1$. The significance of the sample effect can principally be guaranteed also in the case that both testers and days have significant influence (SHARAF et al. [1986]).

In contrast to common statistical techniques, by modern experimental design influencing factors are studied *simultaneously* (*multifactorial design*, MFD). The aim of MFD consists in an arrangement of factors in such a way that their influences can be quantified, compared and separated from random variations.

Frequently the signal intensity of the analyte A is the target quantity, the influences on which are described by Eq. (3.16a). Handling all the influences (interferences and other factors) in the same way and holding x_A at any constant value so that $\alpha_0 = y_{A0} + S_{AA}x_A$, Eq. (3.16a) can be written

Table 5.9. Design matrix for three factors at two levels ($+$ and $-$ stand for $+1$ and -1)

Run	(z_0)	z_1	z_2	z_3	$z_1 z_2$	$z_1 z_3$	$z_2 z_3$	$z_1 z_2 z_3$	Target value
1	$+$	$+$	$+$	$+$	$+$	$+$	$+$	$+$	y_1
2	$+$	$+$	$+$	$-$	$+$	$-$	$-$	$-$	y_2
3	$+$	$+$	$-$	$+$	$-$	$+$	$-$	$-$	y_3
4	$+$	$+$	$-$	$-$	$-$	$-$	$+$	$+$	y_4
5	$+$	$-$	$+$	$+$	$-$	$-$	$+$	$-$	y_5
6	$+$	$-$	$+$	$-$	$-$	$+$	$-$	$+$	y_6
7	$+$	$-$	$-$	$+$	$+$	$-$	$-$	$+$	y_7
8	$+$	$-$	$-$	$-$	$+$	$+$	$+$	$-$	y_8

$$y_A = \alpha_0 + \alpha_1 x_1 + \alpha_2 x_2 + \alpha_3 x_3 + \alpha_{12} x_1 x_2 + \alpha_{13} x_1 x_3 + \alpha_{23} x_2 x_3 + \alpha_{123} x_1 x_2 x_3 \quad (5.6)$$

in case of three influence factors. From the various types of treatment and design, *two-level factorial design* is mostly applied. That means that the influence factors are varied between a higher and a lower level, x_{max} and x_{min}. Using *complete factorial design* (CFD) the number of experiments is $N = 2^m$ for m factors. In the case of $m = 3$ as given in Eq. (5.6), $N = 8$ experiments have to be carried out.

Expediently, factorial design is done on the basis of transformed factors z_i, calculated from the x_i by

$$z_i = \frac{x_i - \overline{x}}{\frac{1}{2}(x_{max} - x_{min})} \quad (5.7)$$

where $\overline{x} = \frac{1}{2}(x_{max} + x_{min})$ so that $z_{max} = \frac{x_{max} - x_{min}}{x_{max} - x_{min}} = +1$, $z_{min} = \frac{x_{min} - x_{max}}{x_{max} - x_{min}} = -1$, and $\overline{z} = 0$.

With the transformation at Eq. (5.7), Eq. (5.6) becomes

$$y = a_0(z_0) + a_1 z_1 + a_2 z_2 + a_3 z_3 + a_{12} z_1 z_2 + a_{13} z_1 z_3 + a_{23} z_2 z_3 + a_{123} z_1 z_2 z_3 \quad (5.8)^2$$

The coefficients a_i are estimated from the results of experiments carried out according to a design matrix such as Table 5.9 which shows a 2^3 plan matrix. The significance of the several factors are tested by comparing the coefficients with the experimental error, to be exact, by testing whether the confidence intervals Δa_i include 0 or not. The experimental error can be estimated by repeated measurements of each experiment or – as it is done frequently in a more effective way – by replications at the centre of the plan (so-called "zero replications"), see Fig. 5.2.

The coefficients are estimated according to

$$a_i = \frac{1}{N} z_i^T y \quad (5.9)$$

---

[2]   the target value y_A is symbolized here y

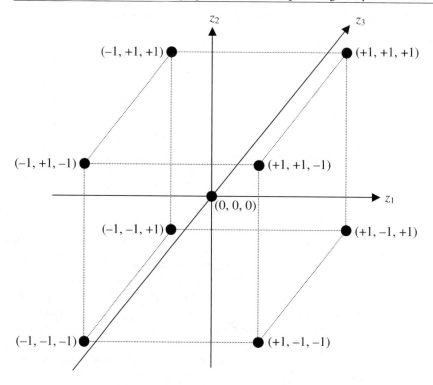

Fig. 5.2. Geometrical representation of a complete two level factorial matrix (three influence factors) with experiments in the centre point (0,0,0)

with z_i^T the corresponding transposed z vector and y the vector of target values obtained as the result of the runs. N is the number of experiments ($N = 2^m$, here $N = 8$). As an example, the coefficients a_i in Eq. (5.8) corresponding to the design matrix at Eq. (5.9) are estimated by

$$a_0 = \frac{1}{8}(+y_1 + y_2 + y_3 + y_4 + y_5 + y_6 + y_7 + y_8)$$

$$a_1 = \frac{1}{8}(+y_1 + y_2 + y_3 + y_4 - y_5 - y_6 - y_7 - y_8)$$

$$a_2 = \frac{1}{8}(+y_1 + y_2 - y_3 - y_4 + y_5 + y_6 - y_7 - y_8)$$

$$a_3 = \frac{1}{8}(+y_1 - y_2 + y_3 - y_4 + y_5 - y_6 + y_7 - y_8)$$

$$a_{12} = \frac{1}{8}(+y_1 + y_2 - y_3 - y_4 - y_5 - y_6 + y_7 + y_8)$$

$$\vdots$$

$$a_{123} = \frac{1}{8}(+y_1 - y_2 - y_3 + y_4 - y_5 + y_6 + y_7 - y_8)$$

according to the vector expression

$$a = Z^T y \tag{5.10}$$

with Z^T being the transposed z matrix.

Because the experimental expenditure increases strongly with the increasing number of influence factors, *fractional factorial design FFD (partial factorial design)* is applied in such cases. It is not possible to evaluate all the interactions by FFDs but only the main effects.

PLACKETT and BURMAN [1946] have developed a special fractional design which is widely applied in analytical optimization. By means of N runs up to $m = N - 1$ variables (where some of them may be dummy variables which can help to estimate the experimental error) can be studied under the following prerequisites and rules:

- The number of experiments (runs) must be a multiple of l^2 (l is the number of levels), that is $N = 8, 12, 16, 24, \ldots$ in case of two-level experiments ($l = 2$).
- The first rows of the design matrixes are

  $N = 8: \quad + + + - + - -$
  $N = 12: \quad + + - + + + - - - + -$
  $N = 16: \quad + + + + - + - + + - - + - - -$ etc.

  in case of two-level design.
- The following rows of the design matrix is generated by shifting the first row cyclically one place ($N - 2$ times).
- The last row has minus in all factors.
- The procedure can be controlled as follows: each row contains $m/2$ times the higher level ($+$) and ($m/2 - 1$) times the lower ($-$), the columns contain each $m/2$ times $+$ and $-$.

Fractional factorial design is especially useful in case of a high number of influence variables from which the insignificant one have to be screened.

An example of a PLACKETT BURMAN plan for $l = 2$ levels, $m = 7$ influence factors (including dummy variables) and, therefore, $N = 8$ runs is given in Table 5.10.

The coefficients a_i of the main effects of the model

$$y = a_0 z_0 + a_1 z_1 + a_2 z_2 + a_3 z_3 + a_4 z_4 + a_5 z_5 + a_6 z_6 + a_7 z_7 \tag{5.11}$$

are obtained by the vector equation

$$a = (Z^T Z)^{-1} Z^T y \tag{5.12}$$

The coefficients characterize the effect of the belonging factor. The influence is significant in the case that

$$|a_i| \geq a_{crit} = \Delta a = s_a \, t_{1-\alpha, v} \tag{5.13}$$

Table 5.10. PLACKETT-BURMAN design matrix for $N = 8$ experiments and consequently $m = 7$ factors (including dummy variables) at two levels

Run	z_1	z_2	z_3	z_4	z_5	z_6	z_7	Target value
1	+	+	+	−	+	−	−	y_1
2	+	+	−	+	−	−	+	y_2
3	+	−	+	−	−	+	+	y_3
4	−	+	−	−	+	+	+	y_4
5	+	−	−	+	+	+	−	y_5
6	−	−	+	+	+	−	+	y_6
7	−	+	+	+	−	+	−	y_7
8	−	−	−	−	−	−	−	y_8

i.e., the influence of z_i (and therefore x_i) is insignificant if the confidence interval $a_i \pm \Delta a$ includes zero. Multifactorial experiments with a low number of factors can also be evaluated by ANOVA (see DANZER et al. [2001], Sect. 5.1.1).

If nonlinear effects are expected the variables must be varied at more than two levels. A screening plan comparable to the PLACKETT BURMAN design but on three levels is that of Box and BEHNKEN [1960].

In case of special conditions, viz. internal correlations, interactions can be estimated in addition to the main effects by means of a 2^{m-1} design.

Multifactorial experiments are used in analytical chemistry for diverse applications, e.g., checking up *significant influences* before optimization procedures, recognizing *matrix effects*, and testing the *robustness* of analytical procedures (WEGSCHEIDER [1996]).

5.2
Optimization of Analytical Procedures

Analytical procedures should always run under optimum conditions. That means that for Eq. (5.6), which is here used only with two factors, the coefficients have to be chosen in such a way that y becomes an optimum

$$y = \alpha_0 + \alpha_1 x_1 + \alpha_2 x_2 \overset{!}{=} opt \tag{5.14}$$

In analytical chemistry the target quantity y which has to be optimized is frequently the signal intensity, absolute or relative (signal-to-noise ratio), but occasionally other parameters like yields of extractions or chemical reactions, too. The classical way to optimize influences, e.g., in an optimization space as shown in Fig. 5.3a is to study the factors independently one after the other. In Fig. 5.3b,c it can be seen that an individual optimum will be found in this way.

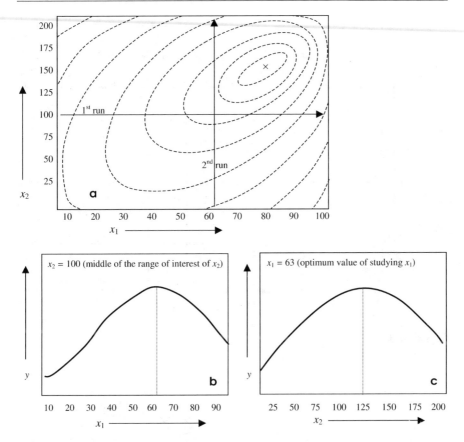

Fig. 5.3. Contour plot of the response surface $y = f(x_1, x_2)$; the optimum is situated at point $\times$ (**a**); Response curves of y in dependence of x_1 as result of a first run (**b**) and x_2 as result of a second run (**c**)

However, the optima of x_1 and x_2 found in this way do not meet the global optimum of the response surface which is situated at $x_1 = 80$ and $x_2 = 150$. Because the global optimum is rarely found by such an obsolete proceeding, multivariate techniques of optimization should be applied.

The most reliable technique to find the global optimum by means of common methods is the transition from the quasi-two-dimensional approach (Fig. 5.3b,c) to a complete two-dimensional one. It consists of a certain number of experiments as shown in Fig. 5.4.

On the basis of the grid experiments a mathematical function $y = f(x_1, x_2, x_3, \ldots)$, called the *response surface*, is estimated that characterizes the response as a function of the factors. In case of only two factors the response surface can be visualized by plots like that in Fig. 5.5.

Response surfaces are mostly described mathematically by polynomial approximations of 1st and 2nd degree. Grid search corresponds to a com-

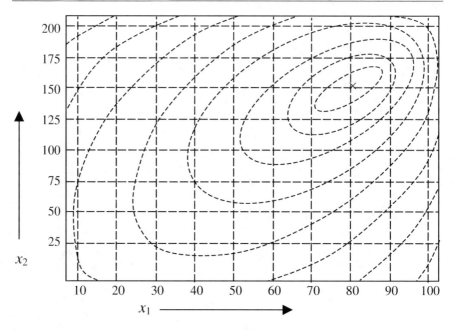

Fig. 5.4. Grid experiments for estimating the response surface $y = f(x_1, x_2)$

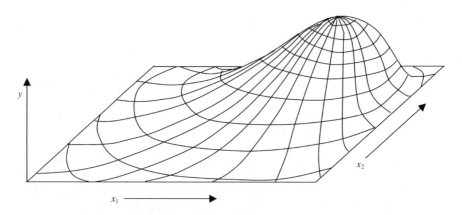

Fig. 5.5. Response surface as a result of grid experiments according to Fig. 5.4

plete factorial design. When the resolution of the variables is high enough, each optimum – the global one and all the local maxima and minima – can be found. But the high number of experiments imposes limits in this regard how it is generally in response surface technique.

The number of experiments can considerably be decreased by iterative optimization methods which starts at an area that can be selected by experience, supposition or randomly. This start area is moved step by

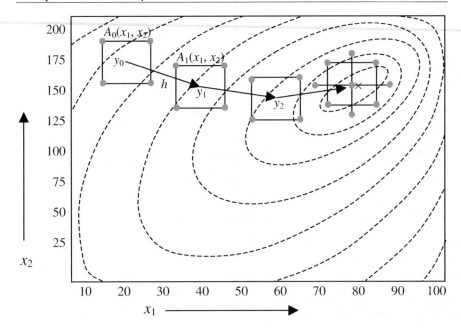

Fig. 5.6. Schematic representation of the Box-Wilson optimization with step width h

step in direction to the optimum following the gradient of the function $y = f(x_1, x_2, x_3, \ldots)$, i.e. the steepest ascent[3].

The well-known Box-Wilson optimization method (Box and Wilson [1951]; Box [1954, 1957]; Box and Draper [1969]) is based on a linear model (Fig. 5.6). For a selected start hyperplane, in the given case an area $A_0(x_1, x_2)$, described by a polynomial of first order, with the starting point y_0, the gradient $grad[y_0]$ is estimated. Then one moves to the next area in direction of the steepest ascent (the gradient) by a step width of h, in general

$$y_{i+1} = y_i + h\ grad[y_i] \tag{5.15}$$

Near the optimum both the step width and the model of the hyperplane are changed, the latter mostly from a first order model to a second order model. The vicinity of the optimum can be recognized by the coefficients $a_1, a_2, \ldots$ of Eq. (5.14) which approximate to zero or change their sign, respectively. For the second order model mostly a Box-Behnken design is used.

Because this proceeding is relatively expensive, an effective semi-quantitative method is widely used in optimization, the *sequential simplex optimization*. Simplex optimization is done without estimation of gradients and setting step widths. Instead of this, the progress of the optimization

---

[3]   steepest descent in case of minima

Table 5.11. Basic simplex operations

Operation	Movement	Condition
Reflexion		$y_1 < y_2 \approx y_3$
Expansion		$y_4 > y_2 \approx y_3$
Contraction		$y_4 \lesssim y_2 \approx y_3$
Strong contraction		$y_4 < y_2 \approx y_3$

procedure results directly from the quality of the preceded experimental values.

A simplex is a geometric figure formed by $p+1$ points in a p-dimensional space. In the two-dimensional case $y = f(x_1, x_2)$ the simplex is a triangle, in the three-dimensional case a tetrahedron etc. With regard to its form, the simplex may be regular, rectangular, or irregular. The simplex optimization starts with a set of $p + 1$ parameters (here $p + 1 = 3$). The movement of the simplex takes place according to the rules given in Table 5.11.

As an example, in Fig. 5.7 a simplex optimization is shown in a simplified way, i.e., only by reflexions and with simplexes of invariable size. The approach to the optimum is indicated by rotation or oscillation of the simplex. Then contractions should be included into the operations.

The optimum found by sequential proceeding, both by BOX-WILSON and simplex technique, is that local optimum situated nearest the starting point. It must not inevitably be identical with the global optimum. Therefore, it may be useful to repeat the optimization procedure one or several times.

5.3
Global Optimization by Natural Design

Some natural processes and principles have stimulated researchers to develop algorithms that imitate concepts of nature and are, therefore, summarized under the name *natural computation* (KATEMAN [1990]; LUCASIUS [1994]). The most prominent methods are:

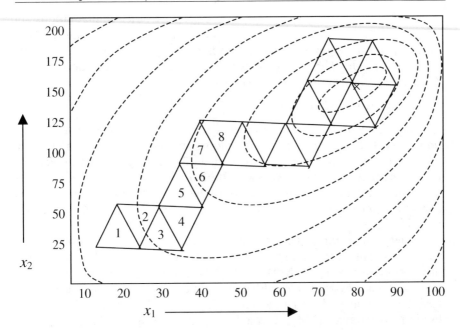

Fig. 5.7. Simplex optimization within a response surface of two factors x_1 and x_2 with simplexes of invariable size

- *Thermal computation* which comprises algorithms inspired by multi-particle systems like BROWNian motion (BROWNian search, Monte Carlo search) on the one hand and *simulated annealing* which is inspired by BOLTZMANNs statistics on the other hand.
- *Evolutionary computation* which is learned by watching population dynamics; the most important programming are *genetic algorithms* which are inspired by the evolutionary processes of mutation, recombination, and natural selection in biology.
- *Connectionist computation* which is inspired by multi-cellular systems like artificial neural networks that mimic the way of working of the human brain.

Simulated annealing is a category of global optimization procedures the name of which is derived from the statistical mechanics simulating the atomic equilibrium according to the BOLTZMANN statistics. By means of simulated annealing it is searched for the most probable configuration of parameters on the basis of simulating the evolution of a substance to thermal equilibrium (KIRKPATRICK et al. [1983]; FRANK and TODESCHINI [1994]). The distribution of configurations x can be described by the BOLTZMANN distribution

$$P(x) = \frac{\exp\left(-C(x)/c\right)}{\sum\limits_{w} \exp\left(-C(x_w)/c\right)} \tag{5.16}$$

where $C(x)$ is the function to be optimized, x_w are other configurations and c is a control parameter. From the initial configuration x another configuration x_r in the neighborhood of x is generated by modifying one randomly selected variable. The new configuration is accepted when the difference $\Delta C(x_r, x) \leq 0$, otherwise the probability

$$P = \exp(-\Delta C(x_r, x)/c) \tag{5.17}$$

is compared with a random number generated from a uniform distribution $[0, 1]$. If P is larger than that random number, the new configuration is also accepted, otherwise it is declined. The iteration is continued until convergence is reached. Afterwards the optimization runs are continued with lowered control parameter c. More detailed information on simulated annealing can be found in VAN LAARHOVEN and AARTS [1987]; KALIVAS [1992].

Genetic Algorithms (GA) are the most important global optimization techniques. GA base on mimicking the evolution process by variation of populations according to the DARWIN rules [DARWIN 1859] such as *selection*, *reproduction*, *crossover* (*recombination*), and *mutation*. Genetic algorithms have been pioneered by HOLLAND [1975], detailed representations can be found in GOLDBERG [1989]; DAVIS [1991], RECHENBERG [1973].

The initial data are binary coded in form of a bit sequence (*bit string*[4]). Start values of the variables $x_1, x_2, \ldots, x_m$ could be, e.g., 011001011, 100101100, $\ldots, 010010101$. This initial population is undertaken an evolution process such as schematically represented in Fig. 5.8.

In the course of each run corresponding to Fig. 5.8 the fitness of the members of the population is tested by means of an objective criterion (e.g., maximum correlation of a regression model or minimum random deviation of a response surface) that is compared with a break-off criterion fixed in advance. According to their fitness, the members from the present population (generation) are selected and reproduced by doubling. On the other hand, less fit members are omitted from the population. In the recombination step, parts of the bit-string are exchanged, namely by single-, two- or three-point-, uniform-, or circular crossover. In this way (by "mating"), from two parent bit strings two offspring are generated. Finally, by mutation only a small number of genes from the whole population is changed by flipping to the opposite value ($0 \to 1$ and $1 \to 0$, respectively).

For the selection and reproduction step the idea of "élitism" plays a role in so far as individuals of high quality should not become extinct. On the other hand, a larger number of élitists produces untimely a homogenisation of the population.

The advantage of Genetic Algorithms, in contrast to the traditional optimization methods, is the fact that a large number of variables can be included into the process. Also in the presence of local optima, GA can find rapidly the global optimum.

---

[4]   Also called "*chromosome*"; a bit in this chromosome is then called "*gen*".

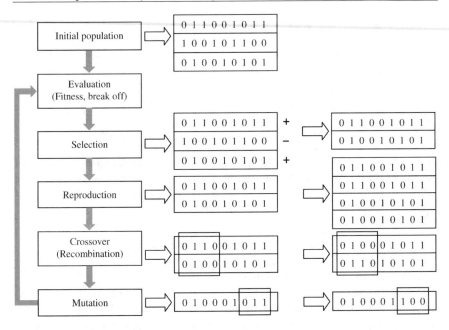

Fig. 5.8. Flow chart of a Genetic Algorithm

In many practical problems, interactions between the variables appear so that the absolute global optimum can be found heavily. As an example, wavelength selection in NIR determination of blood glucose (see Sect. 6.2.6) is considered. The aim of the selection is to find such combinations of wavelengths with which calibration models are obtained their prediction quality is as near at the global optimum as possible (DANZER et al. [2001], p 174). The number of combinations C for the selection of k wavelengths from n channels of the spectrometer is given by

$$C = \binom{n}{k} = \frac{n!}{k!(n-k)!} \tag{5.18}$$

So, for the selection of 15 wavelengths out of 86 the number of 2.1784×10^{16} combinations can be generated. Because of the existence of background, noise and strong multicollinearities in the NIR spectra, a large number of quasi-optimum solutions are available between which cannot be differentiated significantly. In Fig. 5.9 the wavelength selection 15 out of 86 is shown, carried out to improve the quality of the calibration model. Whereas the cross validated *PRESS* value $s^2_{(cv)res}$ (see Eq. 6.105) is 295.6 mmol^2/L^2 when 86 equidistant wavelengths are used, $PRESS_{CV}$ improves to 147.0 mmol^2/L^2 for 15 GA-selected wavelengths.

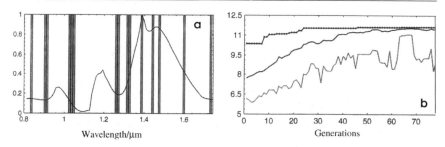

Fig. 5.9. Wavelength selection by Genetic Algorithm: **a** 15 optimum wavelength selected from 86 of the full spectrum; **b** relative fitness ($rf/10^4$) of the run in dependence of the number of generations, above: fitness of the best solution, middle: mean, below: fitness of the worst solution (Fɪsᴄʜʙᴀᴄʜᴇʀ et al. [1994/96;1995])

In connection with the NIR determination of blood glucose, GA has been used also to select spectra according to the quality of recordings (Fɪsᴄʜʙᴀᴄʜᴇʀ et al. [1994/96;1995]).

Optimization problems in general and, in particular in analytical chemistry, have been reported by Gᴏʟᴅʙᴇʀɢ [1989]; Dᴀᴠɪs [1991]; Lᴜᴄᴀsɪᴜs et al. [1991]; Lᴜᴄᴀsɪᴜs and Kᴀᴛᴇᴍᴀɴ [1993]; Wɪᴇɴᴋᴇ et al. [1992, 1993], and others.

In analytical chemistry, *Artificial Neural Networks* (ANN) are mostly used for calibration, see Sect. 6.5, and classification problems. On the other hand, feedback networks are usefully to apply for optimization problems, especially nets of Hᴏᴘꜰɪᴇʟᴅ type (Hᴏᴘꜰɪᴇʟᴅ [1982]; Lᴇᴇ and Sʜᴇᴜ [1990]).

References

Ahrens H (1967) *Varianzanalyse.* Akademie-Verlag, Berlin

Box GPE (1954) *The exploration and exploitation of response surfaces: some general considerations and examples.* Biometrics 10:16

Box GEP (1957) *Evolutionary operation: a method for increasing industrial productivity.* Appl Statist 6:81

Box GEP, Behnken DW (1960) *Some new three level designs for the study of quantitative variables.* Technometrics 2:445

Box GEP, Draper NR (1969) *Evolutionary operation.* Wiley, New York

Box GPE, Wilson KB (1951) *On the experimental attainment of optimum conditions.* J R Statist Soc B 13:1

Danzer K, Marx G (1979) *Application of the two-dimensional variance analysis for the investigation of homogeneity of solids.* Anal Chim Acta 110:145

Danzer K, Hobert H, Fischbacher C, Jagemann K-U (2001) *Chemometrik – Grundlagen und Anwendungen.* Springer, Berlin Heidelberg New York

Darwin C (1859) *On the origin of species by means of natural selection, or the preservation of favoured races in the struggle for life.* John Murray, London

Davis L (ed) (1991) *Handbook of genetic algorithms. Part I. A genetic algorithms tutorial.* Van Nostrand Reinhold, New York

Dixon W, Massey FJ (1983) *Introduction to statistical analysis.* McGraw-Hill, New York

Dunn OJ, Clark VA (1974) *Applied statistics - analysis of variance and regression.* Wiley, New York

Eisenhart C (1947) *The assumptions underlying the analysis of variance.* Biometrics 3:1

Fischbacher C, Jagemann K-U, Danzer K, Müller UA, Mertes B, Papenkorth L, Schüler J (1994/96) *Unpublished results*

Fischbacher C, Jagemann K-U, Danzer K, Müller UA, Mertes B (1995) *Non-invasive blood glucose monitoring by chemometrical evaluation of NIR-spectra.* 24th EAS (Eastern Analytical Symposium), Somerset, NJ, November 12-16

Fisher RA (1925) *Theory of statistical estimation.* Proc Cambridge Phil Soc 22:700

Fisher RA (1935) *The design of experiments.* Oliver & Boyd, Edinburgh (7th edn 1960; 9th edn 1971, Hafner Press, New York)

Frank IE, Todeschini R (1994) *The data analysis handbook.* Elsevier, Amsterdam

Goldberg DE (1989) *Genetic algorithms in search, optimization and machine learning.* Addison-Wesley, New York

Graf U, Henning H-U, Stange K, Wilrich P-Th (1987) *Formeln und Tabellen der angewandten mathematischen Statistik.* Springer, Berlin Heidelberg New York (3. Aufl)

Hald A (1960) *Statistical tables and formulas.* Wiley, New York

Holland JH (1975) *Adaption in natural and artificial systems.* University of Michigan Press, Ann Arbor, MI. Revised ed (1992) MIT Press, Cambridge, MA

Hopfield JJ (1982) *Neural networks and physical systems with emergent collective computational abilities.* Proc Nat Acad Sci USA 79:2554

Kalivas JH (1992) *Optimization using variations of simulated annealing.* Chemom Intell Lab Syst 15:1

Kateman G (1990) *Evolutions in chemometrics.* Analyst 115:487

Kirkpatrick S, Gelatt CD, Vecchi MP (1983) *Optimization by simulated annealing.* Science 220:671

Lee BW, Sheu BJ (1990) *Combinatorial optimization using competitive Hopfield neural network.* Proc Internat Joint Conf Neural Networks II: 627, Washington, DC

Lucasius, CB (1994) *Evolutionary computation: a distinctive form of natural computation with chemometric potential.* Chapter 9 in: Buydens LM , Melssen WJ: *Chemometrics. Exploring and exploiting chemical information.* Katholieke Universiteit Nijmegen

Lucasius CB, Kateman G (1993) *Understanding and using genetic algorithms. Part 1. Concepts, properties and context.* Chemom Intell Lab Syst 19:1

Lucasius CB, Blommers MJJ, Buydens LMC, Kateman G (1991) *A genetic algotithm for conformational analysis of DNA.* In: Davis L (ed) *Handbook of genetic algorithms.* Van Nostrand Reinhold, New York

Mandel J (1961) *Non-additivity in two-way analysis of variance.* J Am Statist Assoc 56:878

Neave HR (1981) *Elementary statistical tables.* Allen Unwin, London

Plackett RL, Burman JP (1946) *The design of optimum multifactorial experiments.* Biometrika 33:305

Rechenberg I (1973) *Evolutionsstrategie. Optimierung technischer Systeme nach Prinzipien der biologischen Evolution.* Frommann-Holzboog, Stuttgart

Sachs L (1992) *Angewandte Statistik, 7th edn.* Springer, Berlin Heidelberg New York

Scheffé H (1961) *The analysis of variance.* Wiley, New York

Sharaf MA, Illman DL, Kowalski BR (1986) *Chemometrics.* Wiley, New York

van Laarhoven PJM, Aarts EHL (1987) *Simulated annealing: theory and applications.* Reidel, Dordrecht

Wegscheider W (1996) *Validation of analytical methods.* In: H. Günzler (ed) *Accreditation and quality assurance in analytical chemistry.* Springer, Berlin Heidelberg New York

Wienke D, Lucasius C, Kateman G (1992) *Multicriteria target vector optimization of analytical procedures using a genetic algorithm. Part I. Theory, numerical simulations and application to atomic emission spectroscopy.* Anal Chim Acta 265:211

Wienke D, Lucasius C, Ehrlich M, Kateman G (1993) *Multicriteria target vector optimization of analytical procedures using a genetic algorithm. Part II. Polyoptimization of the photometric calibration graph of dry glucose sensors for quantitative clinical analysis.* Anal Chim Acta 271:253

6 Calibration in Analytical Chemistry

In measurement sciences, calibration is an operation that establish a relationship between an output quantity, q_{out}, with an input quantity, q_{in}, for a measuring system under specified conditions ($q_{in} \rightarrow q_{out}$). The result of calibration is a model that may have the form of a conversion factor, a mathematical equation, or a graph. By means of this model, then it is possible to estimate q_{in}-values from measured q_{out}-values ($q_{out} \rightarrow q_{in}$) as can be seen in an abstracted form in Fig. 6.1.

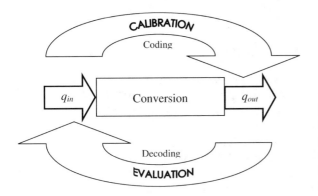

Fig. 6.1. Abstract representation of the calibration procedure by conversion of q_{in} into q_{out} and the reverse evaluation procedure

From the information-theoretical point of view, calibration corresponds to the *coding* of the input quantity into the output quantity and, vice versa, the evaluation process corresponds to *decoding* of output data. From the mathematical viewpoint, q_{in} is the independent quantity in the calibration step and q_{out} the dependent one. In the evaluation step, the situation is reverse: q_{out} is the independent, and q_{in} the dependent quantity. From the statistical standpoint, q_{out} is a random variable both in calibration and evaluation whereas q_{in} is a fixed variable in the calibration step and a random variable in the evaluation step. This rather complicated situation has some consequences on which will be returned in Sect. 6.1.2.

Table 6.1. Input quantities, q_{in}, and output quantities, q_{out}, in analytical calibration

q_{in}	q_{out}	Analytical technique	Calibration function
x	y	Quantitative analysis	$y = f(x)$ (see Sect. 6.2)
Q	z	Identification, qualitative analysis	$z = f(Q)$ (see Sect. 2.4)
$x(Q)$	$y(z)$	Quantitative multicomponent analysis	$Y = f(X)$ (see Sect. 6.4)
I, S	$y(z)$	Identity, structure analysis	Comparison with reference signal functions (see Sect. 2.4)

6.1
General Fundamentals of Calibration

In analytical chemistry, calibration represents a set of operations that connects quantities in the sample domain with quantities in the signal domain (see Sect. 2.3, Fig. 2.12). In Table 6.1 the real analytical quantities and properties behind the abstract input and output quantities are listed.

Calibration in analytical chemistry relates mostly to quantitative analysis of selected species and, therefore, to calibration functions of the kind $y = f(x)$.

6.1.1
Fundamental and Experimental Calibration

Depending on the type of relationships between the measured quantity and the measurand (analytical quantity) it can be distinguished (DANZER and CURRIE [1998]) between calibrations based on *absolute* measurements (one calibration is valid for all[1] on the basis of the simple proportion $y = b \cdot x$, where the *sensitivity factor* b is a fundamental quantity; see Sect. 2.4; HULA-NICKI [1995]; IUPAC ORANGE BOOK [1997, 2000]), *definitive* measurements (b is given either by a fundamental quantity complemented by an empirical factor or a well-known empirical (transferable) constant like molar absorption coefficient and NERNST factor), and *experimental calibration*.

In the simplest case, experimental calibration can be carried out by *direct reference* measurements where the sensitivity factor b is given by the relation of measured value to concentration of a reference material (RM), $b = y_{RM}/x_{RM}$. Direct reference calibration is frequently used in NAA and X-ray analytical techniques (XRF, EPMA, TXRF).

On the other hand, *indirect reference* measurements which result in an empirical calibration function, frequently based on a linear model, $y = a + b \cdot x + e_y$; see Sect. 2.4, Eq. (2.22), where the intercept a corresponds to

---

[1]   Apart from the fact that both the fundamental constants b and the conditions under which they are valid (e.g. under which a reaction proceeds quantitatively) sometime were determined and have to be updated from time to time (e.g., on the basis of IUPAC documents).

the experimental *blank* and the slope b to the experimental *sensitivity*; e_y is the error of the y-measurement.

In analytical practice, some methods using definitive measurements, in principle, are also calibrated by indirect reference measurements using least squares estimating to provide reliable estimates of b (spectrophotometry, potentiometry, ISE, polarography).

Because calibration is the prerequisite of reliable evaluations and, therefore, of analytical results which are both accurate and precise, calibration itself has to be carried out in a very reliable way. For this reason, the following experimental and fundamental conditions have to be realized:

(i)   The *standards* used for the calibration measurements should be both reliable and *traceable*. *Traceability* is the *property of the result of a measurement or the value of a standard whereby it can be related to stated references, usually national or international standards, through an unbroken chain of comparisons all having stated uncertainties* (ISO [1993]; HÄSSELBARTH [1995]). The connection to the *International System of Units* (*SI*, ISO [1993]) is given in particular by the amount of substance (mole) and mass (kg) and is realized in detail by *certified reference materials* (CRMs). If laboratory standards are used for calibration then these have to be validated by CRMs.

(ii) The *measurement strategy* of experimental calibration have to be fixed. Specifically:
      – The independence of the diverse calibration samples have to be guaranteed, that means that each sample must be prepared separately and not by down-diluting of a master sample.
      – The number of calibration points, p, their distance and measure at the concentration scale, the number of replicate measurements, n_i, in each point carried out with independent calibration samples (and not by repetition measurements on one and the same sample), and their distribution at the calibration points (equally distributed, $n_1 = n_2 = \ldots = n_p$, or in large numbers at the ends of the calibration range, $n_1 > n_2 = n_3 = \ldots = n_{p-1} < n_p$).

(iii) The type of the *calibration model*, linear or nonlinear, univariate, bivariate, or multi-variate, respectively.

(iv) The *statistical character* of the variables and, therefore, the type of the regression model (DANZER et al. [2004]), classical (reverse calibration, direct or indirect) or inverse, respectively, unweighted or weighted.

As mentioned above, the random character of the input and output variables are of importance with regard to the calibration model and its estimation by calculus of regression. Because of the different character of the analytical quantity x in the calibration step (no random variables but fixed variables which are selected deliberately) and in the evaluation step (random variables like the measured values), the closed loop of Fig. 6.1 does not correctly describe the situation. Instead of this, a linear progress as shown in Fig. 6.2 takes place.

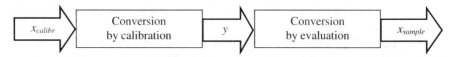

Fig. 6.2. Real representation of the calibration procedure in analytical chemistry

As a consequence, the relationship between y on the one hand and x_{calibr} and x_{sample} on the other hand has to be represented in a three-dimensional way (DANZER [1995]).

6.1.2
The General Three-Dimensional Calibration Model

The three-dimensional model of calibration and the spatial relationships of *calibration*, *evaluation*, and *recovery* is given in Fig. 6.3.

It should be noted that the spatial model contains additionally to the *calibration* and *evaluation* functions the *recovery* function, too. These three functions represent the projection of the *spatial C-E-R function* onto the corresponding planes, as shown in Fig. 6.3.

From the different character of $x_{standard}$ and x_{sample} (Figs. 6.2 and 6.3) it can be seen that the relationship between the calibration function

$$y = f(x) = a + b \cdot x + e_y \tag{6.1}$$

and the evaluation function

$$x = f^{-1}(y) = \frac{y - a}{b} + e_x \tag{6.2}$$

must not be necessarily reversible. This fact has to be checked by validation. Commonly the equivalence is tested by recovery experiments to rule out systematic deviations (see Sect. 4.1.1). The confirmation that $x_{test} = x_{true}$, i.e. $x_{standard} = x_{sample}$ in the calibration procedure, is an important step to ensure the *traceability* of analytical results on the basis of the given calibration. Only in the case when the mutual equivalence of Eqs. (6.1) and (6.2) in the three-dimensional model have been proved, it can be used – as usual – in a two-dimensional way. Pictorially this may be imagined by folding up the planes of the calibration function and of the evaluation function (B and C in Fig. 6.3) to a single plane containing a common calibration-evaluation function as it is usual in analytical chemistry.

Consequences of various types of systematic deviations between x_{test} and x_{true}, and therefore, $x_{standard}$ and x_{sample} are discussed in DANZER [1995].

The different statistical character of the three variables becomes most clear in the different uncertainties of the calibration and evaluation lines. Notwithstanding the fundamental differences between $x_{standard}$ and x_{sample}, the calculation of the calibration coefficients is carried out by regression calculus.

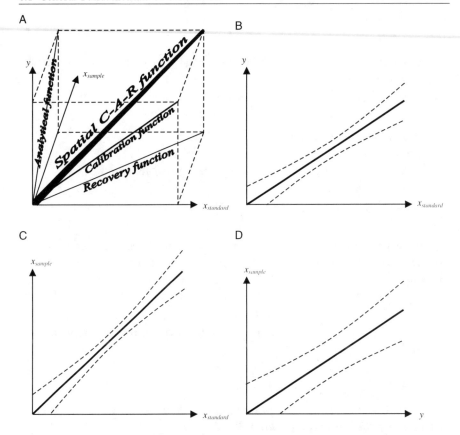

Fig. 6.3. Three-dimensional model of calibration, analytical evaluation and recovery: spatial model (**A**); the three relevant planes are given separately in (**B**) as the calibration function with confidence interval, in (**C**) as the recovery function with confidence interval, and in (**C**) as the evaluation function with prediction interval (**D**)

6.1.3
Regression and Calibration

Having two random variables, x and y, e.g. values measured independently from each other and each of them being normally distributed, then the following questions may be of interest:

(i) Does a relation exist between y and x and how strong is it?

(ii) How the relation can functionally be described, i.e. how can y be estimated from x and vice versa?

Whereas the first question is answered by *correlation analysis*, the second one is subject of *regression analysis*.

   Correlation analysis investigates stochastic relationships between *random* variables on the basis of samples. The interdependence of two variables x

and y is characterized by the *correlation coefficient* r_{xy} (PEARSON's *correlation coefficient*)[2].

For the calculation of measures in correlation and regression analysis the following sums are of relevance:

$$S_{xx} = \sum_{i=1}^{n} (x_i - \bar{x})^2, \qquad S_{yy} = \sum_{i=1}^{n} (y_i - \bar{y})^2,$$

$$S_{xy} = \sum_{i=1}^{n} (x_i - \bar{x})(y_i - \bar{y}) \tag{6.3'}$$

With this the correlation coefficient of x and y is calculated by

$$r_{xy} = \frac{S_{xy}}{\sqrt{S_{xx} S_{yy}}} = \frac{S_{xy}}{\sqrt{s_x^2 s_y^2}} = \frac{S_{xy}}{s_x s_y} \tag{6.3}$$

As the last term in Eq. (6.3) shows, the correlation coefficient corresponds to the covariance of x and y, $\mathrm{cov}(x, y) = s_{xy}$, divided by the standard deviations s_x and s_y.

Concrete values of correlation coefficients indicate the following situations:

- $r_{xy} = +1$, $r_{xy} = -1$ perfect (positive and reverse, respectively) interdependence of x and y
- $0 < |r_{xy}| < 1$ stochastic interdependence
- $r_{xy} = 0$ missing dependence.

The pairwise correlation of more than two variables $x_1, x_2, \ldots, x_m$ is characterized by the *correlation matrix* R

$$R = \begin{pmatrix} 1 & r_{x_1 x_2} & \cdots & r_{x_1 x_m} \\ r_{x_2 x_1} & 1 & \cdots & r_{x_2 x_m} \\ \vdots & \vdots & & \vdots \\ r_{x_m x_1} & r_{x_m x_2} & \cdots & 1 \end{pmatrix} \tag{6.4}$$

In multivariate data analysis frequently the *covariance matrix* S is used

$$S = \begin{pmatrix} \mathrm{var}(x_1) & \mathrm{cov}(x_1, x_2) & \cdots & \mathrm{cov}(x_1, x_m) \\ \mathrm{cov}(x_2, x_1) & \mathrm{var}(x_2) & \cdots & \mathrm{cov}(x_2, x_m) \\ \vdots & \vdots & & \vdots \\ \mathrm{cov}(x_m, x_1) & \mathrm{cov}(x_m, x_2) & \cdots & \mathrm{var}(x_m) \end{pmatrix}$$

$$= \begin{pmatrix} s_{x_1}^2 & s_{x_1 x_2} & \cdots & s_{x_1 x_m} \\ s_{x_2 x_1} & s_{x_2}^2 & \cdots & s_{x_2 x_m} \\ \vdots & \vdots & & \vdots \\ s_{x_1 x_2} & s_{x_1 x_2} & \cdots & s_{x_m}^2 \end{pmatrix} \tag{6.5}$$

---

[2] The correlation coefficient r_{xy} of a sample is an estimate of the correlation coefficient ρ_{xy} of the population.

In contrast to correlation matrix the covariance matrix is scale-dependent. In case of autoscaled variables the covariance matrix equals the correlation matrix.

In analytical calibration, there exists – strictly speaking – no correlation problem for the following two reasons:

(1) The interdependence of the measured values, y, and the analytical values, x, is well-known a priori – mostly by natural laws – and is, therefore, not subject of verification as a rule

(2) The analytical values of the calibration standards, $x_{standard}$, are *no random variables* but fixed one and carefully selected.

Two consequences result from this fact:

- The correlation coefficient, which is a characteristic for the relationship between random variables, is not meaningful in calibration (CURRIE [1995]; DANZER and CURRIE [1998]) and should, therefore, not be used to characterize the quality of calibration (instead of r_{xy} the residual standard deviation $s_{y.x}$ should be applied; see Eq. (6.19)).

- From the diverse possible regression calculi, a certain algorithm has to be selected, namely that of the regression of y onto x; see Eq. (6.6).

The general pre-conditions for the estimation of regression parameters (regression coefficients a and b as well as the uncertainties of the model and the relevant estimates) are the following:

(i) Real replication measurements must be carried out (runs with the same treatment), not only repeated measurements at the same samples

(ii) Normal distribution of the x- and y-values

(iii) Homoscedasticity, i.e. homogeneity of variances in the diverse calibration points

Depending on whether x can be considered to be free or almost free of errors ($s_x \ll s_y/b$) or vice versa[3] ($s_y/b \ll s_x$), a regression of y onto x

$$\hat{y} = a_x + b_x x \qquad (6.6)$$

by minimizing the y-deviations according to the GAUSSian least squares criterion or a regression of x onto y

$$\hat{x} = a_y + b_y y \qquad (6.7)$$

by minimizing the x−deviations have to be carried out.

The situation is illustrated in Fig. 6.4 from which can be seen that the reversion of the dependent and independent variables gives different estimates $\hat{y}$ and $\hat{x}$.

The regression coefficients in Eq. (6.6), b_x and a_x, for the estimation of y from x are estimated by the following expressions:

---

[3]   This model usually does not have any relevance in analytical calibration.

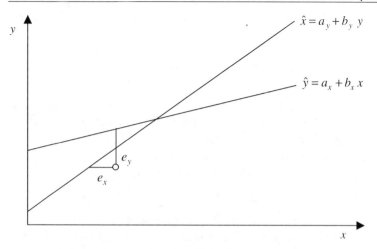

Fig. 6.4. Different regression models as a result of the minimization of the x- and the y-deviations, respectively

$$b_x = \frac{S_{xy}}{S_{xx}} \qquad\qquad (6.8)$$

$$a_x = \frac{\sum y - b_x \sum x}{n} \quad . \qquad\qquad (6.9)$$

Because of the irrelevant condition $s_y/b \ll s_x$), the reverse model, viz. the estimation of x from y, according to Eq. (6.7) and therefore the coefficients b_y and a_y, are not of direct relevance in analytical calibration. Notwithstanding, their estimates will be given here for completion and as auxiliary quantities for further calculations:

$$b_y = \frac{S_{xy}}{S_{yy}} \quad , \qquad a_y = \frac{\sum x - b_y \sum y}{n} \quad . \qquad\qquad (6.10)$$

All the special cases that may be appear in analytical calibration in dependence from the specific conditions as well as the uncertainties of the calibration measures will be given in the following.

6.2
Single Component Calibration

6.2.1
Linear Calibration Model

On the condition that the errors of the measurement have a zero mean and are uncorrelated, the linear function (Eq. 6.1) can be fitted to the measured values by means of the GAUSS*ian least squares estimation* (GLS, or *ordinary least square estimation,* OLS, respectively; see DANZER and CURRIE [1998]).

With the fundamental relations

- Model: $\quad\quad y_i = a + b\,x_i + e_{y_i}$ $\hfill$ (6.11a)

- Estimate: $\quad\quad \hat{y}_i = \hat{a} + \hat{b}\,x_i$ $\hfill$ (6.11b)

- Residual: $\quad\quad e_{y_i} = y_i - \hat{y}_i = y_i - \hat{a} - \hat{b}\,x_i$ $\hfill$ (6.11c)

the general least squares criterion expressed by the *sum of squares of residuals,* ssr(y), reads (DANZER [1990])

$$ssr(y) = \sum_{i=1}^{n}\left(\frac{y_i - \hat{y}_i}{s_i}\right)^2 = \sum_{i=1}^{n}\left(\frac{e_{y_i}}{s_i}\right)^2 \hspace{2cm} (6.12)$$

where s_i is the estimated standard deviation at the given point i and n the total number of calibration measurements which are equal to the sum of measurements n_j in each calibration point p

$$n = \sum_{j=1}^{p} n_j \hspace{2cm} (6.13a)$$

which is simplified to

$$n = p \cdot n_j \hspace{2cm} (6.13b)$$

if an equal number n_j of repetitions in each calibration point is carried out.

The sum of squares of residuals has to be minimized according to the general least squares (LS) criterion

$$\sum_{i=1}^{n}\left(\frac{e_{y_i}}{s_i}\right)^2 \overset{!}{=} \min \hspace{2cm} (6.14)$$

Depending on the fulfilment of the conditions mentioned above, namely (i) to (iii) and following, the least squares criterion has to be modified as follows:

1. The errors are only or essentially in the measured values y as the dependent variable ($bs_x \ll s_y$) and in addition, the errors s_y are constant in the several calibration points (*Homoscedasticity*):

$$s_{y_1}^2 \overset{\alpha}{=} s_{y_2}^2 \overset{\alpha}{=} \cdots \overset{\alpha}{=} s_{y_p}^2 \overset{\alpha}{=} s_y^2 \hspace{2cm} (6.15)$$

where $\overset{\alpha}{=}$ means equality for a given risk of error α. Only in this homoscedastic case and if errors in x can be neglected, the LS criterion is reduced to

$$\sum_{i=1}^{n} e_{y_i}^2 \overset{!}{=} \min \hspace{2cm} (6.16)$$

and the ordinary least squares estimation of the calibration model according to Eqs. (6.6), (6.8) and (6.9) can be applied. The evaluation of analyses carried out on the basis of the inverted calibration function (Eq. 6.6)

$$\hat{x} = \frac{\hat{y} - \hat{a}}{\hat{b}} \qquad (6.17)$$

and not according to Eq. (6.7) that holds for other statistical assumptions. On the other hand, CENTNER et al. [1998] found that predictions on the basis of inverse calibration (Eq. 6.7) may be more reliable, especially in the case of less precise measurements.

2. In case that the measuring errors s_y vary and therefore *heteroscedasticity* must be assumed, the original LS criterion (Eq. 6.14) must be applied and the model of *weighted least squares* (WLS) results from this criterion as will be shown in Sect. 6.2.3.

3. In the most general case, if both variables are subject to error and, therefore, neither the condition $s_x \ll s_y/b$ nor $s_y/b \ll s_x$ is fulfilled, the following total error results:

$$s_i^2 = s_{y_i}^2 + b^2\, s_{x_i}^2 \quad . \qquad (6.18)$$

In this case, in which there are errors in both variables, y and x, the resulting error e_{x+y}^2 has to be minimized (see Fig. 6.5) and *orthogonal least squares fitting* must be carried out as will be shown in Sect. 6.2.4.

4. Sometimes the basic conditions for the use of LS are not fulfilled, either by measurement values being not-normally distributed or by strongly

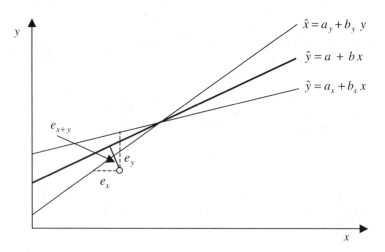

Fig. 6.5. Regression model of errors in both variables in comparison to the model given in Fig. 6.4

deviating calibration points (*leverage points*, see Rousseeuw and Leroy [1987]). In such cases, methods of *robust calibration* may be used; see Sect. 6.2.6.

5. In addition to statistical peculiarities, special features may also result from certain properties of samples and standards which make it necessary to apply special calibration techniques. In cases when matrix effects appear and matrix-matched calibration standards are not available, the *standard addition method* (SAM, see Sect. 6.2.6) can be used.

Measurements in living things, e.g. patients, is occasionally done by sampling in a non-invasive way. Because no "standard patients" are available which can be selected for calibration, another principle of calibration must be applied, e.g., by a *reference method* that measures parallel to the actual measuring method. So, the calibration consists in a comparison of the measured results with that of the reference method which are considered to be true (that have been validated beforehand). An example of a *reference calibration* is represented by non-invasive blood glucose determination by means of NIR spectroscopy (Müller et al. [1997]; Fischbacher et al. [1997]; Danzer et al. [1998], see Sect. 6.2.6).

The several variants deriving from the items 1 to 4 are represented in the flow sheet given in Fig. 6.6. Common calibration by Gaussian least squares estimation (OLS) can only be applied if the measured values are independent and normal-distributed, free from outliers and leverage points and are characterized by homoscedastic errors. Additionally, the error of the values in the analytical quantity x (measurand) must be negligible compared with the errors of the measured values y.

From the chemical point of view, in cases where matrix effects appear and no suitable certified reference materials are available, the calibration may be performed in the sample matrix itself by means of standard addition.

In analytical practice, linear calibration by ordinary least squares is mostly used. Therefore, the estimates are summarized before the uncertainties of the estimates will be given:

OLS estimates:
$$\hat{y} = a_x + b_x\, x \tag{6.6}$$

Evaluation estimate:
$$\hat{x} = \frac{\hat{y} - a_x}{b_x} \tag{6.17}$$

Slope (sensitivity):
$$b_x = \frac{S_{xy}}{S_{xx}} \tag{6.8}$$

Intercept ("blank"):
$$a_x = \frac{\sum y - b_x \sum x}{n} \tag{6.9}$$

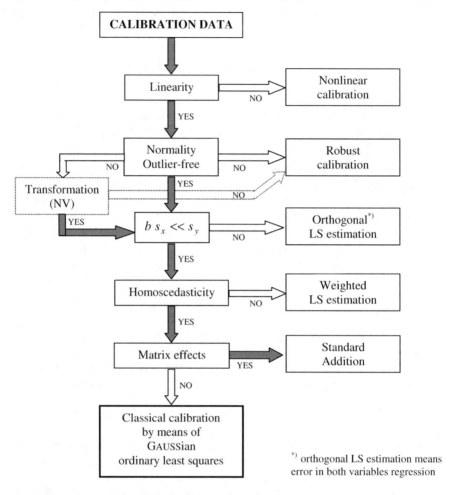

Fig. 6.6. Flow sheet of several calibration models in dependence of the fulfilment of certain statistical and chemical conditions

6.2.2
Errors in Linear Calibration

Fundamentally, the uncertainties of measured values $\hat{y}$ estimated by calibration, e.g. according to Eq. (6.6), on the one hand and of analytical results $\hat{x}$ (analyte contents, concentrations) estimated by means of a calibration function, e.g. according to Eq. (6.17), on the other hand differ from one another as can be seen from Fig. 6.3B,C, and Fig. 6.7. Whereas the uncertainty of y values in calibration is characterized by the *confidence interval cnf($\hat{y}$)*, the uncertainty of estimated x values is characterized by the *prediction interval prd($\hat{x}$)*.

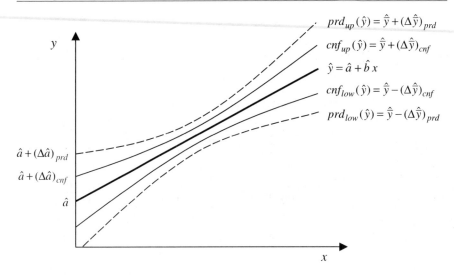

Fig. 6.7. Calibration straight line with relevant upper and lower confidence and prediction bands

The uncertainty of OLS calibration is characterized by the following special standard deviations and uncertainty intervals:

(1)   *Residual standard deviation*

$$s_{y.x} = \sqrt{\frac{\sum\limits_{i=1}^{n}(y_i - \hat{y}_i)^2}{n-2}} = \sqrt{\frac{\sum\limits_{i=1}^{n}(y_i - \hat{a}_x - \hat{b}_x x_i)^2}{n-2}} \qquad (6.19)$$

In calibration, the number of degrees of freedom depends on the number of parameters estimated by the given model. In case of the two-parametric model (Eq. 6.6) $v = n - 2$, in case of linear calibration through the coordinate origin $(a = 0)$ $v = n - 1$, and in case of a three-parametric nonlinear calibration model $(y = a + bx + cx^2)$ $v = n - 3$.

(2)   *Estimated standard deviation (ESD) of the estimated intercept (blank) â*

$$s_{\hat{a}} = s_{y.x}\sqrt{\frac{1}{n} + \frac{\bar{x}^2}{S_{xx}}} \qquad (6.20)$$

with n being the total number of calibration measurements, $\bar{x}$ the mean of all the x values of the calibration experiment, and S_{xx} the sum of the squared x deviations.

(3)   *ESD of the estimated slope* $\hat{b}$

$$s_{\hat{b}} = \frac{s_{y.x}}{\sqrt{S_{xx}}} \tag{6.21}$$

(4)   *ESD of an estimated mean* $\hat{\bar{y}}$ *at position* x_i

$$s_{\hat{\bar{y}}} = s_{y.x} \sqrt{\frac{1}{n} + \frac{(x_i - \bar{x})^2}{S_{xx}}} \tag{6.22}$$

(5)   *ESD of a predicted single value* $\hat{y}_{prd}$ *at position* x_i

$$s_{\hat{y}_{prd}} = s_{y.x} \sqrt{1 + \frac{1}{n} + \frac{(x_i - \bar{x})^2}{S_{xx}}} \tag{6.23}$$

(6)   *ESD of a predicted mean* $\hat{\bar{y}}_{prd}$ *from N repetitions in position* x_i

$$s_{\hat{\bar{y}}_{prd}} = s_{y.x} \sqrt{\frac{1}{N} + \frac{1}{n} + \frac{(x_i - \bar{x})^2}{S_{xx}}} \tag{6.24}$$

(7)   *ESD of a predicted mean* $\hat{\bar{x}}_{prd}$ *from N repetitions for a measured value* y_i

$$s_{\hat{\bar{x}}_{prd}} = \frac{s_{y.x}}{b} \sqrt{\frac{1}{N} + \frac{1}{n} + \frac{(y_i - \bar{y})^2}{b^2 S_{xx}}} \tag{6.25}$$

(8)   The *confidence band, CB, of the entire calibration straight line* as shown in Fig. 6.7 is given by

$$CB = \hat{y} \pm s_{\hat{y}} \sqrt{2F_{1-\alpha, v_1=2, v_2=n-2}} \tag{6.26a}$$

and the *prediction band, PB*

$$PB = \hat{y}_{prd} \pm s_{\hat{y}_{prd}} \sqrt{2F_{1-\alpha, v_1=2, v_2=n-2}} \tag{6.26b}$$

The following uncertainty intervals resulting from Eqs. (6.19) to (6.25) are of practical interest:

(9)   *Confidence interval of the intercept* (blank) $\hat{a}$

$$cnf(\hat{a}) = \hat{a} \pm s_{\hat{a}} \, t_{1-\alpha, v=n-2} \tag{6.27}$$

(10) *Prediction interval prd(a) of a single a-value* by Eq. (6.28) and that of *an average ā from N repetition measurements* according to Eq. (6.29)

$$prd(a) = a \pm s_{y.x}\, t_{1-\alpha, v=n-2} \sqrt{1 + \frac{1}{n} + \frac{\bar{x}^2}{S_{xx}}} \qquad (6.28)$$

$$prd(\bar{a}) = \bar{a} \pm s_{y.x}\, t_{1-\alpha, v=n-2} \sqrt{\frac{1}{N} + \frac{1}{n} + \frac{\bar{x}^2}{S_{xx}}} \qquad (6.29)$$

The latter is important with regard to the estimation of the detection limit from blanks

(11) *Confidence interval of an estimated mean $\hat{\bar{y}}_i$ at position x_i*

$$cnf(\hat{\bar{y}}_i) = \hat{\bar{y}}_i \pm s_{y.x}\, t_{1-\alpha, v=n-2} \sqrt{\frac{1}{n} + \frac{(x_i - \bar{x})^2}{S_{xx}}} \qquad (6.30)$$

(12) *Prediction interval of a single value $\hat{y}_{prd}$ at position x_i*

$$prd(\hat{y}_i) = \hat{y}_i \pm s_{y.x}\, t_{1-\alpha, v=n-2} \sqrt{1 + \frac{1}{n} + \frac{(x_i - \bar{x})^2}{S_{xx}}} \qquad (6.31)$$

(13) *Prediction interval of a mean $\hat{\bar{y}}_{prd}$ from N repetitions at position x_i*

$$prd(\hat{\bar{y}}_i) = \hat{\bar{y}}_i \pm s_{y.x}\, t_{1-\alpha, v=n-2} \sqrt{\frac{1}{N} + \frac{1}{n} + \frac{(x_i - \bar{x})^2}{S_{xx}}} \qquad (6.32)$$

(14) *Prediction interval of a mean $\hat{\bar{x}}_{prd}$ from N repetitions for a measured value y_i*

$$prd(\hat{\bar{x}}_i) = \hat{\bar{x}}_i \pm \frac{s_{y.x}}{b}\, t_{1-\alpha, v=n-2} \sqrt{\frac{1}{N} + \frac{1}{n} + \frac{(y_i - \bar{y})^2}{b^2\, S_{xx}}} \qquad (6.33)$$

6.2.3
Weighted Linear Least Squares Estimation (WLS)

In cases in that homoscedasticity according to Eq. (6.15) is not given, the estimated standard deviation is frequently a function of the measured quantity, $s_y = f(y)$. In this case the calibration system is *heteroscedastic* and *weighted least squares fitting* has to be applied (GARDEN et al. [1980]; DRAPER and SMITH [1981]). In general, least squares fitting starts from Eq. (6.14) and only in the case of homoscedasticity Eq. (6.16) can be taken as the basis.

From Eq. (6.14) it suggests itself that squared variance weighting $w_{y_i} \sim 1/s_{y_i}^2$ should be applied. In analytical practice, frequently relative weights

$$w_{y_i} = \frac{\dfrac{1}{s_{y_i}^2}}{\dfrac{1}{p} \sum_{i=1}^{p} \dfrac{1}{s_{y_i}^2}} \tag{6.34}$$

are used for the p calibration points in which each n_j repetitions are made, see Eq. (3.13a,b).

The minimizing criterion (Eq. 6.14) now becomes

$$\sum_{i=1}^{n} w_{y_i} e_{y_i}^2 \overset{!}{=} \text{min} \tag{6.35}$$

and the calibration coefficients are calculated analogous to Eqs. (6.8) and (6.9) by means of the weights ($w_{y_i} = w_i$)

$$b_{x_w} = \frac{n \sum w_i x_i y_i - \sum w_i x_i \sum w_i y_i}{n \sum w_i x_i^2 - \left(\sum w_i x_i \right)^2} \tag{6.36}$$

$$a_{x_w} = \frac{\sum w_i y_i - b_{x_w} \sum w_i x_i}{n} \tag{6.37}$$

As a rule the calibration coefficients do not alter significantly when WLS is applied instead of OLS. On the other hand, the uncertainty in the lower range of calibration is reduced in a remarkable way. The estimate of the residual standard deviation is

$$s_{(y.x)_w} = \sqrt{\frac{\sum w_i (y_i - \hat{y}_i)^2}{n - 2}} \tag{6.38}$$

Other quantities characterizing uncertainties can be estimated in analogy to Eqs. (6.20)–(6.33). Today, software packages for regression analysis usually allow one to enter an estimate of the functional dependence $s_y = f(y)$ and to carry out a suitable weighting with this or other functions offered.

In general, the decision on weighted or unweighted least squares can be reached on the basis of a statistical test, (see Eq. 6.2.5), or on the basis of a theoretical model.

6.2.4
Linear Least Squares Fitting in Case of Errors in Both Variables

In Fig. 6.5 three different calibration functions are given. First, the model to estimate y from (practically) error-free x values. This model is commonly used for analytical calibration in form of OLS, $\hat{y} = a_x + b_x x$ (Eq. 6.6). Another model (usually without relevance in analytical calibration because the measured values y are considered to be practically error-free) has been formulated in Eq. (6.7), $\hat{x} = a_y + b_y y$. The third model represented in Fig. 6.5,

$$\hat{y} = a + b\,x \tag{6.39}$$

is the so-called *error in both variables model* (*EBV model*) and has to be estimated by orthogonal least squares minimizing, that means the errors in both the dependent and the independent variable are minimized simultaneously. So, the general condition at Eq. (6.14) becomes, with Eq. (6.18):

$$\sum_{i=1}^{n} \frac{e_{y_i}^2}{s_{y_i}^2 + b^2 s_{x_i}^2} \stackrel{!}{=} \min \tag{6.40}$$

The EBV model has to be applied if both the measured values y and the analytical value x are error-affected quantities. The calibration coefficients of the model at Eq. (6.39) cannot be determined directly for the general case, but only according to certain assumptions or by approximations (DANZER et al. [1995]) from which three will be given here:

(1) MANDEL's approximation (MA, see MANDEL [1984])

$$\hat{b}_{MA} = \frac{S_{yy} - S_{xx} + \sqrt{(S_{xx} - S_{yy})^2 + 4 S_{xy}^2}}{2\,S_{xy}} \tag{6.41}$$

with S_{xx}, S_{yy}, and S_{xy} as defined in Eq. (6.3').

(2) WALD's approximation (WA, see WALD [1940]; SHARAF et al. [1986])

$$\hat{b}_{WA} = \frac{\displaystyle\sum_{i=1}^{g} y_i - \sum_{j=h}^{n} y_j}{\displaystyle\sum_{i=1}^{g} x_i - \sum_{j=h}^{n} x_j} \tag{6.42}$$

with n being the number of calibration measurements and the index $g = n/2 = h - 1$ if n is even, and $g = \frac{1}{2}(n + 1) = h$ if n is uneven.

(3) Geometrical averaging (GA, see DANZER [1990]; DANZER et al. [1995])

$$\hat{b}_{GA} = \tan\left(\frac{\arctan b_x + \arctan b_y^{-1}}{2}\right) \tag{6.43}$$

with b_x according to Eq. (6.8) and b_y according to Eq. (6.10). The intercepts $\hat{a}_{MA}$, $\hat{a}_{WA}$, and $\hat{a}_{GA}$ are each estimated analogously to Eq. (6.9) using $\hat{b}_{MA}$, $\hat{b}_{WA}$, and $\hat{b}_{GA}$ instead of b_x.

Furthermore it should be mentioned that the first principal component p_1 of a *principal component analysis* (PCA) gives a good approximation of the orthogonal calibration line (DANZER et al. [1995]). Principal component analysis is a useful method to extract new uncorrelated variables from a data matrix X. The mathematical fundamentals of PCA can be found in Sect. 8.3, MALINOWSKI and HOWERY [1980]; FLURY [1988]; MARTENS and NÆS [1989]. The PC principle may be explained graphically as follows: In an m-dimensional data space a new coordinate system is spread by principal components in such a way that the first principle component p_1 stretches in the direction of the largest variation of the data, the second one, p_2, in the direction of the largest remaining variation, orthogonal to p_1, and so on. In case of calibration mean-centered data have to be used and the complete principal component solution is relevant. There exist only two principle components, p_1 and p_2, where the first, p_1, fits very good the EBV calibration straight line and the second, p_2, represents the orthogonal scattering of the data around the straight line.

The complete principal component decomposition of the data matrix X into a score matrix P and a loading matrix P is given by

$$X = P \times L^T. \tag{6.44}$$

By means of the transposed loading matrix

$$L^T = \begin{pmatrix} l_{11} & l_{12} \\ l_{21} & l_{22} \end{pmatrix} \tag{6.45}$$

the slope of the EBV (orthogonal) calibration line can be estimated according to

$$\hat{b}_{PC} = \frac{l_{12}}{l_{11}} \tag{6.46}$$

and $\hat{a}_{PC}$ again analogously to Eq. (6.9) using $\hat{b}_{PC}$ instead of b_x.

Diverse procedures of EBV calibration and OLS have been compared for both simulated and real analytical data in DANZER et al. [1995]. Especially in cases when large errors in the x values (concentration) exist, EBV calibration should be used instead of OLS.

6.2.5
Statistical Tests and Validation of Calibration

A calibration procedure has to be validated with regard to general and specific requirements under which the calibration model has been developed. For this purpose, it is important to test whether the conditions represented in Fig. 6.6 are fulfilled. On the other hand, it is to assure by experimental studies that certain performance features (*accuracy, precision, sensitivity, selectivity, specificity, linearity, working range, limits of detection* and *of quantification, robustness,* and *ruggedness,* see Chap. 7) fulfil the expected requirements.

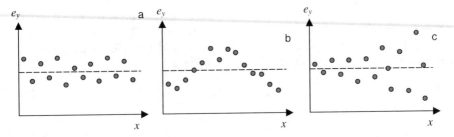

Fig. 6.8. Typical plots of residual deviations: random scattering (**a**), systematic deviations indicating nonlinearity (**b**), and trumpet-like form of heteroscedasticity (**c**)

The underlying calibration procedure of a newly developed analytical method has to be examined by *basic validation* studies to determine the reliability of the method and its efficiency in comparison with traditional methods. In order to ensure long-term stability, it is necessary to perform *revalidations*, which can be combined with the use of quality control charts, over meaningful time periods.

Regarding calibration, it is of special importance to characterize the following features:

(i)   *Accuracy* of analytical results

(ii) *Precision* of calibration and analytical values

(iii) *Calibration model* (linear/nonlinear) and *scope* (working range, sensitivity, limits of application/detection/quantitation)

Commercial software packages are usually able to represent graphically the residual errors (deviations) of a given calibration model which can be examined visually. Typical plots as shown in Fig. 6.8 may give information on the character of the residuals and therefore on the tests that have to be carried out, such as randomness, normality, linearity, homoscedasticity, etc.

Accuracy. In general, the accuracy of analytical results is assured by *recovery studies* (WEGSCHEIDER [1996]; DANZER [1995]; BURNS et al. [2002]). According to the recovery function in the general three-dimensional calibration model (see Fig. 6.3), common studies on systematic deviations (Fig. 4.3), and Eqs. (4.2) and (4.3) the following *recovery formulae*

$$x_{test} = \alpha + \beta\, x_{true} \quad (x_{sample} = \alpha + \beta\, x_{standard}) \tag{6.47a}$$

$$x_{test(I)} = \alpha + \beta\, x_{true(II)} \tag{6.47b}$$

have to be tested. By means of selected *standard samples* (if possible, *certified reference materials*) with known ("*true*") analyte amounts the *recovery function* (*validation function* (see DANZER [1995]; BURNS et al. [2002]) has to be determined by normal linear regression[4]. The measures α and β repre-

---

4   It should be noted that a recovery factor obtained using a single reference material or single spike addition experiment does not indicate the absence of systematic

sent here the *validation coefficients* with the analytical meaning of a constant bias (intercept α) and a proportional bias (slope β). The estimates of α and β can be determined analogous to Eqs. (6.8) and (6.9). By testing the null hypotheses $H_0: \alpha = 0$ and $H_0: \beta = 1$ according to

$$\hat{t}_{(\alpha)} = \frac{\alpha}{s_\alpha} \tag{6.48a}$$

$$\hat{t}_{(\beta)} = \frac{\beta}{s_\beta} \tag{6.48b}$$

the absence of systematic deviations can be verified. For this purpose, the estimated $\hat{t}$-values have to be compared with the corresponding quantiles of the t-distribution $t_{1-\alpha, v=n-2}$.

Systematic deviations are also detected if the corresponding confidence intervals of the validation coefficients do not include 0 or 1, respectively, namely

(i)   An *additive bias*   if $\alpha + \Delta\alpha < 0$ or $\alpha - \Delta\alpha > 0$ $(\alpha > |\Delta\alpha|)$

(ii) A *proportional bias*   if $\beta + \Delta\beta < 1$ or $\beta - \Delta\beta > 1$ $(\beta > |\Delta\beta|)$

Another way to verify the accuracy of a method (I) under validation is the analysis of a given set of test samples with graduated analyte concentrations by both methods, (I), and an independent one (II), which is known to be accurate. The special recovery function in this case is given by Eq. (6.47b).

Because both quantities, $x_{test(I)}$ and $x_{test(II)}$ are subject to error in this processing, EBV fitting according to Eqs. (6.41)–(6.43) or principal component analysis (Eq. 6.46) must be applied. The test on significant deviations from $\alpha = 0$ and $\beta = 1$ are carried out as above.

Precision. The *precision of the calibration* is characterized by the confidence interval $cnf(\hat{y}_i)$ of the estimated y values at position x_i according to Eq. (6.30). In contrast, the *precision of analysis* is expressed by the prediction intervals $prd(\hat{y}_i)$ and $prd(\hat{x}_i)$, respectively, according to Eqs. (6.32) and (6.33). The precision of analytical results on the basis of experimental calibration is closely related to the adequacy of the calibration model.

Linearity. Whether the chosen linear model is adequate can be seen from the residuals e_y over the x values. In Fig. 6.8a the deviations scatter randomly around the zero line indicating that the model is suitable. On the other hand, in Fig. 6.8b it can be seen that the errors show systematic deviations and even in the given case where the deviations alternate in the real way, it is indicated that the linear model is inadequate and a nonlinear model must be chosen. The hypothesis of linearity can be tested:

---

error or that an analytical procedure has successfully been validated (WEGSCHEIDER [1996]; BURNS et al. [2002])

(a) *A priori* (no actual nonlinear model is considered) by comparison of the deviations of the means from the calibration line (the residual standard deviations $s_{y.x}$; see Eq. (6.19)) with that of the y values from their means (s_y):

$$\hat{F} = \frac{s_{y.x}^2}{s_y^2} = \frac{\dfrac{1}{p-2} \displaystyle\sum_{i=1}^{p} n_i(\overline{y}_i - \hat{y}_i)^2}{\dfrac{1}{n-p} \displaystyle\sum_{i=1}^{p} \sum_{j=1}^{n_i} (y_{ij} - \overline{y}_i)^2} \qquad (6.49)$$

with n_i being the number of measurements in the p calibration points, see Eq. (6.13a,b). The test is carried out by comparison of the quotient (Eq. 6.49) with the corresponding quantile of the F-distribution. In case that $\hat{F}$ exceeds $F_{1-\alpha,\nu_1=p-2,\nu_2=n-p}$ a linear relationship cannot be assumed. This kind of ANOVA (SHARAF et al. [1986]) is usually implemented into software packages of regression.

(b) *A posteriori* by comparison of the residual standard deviation of the linear model, $s_{y.x\,lin}^2$, with that of a certain nonlinear model, $s_{y.x\,non}^2$:

$$\hat{F} = \frac{s_{y.x_{lin}}^2}{s_{y.x_{non}}^2} = \frac{\dfrac{1}{\nu_{lin}} \displaystyle\sum_{i=1}^{n} (y_i - \hat{y}_i)^2}{\dfrac{1}{\nu_{non}} \displaystyle\sum_{i=1}^{n} (y_i - \hat{y}_i)^2} \qquad (6.50)$$

The number of degrees of freedom in case of linear models is $\nu_{lin} = n - 2$ or $n - 1$, respectively, depending on whether two parameters are estimated according to Eq. (6.6) or (6.39) or only one parameter according to $\hat{y} = b\,x$. In the nonlinear case, ν_{non} results from the actual model (e.g., for a quadratic equation, $\hat{y} = a + b\,x + c\,x^2$, $\nu_{non} = n - 3$). A suitable test for Eq. (6.50) can also be carried out according to MANDEL [1964]:

$$\hat{F} = \frac{s_{y.x_{lin}}^2 - s_{y.x_{non}}^2}{s_{y.x_{non}}^2} \qquad (6.51)$$

by comparison with $F_{1-\alpha,\nu_1=1,\nu_2=\nu_{non}}$. In each case where $\hat{F} \geq F_{1-\alpha,\nu_1,\nu_2}$, the linear model cannot be applied.

Homoscedasticity. Unequal variances are recognizable from residual plots as in Fig. 6.8c where frequently e_y is a function of x in the given trumpet-like form. In such a case, the test of homoscedasticity can be carried out in a simple way by means of the HARTLEY test (F_{max} test), $\hat{F}_{max} = s_{max}^2/s_{min}^2$, see Sect. 4.3.4 (1).

In cases in which the situation is more obscure as represented in Fig. 6.8c, the BARTLETT test of homogeneity of variances (see Sect. 4.3.4 (3)) has to be applied. The test statistic $\hat{\chi}^2$ has to be compared with the critical value and the null hypothesis H_0: $\sigma_1^2 = \sigma_2^2 = \cdots = \sigma_p^2$ must be rejected if $\hat{\chi}^2 \geq \chi_{1-\alpha,v}^2$.

In this case, instead of normal OLS calibration, the weighted calibration has to be chosen. For WLS the dependence of the variance of the measurement values, s_y^2, on the analytical values x (and therefore on the measured values y, too) has to be determined in preliminary studies. Then the resulting *variance function*, $s_y^2 = f(y)$, can be entered for weighting directly in corresponding calibration programs, e.g. in the form $w = 1/s_y^2 = 1 = 1/y$; see Sect. 6.2.3.

Sometimes, as an alternative to the weighted calibration, it is recommended to restrict the working range to those partial ranges in which homoscedasticity can be supposed. Considering the relatively frequent occurrence of heteroscedastic calibration data, however, in analytical practice, the possibility of applying WLS should, in general, be taken into account to a stronger extent. In this way, more reliable analytical results can be obtained with a higher precision which is constant to a large extent and covers the whole working range.

6.2.6
Alternative Calibration Procedures

If the basic conditions for the use of least squares fitting are not fulfilled (Fig. 6.6), especially if strongly deviating calibration points appear (*"outliers"* or, more exactly, *leverage points*), the OLS method fails, i.e., the estimated calibration are biased and, therefore, are not representative for the relation between x and y. Whereas normality of the measured values can be frequently obtained by a suitable transformation, especially in the case of outlying calibration points, robust calibration has to be applied (ROUSSEEUW and LEROY [1987]; DANZER [1989] DANZER and CURRIE [1998]).

Robust calibration. The GAUSSIAN OLS criterion according to Eq. (6.16) is strongly sensitive against outliers. Therefore, robust methods of fitting have been developed following two strategies (ROUSSEEUW and LEROY [1987]):

(i)   Recognition and selection of outliers and fitting of the remaining data by OLS (*outlier diagnostics*)

(ii) Fitting of only the representative data by means of robust techniques (*robust regression* in the stricter sense)

It has been known for a long time (EDGEWORTH [1887]) that minimizing of the linear sums of deviations

$$\sum_{i=1}^{n} e_{y_i} \overset{!}{=} \min \tag{6.52}$$

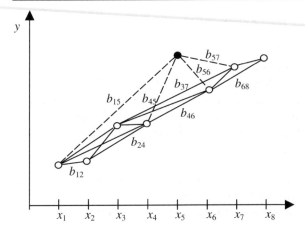

Fig. 6.9. Schematic representation of some of the slopes b_{ij} between eight calibration points[5] including an outlier (*black marked*)

is more insensitive against outliers than that of squared sums according to Eq. (6.16). Robust regression techniques are also based on linear deviation criteria and use special *minimization functions* (HUBER [1981]) and *influence functions* (HAMPEL [1980]; HAMPEL et al. [1986]):

$$\sum_{i=1}^{n} q \cdot e_{y_i} \overset{!}{=} \min \tag{6.53}$$

On the other hand, *least trimmed squares* (LTS) can be used to estimate the calibration parameters: that points which produce the $(n - h)$ largest deviation squares are rejected and the remaining h points are used for the LTS fitting:

$$\sum_{i=1}^{h} e_{y_i}^2 \overset{!}{=} \min \tag{6.54}$$

Another way is the robust parameter estimation on the basis of *median statistics* (see Sect. 4.1.2; DANZER [1989]; DANZER and CURRIE [1998]). For this, all possible slopes between all the calibration points $b_{ij} = (y_j - y_i)/(x_j - x_i)$ for $j > i$ are calculated. After arranging the b_{ij} according to increasing values, the average slope can be estimated as the median by

$$\bar{b} = med\{b_{ij}\} \tag{6.55}$$

where $med\{b_{ij}\}$ is determined according to Eq. (4.22). The principle of the procedure is illustrated in Fig. 6.9.

---

[5] The total number n_b of all the single slopes b_{ij} is given by the combinations of n calibration points to the 2nd class: $n_b = C_n^2 = \binom{n}{2} = \frac{1}{2}n(n - 1)$, i.e., in this case $n_b = 28$.

The intercept $\tilde{a}$ will then be obtained by

$$\tilde{a} = med\{y_i - \tilde{b} x_i\} \tag{6.56}$$

Estimates of the variance and uncertainty intervals in robust calibration can be taken from the literature (HUBER [1981]; ROUSSEEUW and LEROY [1987]).

Robust calibration corresponds in most cases to the problem of outlying calibration points (leverage points). In consideration of that, attention must be directed to the linearity of the relationship in general and the randomness of the residuals.

Calibration by Standard Addition Method (SAM). When matrix effects appear or are to be expected and matrix-matched calibration samples are not available, the standard addition method (SAM) can be the calibration method of choice.

Especially in the case of biochemical and environmental systems and generally in ultra trace analysis, SAM is frequently applied. By addition of standard solutions to the sample a similar behaviour of the calibration set and the sample is created provided that the analyte is added in form of *same species*.

The model of standard addition is based on the prerequisite that blanks do not appear, $y = bx$, or can be eliminated, $y - a = y_{net} = bx$:

$$y_0 = bx_0 \tag{6.57}$$

where y_0 is the measured value of the unspiked test sample and x_0 the initial content of the analyte. Known amounts x_i of the analyte are added to the sample; in doing so it is recommendable to use equimolar amounts of x_i in the range of

$$x_1 \approx x_0, x_2 = 2x_1, \ldots, x_p = px_1$$

where frequently $p = 3$ or 4 is used. Therefore, some ideas about the initial concentration x_0 should exist. The principle of SAM is illustrated in Fig. 6.10.

The standard addition calibration function is estimated by least squares fitting. The slope b (sensitivity of the SAM) will be obtained according to

$$b = \frac{\hat{y}_p - \hat{y}_0}{x_p} \tag{6.58}$$

This procedure is justified if the sensitivity of the determination of the species in the sample is the same as of the species added:

$$b_{(AD)} = \frac{\Delta y}{\Delta x} = \frac{y_0}{x_0} \overset{!}{=} b_{(SA)} = \frac{\hat{y}_p - \hat{y}_0}{x_p} \tag{6.58'}$$

The proceeding is simplified for such cases in which the volume of the added analyte amounts can be neglected in comparison with that of the initial sample (solution). Otherwise, a procedure using volume-corrected measurement values has to be applied (see SHARAF et al. [1986]).

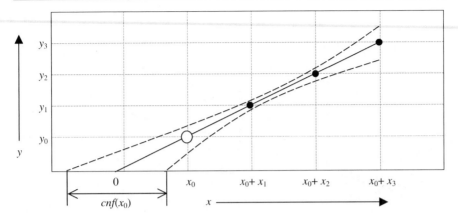

Fig. 6.10. Calibration by the standard addition method (SAM)

The estimation of the analyte content x is given by the *SA analytical function* (the inverse function of Eq. (6.57)) whose extrapolation for $y = 0$ yields

$$x_0 = \frac{\hat{y}_0}{b} = \frac{x_p \, \hat{y}_0}{\hat{y}_p - \hat{y}_0} \tag{6.59}$$

with the confidence interval

$$cnf(x_0) = x_0 \pm \frac{s_{y.x} \, t_{1-\alpha,v}}{b} \sqrt{\frac{1}{n} + \frac{\left(-x_0 - \frac{x_p}{2}\right)^2}{S_{xx}}} \tag{6.60}$$

This confidence interval is wider as compared to normal calibration (see Eq. 6.33) because of the extrapolation to $x_i = -x_0$. The number of calibration measurements n results from $n_0 + p \cdot n_i$ or $n_0 + \sum^p n_i$, respectively, where n_0 is the number of measurements of y_0.

When a blank appears, it has to be estimated from a sufficiently large number of blank measurements and the measured values must be corrected in this respect. To ensure the adequateness of the SA calibration model, $p \geq 2$ additions should be carried out. Only in the case when it is definitely known that the linear model holds true, then one single addition (n_i times repeated) may be carried out. In general, linearity can be tested according to Eqs. (6.49)–(6.51).

Although standard addition calibration is an unreliable method if linearity in the range $x < x_0$ is not experimentally verified but only supposed, there is scarcely an alternative in trace and ultra trace analysis when matrix effects are seriously suspected.

Individual Three-Dimensional Calibration. In general, the intensity y of a signal is determined by both the analyte content x and the sample weight w. This is because the direct quantity that causes the value of the measuring quantity is the number of analyte species, N, in the measuring sample

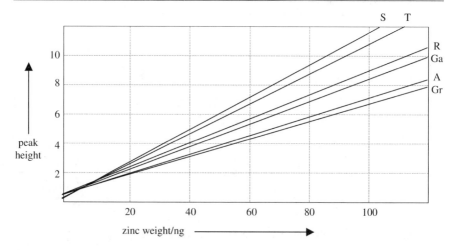

Fig. 6.11. Calibration lines by using several CRM samples (S: soil BHB-1, T: Tibet soil GBW 8302, R: river sediment CRM 320, Ga: Gabbro MRG-1, A: Andesite AGV-1, Gr: Granodiorite GSP-1) and varying the sample weight (Zn 307.6 nm)

(sample solution, sample vapour, plasma etc), as it is expressed, e.g., by the general law of absorption, $y = y_0 \exp(-kN)$. In most cases, conditions of the analytical procedures are fixed which provide the direct connection between the signal intensity and the analyte content (e.g. constant measuring volumina or sample weights).

Not in all cases can such constant conditions be realized. In Solid Sampling AAS (SS-AAS), calibration is commonly carried out by using only one suitable calibration sample (frequently a certified reference material, CRM) and varying its weight. The absolute analyte amount of the analyte is then used as the independent quantity and calibrated vs the dependent signal intensity (peak height or area, respectively). The second variant, the use of several CRMs with varying analyte content, can only rarely applied because of high costs of CRMs and matrix influences that will be distinct under the given conditions.

By combination of both calibration variants it can be shown (DANZER et al. [1998]) that strong matrix effects appear which are represented by the different sensitivities in Fig. 6.11.

To overcome these matrix influences, other calibration strategies can be used (DANZER et al. [1998]). According to one of them, a specialized three-dimensional calibration model, the relation between signal intensity and both sample weight and content is evaluated. The relationship between the three quantities, $y = f(x, w)$ is demonstrated in Fig. 6.12 by the example of Zn determination represented above.

Both planar and curved surfaces can be fitted by statistical software packages like STATISTICA [1993]. In the general case, the 3-D calibration surface is given by

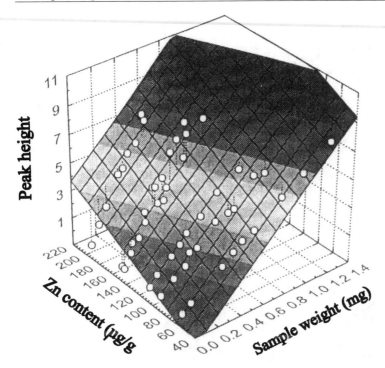

Fig. 6.12. Three-dimensional calibration plot for the zinc determination using a plain model for the relationship peak height vs zinc content and sample weight

$$y = a + b_1 x + b_2 w + b_3 x^2 + b_4 xw + b_5 w^2 \qquad (6.61)$$

where y is the signal intensity (peak height or area, respectively), x the analyte content, w the sample weight, and a as well as the b_i are the 3D calibration coefficients. Upper and lower confidence planes can be estimated by *bootstrapping* (EFRON [1982]; EFRON and TIBSHIRANI [1993]) as described in detail by NIMMERFALL and SCHRÖN [2001].

Calibration by means of results of another method. In life science and other fields there can exist cases in which no real calibration samples are available. That may concern substances of metabolism in living organisms like cholesterol and glucose in blood.

Intensified metabolic control, especially in case of diabetes, demands minimal-invasive or non-invasive methods of analytical measurement. For this goal, a method has been developed to measure the blood glucose content in vivo, in direct contact with the skin, by means of diffuse reflection near infrared (NIR) spectroscopy on the basis of multivariate calibration and neural networks (MÜLLER et al. [1997]; FISCHBACHER et al. [1997]; DANZER et al. [1998]). Because no patients with any "standard blood glucose value" are available in principle, a method of indirect calibration has

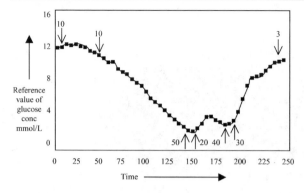

Fig. 6.13. Profile of blood glucose concentration for a single test person; ↓ taking of carbohydrates/g, ↑ insulin injection/IU

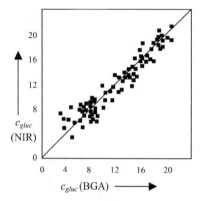

Fig. 6.14. Indirect calibration of NIR measurements vs BGA measurements for several test persons

to be applied. For this aim, type-1 diabetes patients (involved into an intensified conventional insulin therapy) have been selected in which variations of blood glucose values were provoked by subcutaneous or intramuscular injections of 2–5 units of regular insulin and by carbohydrate supply. While recording the NIR spectra at the right hand finger, blood samples were taken from a cannula placed in the dorsum of the left hand. The reference concentrations were measured by a BECKMAN Glucose Analyzer (BGA).

A typical blood glucose profile obtained in this way is represented in Fig. 6.13 (MÜLLER et al. [1997]; FISCHBACHER et al. [1997]). On the other hand, Fig. 6.14 shows the calibration of the NIR measurements vs the BGA reference measurements.

In contrast to the common calibration procedure measuring $y = f(x_{standard})$ as shown in the front plane of Fig. (6.3a,b), the glucose reference calibration takes place in the base area of the three-dimensional model, $x_{sample} = f(x_{standard})$; see Sect. 6.1.2.

Calibration with imprecise signals and analytical values (concentrations) based on fuzzy theory has been dealt with by OTTO and BANDEMER [1986].

6.2.7
Nonlinear Calibration

Nonlinear relationships between analytical quantities and measured values results from different reasons. In most cases, the relating natural laws can determine the nonlinearity. On the other hand, factors and conditions can influence an analytical system in such a way that relationships which are originally linear can become curved. Absorption of electro-magnetic radiation is an example that can illustrate both phenomena. The LAMBERT-BEER law of absorption describes a nonlinear dependence of the radiation intensity after the absorption step, I, on the number of absorbing species N, namely $I = I_0 \exp(-kN)$. I_0 is the incident intensity and k an absorption coefficient. It is well-known that this relationship mostly is used in its linearized form, the so-called BEER law, $A = -\log(I_0/I)) = k \cdot N = \varepsilon \cdot l \cdot c$ with A absorbance, k absorption coefficient, ε molar absorptivity, l length of the absorbing cell, and c concentration of the absorbing species. BEER's law usually is only valid for dilute solutions ($c \leq 0.1$ mol/L). At higher concentrations deviations from the linear calibration curve may occur.

Because nonlinear calibration needs higher expense both in experimental and computational respects, linear models are mostly preferred.

> Nonlinear calibration is carried out by nonlinear regression where two types have to be distinguished: (1) real (intrinsic) nonlinear regression and (2) quasilinear (intrinsic linear) regression. The latter is characterized by the fact that only the data but not the regression parameters are nonlinear. Typical examples are polynomials and trigonometric functions.

When nonlinearity is detected, either by visual inspection, test on linearity, or logical reasons, the following possibilities can be used:

(i) *Transformation* of the analytical values and/or measured values

$$x_{trans} = f(x), \qquad y_{trans} = f(y) \tag{6.62}$$

and modelling of the linear function

$$y_{trans} = a^* + b^* x_{trans} \tag{6.63}$$

(ii) *Modelling of a nonlinear relationship* between y and x by means of a suitable function, some of them are given in Table 6.2.

(iii) *Multi-range calibration* corresponding to a multi-range regression; see, e.g., SACHS [1992]. Several working ranges are fitted by diverse calibration straight lines where in analytical practice the use of only two ranges is customary.

(iv) *Spline functions* fit small intervals of the calibration function by polynomials of low (2nd or 3rd) order under the condition that the resulting overall curve represents a continuous function (WOLD [1974]).

Table 6.2. Suitable nonlinear calibration functions, their normal equations and sensitivity functions

Calibration function	Normal equations	Sensitivity function $S(x)$
$y = a_0 + a_1 x + a_2 x^2$	$a_0 n + a_1 \sum x + a_2 \sum x^2 = \sum y$ $a_0 \sum x + a_1 \sum x^2 + a_2 \sum x^3 = \sum xy$ $a_0 \sum x^2 + a_1 \sum x^3 + a_2 \sum x^4 = \sum x^2 y$	$\dfrac{dy}{dx} = a_1 + 2a_2 x$
$y = a_0 + a_1 x + a_2 x^2 + a_3 x^3$	$a_0 n + a_1 \sum x + a_2 \sum x^2 + a_3 \sum x^3 = \sum y$ $a_0 \sum x + a_1 \sum x^2 + a_2 \sum x^3 + a_3 \sum x^4 = \sum xy$ $a_0 \sum x^2 + a_1 \sum x^3 + a_2 \sum x^4 + a_3 \sum x^5 = \sum x^2 y$ $a_0 \sum x^3 + a_1 \sum x^4 + a_2 \sum x^5 + a_3 \sum x^6 = \sum x^3 y$	$\dfrac{dy}{dx} = a_1 + 2a_2 x + 3a_3 x^2$
$y = a_0 + a_1 \lg x$	$a_0 n + a_1 \sum \lg x = \sum y$ $a_0 \sum \lg x + a_1 \sum (\lg x)^2 = \sum (y \lg x)$	$\dfrac{dy}{dx} = \dfrac{1}{x} \lg e = \dfrac{1}{x \ln 10}$
$\lg y = a_0 + a_1 \lg x$	$a_0 n + a_1 \sum \lg x = \sum \lg y$ $a_0 \sum \lg x + a_1 \sum (\lg x)^2 = \sum (\lg x \lg y)$	$\dfrac{d \lg y}{d \lg x} = a_1$
$y = a \cdot b^x$ corresponding to $\lg y = \lg a + x \lg b$	$n \lg a + \lg b \sum x = \sum \lg y$ $\lg a \sum x + \lg b \sum x^2 = \sum (x \lg y)$	$\dfrac{dy}{dx} = b^x \ln b$

In general, from nonlinear calibrations result variable sensitivities expressed by *sensitivity functions* $S(x)$:

$$S(x) = y' = f'(x) = \frac{dy}{dx} = \frac{df(x)}{dx} \tag{6.64}$$

For the purpose of comparisons, the sensitivity of the centre of the calibration curve $S(\overline{x})$ having the character of an average sensitivity can be used. Sensitivity functions corresponding to diverse calibration functions are given in Table 6.2.

6.3
Multisignal Calibration

By diverse spectroscopic techniques not only a single signal is generated by given species but several their number can be lower (XRF, MS) or higher (OES). Whereas especially in OES several spectral lines are used for identification and qualitative analysis, quantitative analysis of a given analyte is carried out by means of only one single line. In doing so, mostly the most sensitive, undisturbed line is selected whereas all the other signals are ignored.

It has been shown (DANZER and WAGNER [1993]; DANZER and VENTH [1994]; VENTH et al. [1996]) that the reliability of quantitative analyses can be increased when several signals are used for calibration and evaluation. From theoretical considerations it is expected that a multivariate evaluation increases both the sensitivity and the precision and, therefore, the detection power, too.

Multisignal evaluation is carried out by means of *Principal Component Analysis* (PCA) or (*Partial Least Squares* (PLS) regression. The fundamentals

A

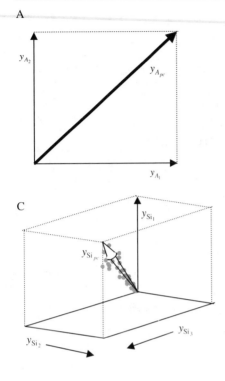

B

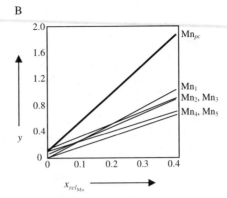

Fig. 6.15. A Principle of the derivation of a latent intensity $y_{A_{pc}}$ from original intensities y_{A_1} and y_{A_2}. **B** Sensitivity of the pc intensity in comparison to the five original intensities of the lines Mn 279.8 nm, Mn 280.1 nm, Mn 293.3 nm, Mn 403.3 nm, and Mn 403.4 nm (DANZER and VENTH [1994]). **C** pc intensity of Si resulting from Si 251.5 nm, Si 252.9 nm and Si 288.2 nm (DANZER et al. [1991])

of these chemometric methods can be found in Chapter 8: DILLON and GOLDSTEIN [1984] MARTENS and NÆS [1989]; DANZER and VENTH [1994]; VENTH et al. [1996].

The principle of multisignal calibration will be represented here in a more illustrative way. Starting with two signal intensities of one and the same analyte A, namely y_{A_1} and y_{A_2}. Plotting y_{A_1} vs y_{A_2} as done in Fig. 6.15A, it can be seen that the resulting (latent) intensity $y_{A_{pc}}$ is larger then the both original values of the same species. Figure (6.15B,C) shows the results of experimental studies that confirm the theoretical expectations.

Because the first principal component is always situated in direction of the largest variation of the data, the intensity of y_{pc} should be increased compared with the original intensities from which it was calculated, namely theoretically according to the n-dimensional PYTHAGORAS' principle

$$y_{pc_{\max}} = \sqrt{\sum_{i=1}^{n} I_n^2} \qquad (6.65)$$

According to the fundamentals of PCA, simultaneously the precision should be improved because the noise is preferably to find in the higher pcs as shown by simulations and experimentally in DANZER and WAGNER [1993], OES determination of Mn, Ni, and Cr in steel (DANZER and VENTH [1994])

Table 6.3. Spectral interferences in ICP-MS of the five most abundant cadmium isotopes relevant for the cadmium determination in Mo-Zr alloys; additional Cd isotopes: ^{106}Cd 1.25%, ^{108}Cd 0.89%, ^{116}Cd 7.49%

Mass	Abundance (%)	Molecule ion interferences	Abundance (%)	Isobaric interferences	Abundance (%)
		ZrO	17.49		
110	12.49	ZrF	11.20	Pd	11.72
		MoO	9.31		
		ZrOH	17.31		
		ZrF	17.10		
111	12.80	MoO	15.86		
		MoF	14.80		
		MoOH	9.10		
		MoO	16.68		
112	24.13	MoOH	15.66	Sn	0.97
		ZrO	2.83		
		ZrF	17.50		
		MoOH	16.46	In	4.30
113	12.22	MoO	9.58		
		MoF	9.3		
		ZrOH	2,79		
		MoO	24.07		
114	28.73	MoF	15.90	Sn	0.65
		MoOH	9.43		

and ICP-MS determination of Sb, Cd, In, Pd, Ag, and Sn in Mo-Zr alloys (VENTH et al. [1996]). The advantage of multisignal calibration was shown impressively in case of cadmium. Its determination on the basis of single signal evaluation is practically impossible because each signal of the Cd isotopes is disturbed as shown in Table 6.3.

By means of the most sensitive signal Cd-114 only a relative uncertainty of 22% could be observed. In contrast, PCA calibration including the five isotopes given in Table 6.3 yields an uncertainty of 0.6% with nearly 100% recovery. This example demonstrates that not only undisturbed signals can be included in the calibration but also disturbed ones.

6.4
Multicomponent Calibration

The most important techniques in multicomponent analysis are spectroscopy (of radiation of various wavelengths as well as of particles) and chromatography. Spectroscopic and chromatographic methods are able to analyze diverse species in a more or less selective way. For the determination of n species Q_i $(i = 1, 2, \ldots, n)$, see 2.3, at least n signals must be measured, which should be well-separated in the ideal case.

In analytical practice, the situation can be different as shown in Fig. 6.16; see ECKSCHLAGER and DANZER [1994]; DANZER et al. [2004]. The given detail of a spectrum may show either well-separated signals as represented in (a) or signals that are overlapped to different degree; see (b) and (c).

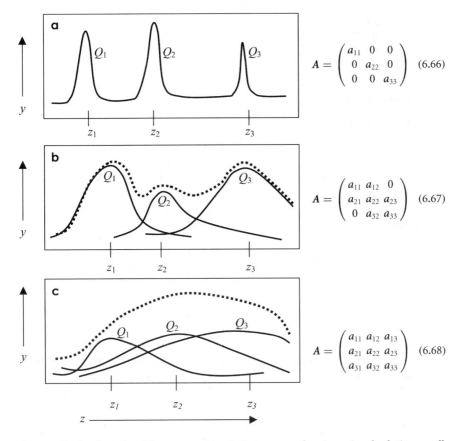

$$A = \begin{pmatrix} a_{11} & 0 & 0 \\ 0 & a_{22} & 0 \\ 0 & 0 & a_{33} \end{pmatrix} \quad (6.66)$$

$$A = \begin{pmatrix} a_{11} & a_{12} & 0 \\ a_{21} & a_{22} & a_{23} \\ 0 & a_{32} & a_{33} \end{pmatrix} \quad (6.67)$$

$$A = \begin{pmatrix} a_{11} & a_{12} & a_{13} \\ a_{21} & a_{22} & a_{23} \\ a_{31} & a_{32} & a_{33} \end{pmatrix} \quad (6.68)$$

Fig. 6.16. Evaluation of multicomponent analysis in cases of various signal relations: well-separated (**a**), moderately overlapped (**b**), and strongly overlapped (**c**) in form of spectra (*left*) and relevant matrices – Eqs. (6.66) to (6.68) (*right*); Q_1, Q_2, Q_3 are different species (analytes), z_1, z_2, z_3 wavelengths at which the intensities y_1, y_2, y_3 are measured, A are the matrixes of absorbances and a_{ij} absorption coefficient of species i at wavelength j

In case (a) each species can be calibrated and evaluated independently from the other. In this fully selective case, the following equation system corresponds to the matrix A in Eq. (6.66):

$$
\begin{aligned}
y_1 &= a_{10} + a_{11}x_1 + e_1 \\
y_2 &= a_{20} + a_{22}x_2 + e_2 \\
&\;\vdots \\
y_m &= a_{m0} + a_{mn}x_n + e_m
\end{aligned}
\tag{6.69}
$$

where the number of measured signal absorbances, m, usually is equal to the number of species, n. On the other hand, case (b), in which neighbouring signals overlap to a certain degree, can be handled by means of multiple linear calibration provided that the absorbances y_i are additive and related signal maxima can be measured for each species:

$$
\begin{aligned}
y_1 &= a_{10} + a_{11}x_1 + a_{12}x_2 + \cdots + a_{1n}x_n + e_1 \\
y_2 &= a_{20} + a_{21}x_1 + a_{22}x_2 + \cdots + a_{2n}x_n + e_2 \\
&\;\vdots \\
y_m &= a_{m0} + a_{m1}x_1 + a_{m2}x_2 + \cdots + a_{mn}x_n + e_m
\end{aligned}
\tag{6.70a}
$$

or, in matrix form,

$$
y = Ax + e_y
\tag{6.70b}
$$

Multicomponent systems of the kind shown in Fig. 6.16b can be calibrated with a high degree of reliability when the preconditions mentioned above are valid. The uncertainties e_i contain both deviations from the model and random errors.

In the case of strongly overlapped signals, see Fig. 6.16c, multiple linear calibration cannot be used for the following reasons:

(i)   In real analytical systems, not all the sample constituents are known. In such cases, an alternative way may be the application of the inverse calibration model of Eq. (6.70a):

$$
x = Ya + e_x
\tag{6.71}
$$

where Y is a matrix of spectra with m given wavelengths for n component mixtures in which other variations like baseline effects must be included. The vector of the sensitivity coefficients a can be estimated by

$$
\hat{a} = (Y^T Y)^{-1} \, Y^T x
\tag{6.72}[6]
$$

(ii) In principle, for spectra like that in Fig. 6.16c, multicollinearities have to be expected. That means that overlapping signal functions and consequently the resulting sum curve are correlated and the measured absorbance values at the respective wavelengths are not independent

---

[6] Y^T is the transposed of the matrix Y.

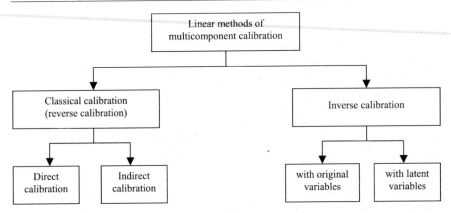

Fig. 6.17. Multivariate calibration methods (according to DANZER et al. [2001, 2004])

from each other. Therefore, Eq. (6.72) becomes unstable and other methods of estimation of the a values must be used. These methods use overdetermined equation systems as their basis, such as

$$A = \begin{pmatrix} a_{11} & a_{12} & a_{13} & \cdots & a_{1m} \\ a_{21} & a_{22} & a_{23} & \cdots & a_{2m} \\ a_{31} & a_{32} & a_{33} & \cdots & a_{nm} \end{pmatrix} \tag{6.73}$$

instead of Eq. (6.68). The number of wavelengths, m (sensors, detecting channels, respectively), is usually much higher than the number of species n (here $n = 3$). The estimation of the calibration coefficients is then carried out by *multivariate calibration*.

Depending on whether the spectra Y are calibrated as dependent on analyte amounts X or conversely, different methods of multicomponent calibration, as represented in Fig. 6.17, can be applied.

6.4.1
Classical Multivariate Calibration

The classical multivariate calibration represents the transition of common single component analysis from one dependent variable y (measured value) to m dependent variables (e.g., wavelengths or sensors) which can be simultaneously included in the calibration model. The classical linear calibration (DANZER and CURRIE [1998]; DANZER et al. [2004]) is therefore represented by the generalized matrix relation

$$Y = XA \tag{6.74}$$

where Y is the $(p \times m)$ matrix of dependent variables (e.g., absorbances at m wavelengths or responses at m sensors), p is the number of calibration standards (calibration mixtures) that is identical with the number of spectra or similar measured records, X is the $(p \times n)$ matrix of independent variables (e.g., the concentration of the n components), and A is the $(n \times m)$ matrix of the calibration coefficients, often called the *sensitivity matrix* (KAISER [1972]; MARTENS and NÆS [1989]; DANZER [2001]). The rows of the matrix A correspond to the spectra of the pure components, which can be directly measured or indirectly estimated.

Direct calibration can be applied when the calibration coefficients are known, otherwise – in case of *indirect calibration* – the calibration coefficients are computed by means of experimentally estimated spectra-concentrations relations.

Classical calibration procedure can only be applied when all the species that contribute to the form of the spectra are known and can be included into the calibration. Additionally, there is the constraint that no interactions between the analytes and other species (e.g. solvent) or effects (e.g. of temperature) should occur.

The analytical values (concentrations) are estimated by

$$\hat{x} = Y A^+ \tag{6.75}$$

where A^+ is the so-called MOORE-PENROSE generalized pseudo inverse (MOORE [1920]; PENROSE [1955])

$$A+ = (A^T A)^{-1} A^T \tag{6.76}$$

with the same dimensions $(m \times p)$ as the transposed matrix.

In case of baseline shift, the sensitivity matrix in Eqs. (6.74) and (6.75) must be complemented by a vector **1**:

$$A = (1\ A) = \begin{pmatrix} 1 & a_{11} & \cdots & a_{1m} \\ \vdots & \vdots & & \vdots \\ 1 & a_{p1} & \cdots & a_{pm} \end{pmatrix} \tag{6.77}$$

Instead of the addition of the 1-vector the calibration data may be centered ($y_i - \bar{y}$ and $x_i - \bar{x}$, respectively). Even if the spectra of the pure species cannot be measured directly then the A-matrix can be estimated indirectly from the spectra *provided that all components of the analytical system are known*:

$$\hat{A} = (X^T X)^{-1} X^T Y \tag{6.78}[7]$$

For inversion of the matrix $X^T X$ it is necessary that a sufficient number of spectra for different concentration steps have been measured. The concentration vectors must vary independently from each other. For this reason, experimental design (see, e.g., DEMING and MORGAN [1993]) can be

---

[7]   Instead of the symbol A and the term sensitivity matrix also the symbol K (matrix of calibration coefficients, matrix of linear response constants etc) is used. Because of the direct metrological and analytical meaning of the sensitivities a_{ij} in the A-matrix the term *sensitivity matrix* is preferred.

used. In the case that the preparation of samples of defined composition is impossible, then the samples should be selected as representative and as uncorrelated as possible (natural design, see MARTENS and NÆS [1989]).

The prediction of analytical values X according to the classical indirect calibration model follows Eq. (6.75):

$$\hat{X} = Y \hat{A}^+ \tag{6.79}$$

The desired independence between the variables of the different analytical signals corresponds directly with the *selectivity* of the analytical system (KAISER [1972]; DANZER [2001], and Sect. 7.3). In case of multivariate calibration, the selectivity is characterized by means of the condition number

$$\text{cond}(A) = \|A\| \cdot \|A^{-1}\| \tag{6.80}$$

where $\|A\|$ is the matrix norm of A and $\|A^{-1}\|$ the norm of the inverse matrix. The matrix norm of A is calculated from $\sqrt{\lambda_{\max}}$, the square root of the largest eigenvalue $\lambda_{\max}$, and the norm of A^{-1} from the reciprocal square root of the lowest eigenvalue $\lambda_{\min}$:

$$\text{cond}(A) = \frac{\sqrt{\lambda_{\max}}}{\sqrt{\lambda_{\min}}} \tag{6.81}$$

Equation (6.79) is valid for exactly determined systems ($m = n$). In case of overdetermined systems, $m > n$, the condition number is given by

$$\text{cond}(A) = \sqrt{\text{cond}(A^{\mathrm{T}}A)} \tag{6.82}$$

If systems are well-conditioned the selectivity is expressed by condition numbers close to 1.

The *uncertainty in multivariate calibration* is characterized with respect to the evaluation functions at Eqs. (6.75) and (6.79). The prediction of a row vector x of dimension n from a row vector y of dimension m results from

$$x = y A^{\mathrm{T}} (A A^{\mathrm{T}})^{-1} \tag{6.83}$$

The relative uncertainty for the prediction of the x-values can be estimated by

$$\frac{\|\delta x\|}{\|x\|} = \text{cond}\, A \left(\frac{\|\delta y\|}{\|y\|} + \frac{\|\delta A\|}{\|A\|} \right) \tag{6.84}$$

where $\|\delta y\|/\|y\|$ is the relative uncertainty of the y-values (error of measurement) and $\|\delta A\|/\|A\|$ the relative uncertainty of the estimation of A (modeling error). The condition number is calculated from Eqs. (6.80)–(6.82).

6.4.2
Inverse Calibration

The classical direct or indirect calibration is carried out by OLS minimization according to GAUSS. Error-free analytical values x are assumed or at least that the errors in x are very small compared with those of the y-values (DANZER and CURRIE [1998]). Additionally all the components in the analytical system must be known and included in the calibration. If these preconditions are not fulfilled the inverse calibration must be applied.

The inverse calibration regresses the analytical values (concentrations), x, on the measured values, y. Although with it a prerequisite of the GAUSSIAN least squares minimization is violated because the y-values are not error-free, it has been proved that predictions with inverse calibration are more precise than those with the classical calibration (CENTNER et al. [1998]). This holds true particularly for multivariate inverse calibration.

In chemometrics, the inverse calibration model is also denoted as the P-matrix model (the dimension of P is $m \times n$):

$$X = YP \tag{6.85}$$

The calibration coefficients are elements of the matrix P which can be estimated by

$$\hat{P} = Y^{+} \cdot X = (Y^{T}Y)^{-1}Y^{T}X \tag{6.86}$$

The analysis of an unknown sample is carried out by multiplication of the measured spectrum y by the P-matrix

$$\hat{x} = y\hat{P} \tag{6.87}$$

In the case that the original variables, the measured values y, are used for inverse calibration, there are no significant advantages of the procedure apart from the fact that no second matrix inversion has to be carried out in the analysis step; see Eq. (6.87). On the contrary, it is disadvantageous that the calibration coefficients (elements of the P-matrix) do not have any physical meaning because they do not reflect the spectra of the single species. In addition, multicollinearities may appear which can make inversion of the Y-matrix difficult; see Eq. (6.86).

On the other hand, when latent variables instead of the original variables are used in inverse calibration then powerful methods of multivariate calibration arise which are frequently used in multispecies analysis and single species analysis in multispecies systems. These so-called "*soft modeling methods*" are based, like the P-matrix, on the inverse calibration model by which the analytical values are regressed on the spectral data:

$$X = YB \tag{6.88}$$

where B is the ($m \times n$)-matrix of calibration coefficients, in concrete terms the matrix of B-coefficients. In contrast to the P-matrix, not all the dimensions of the spectra (the Y-matrix) are used but only those that are significant are realized by certain principal components. Therefore, the estimation of the matrix of B-coefficients can be carried out by PCR (Principal Component Regression) or PLS (Partial Least Squares) Regression.

Both PCR and PLS form latent variables T (principal components, factors) from the original variables, viz., from the matrix of measured values according to

$$Y = TL^T + E_Y \tag{6.89}$$

where T is the factor (score) matrix and L the loading matrix with the dimension $m \times n$; E_Y is the matrix of non-significant factors which is regarded as an error matrix. Additionally, in PLS the matrix of analytical values (e.g. concentrations) is decomposed in the same way:

$$X = TQ^T + E_X \tag{6.90}$$

PCR and PLS have in common the following steps:

(1) Estimation of a weight matrix (eigenvalues) V (from Y in PCR and from Y and X in PLS)

(2) Calculation of the factor matrix $T = ZV$ by means of the standardized variables Z

(3) Calculation of the matrices P and Q according to

$$P^T = T^+ Y \tag{6.91}$$

$$Q^T = T^+ X \tag{6.92}$$

In PCR the calibration coefficients (B-matrix) are estimated column by column according to

$$\hat{b} = VQ^T \tag{6.93}$$

and

$$\hat{b}_0 = \bar{x} - \bar{Y}\hat{b} \tag{6.94}$$

The prediction then is carried out by

$$\hat{x} = Y\hat{b} + \hat{b}_0 \tag{6.95}$$

The significance and non-significance of principal components are decided on the basis of the variance that is explained by each them. Normally, in analytical methods the main variance is caused by the analyte concentration. But sometimes properties of the sample, such as moisture or surface roughness, or effects of the measuring procedure such as spectral baselines or scattered light, can exceed the effect of analyte concentration. Therefore, additional tests should be made as to what degree the principal components postulated to be non-significant by the software are correlated with the analytical values. Principal components which are highly correlated with the variable of interest (e.g., concentration) should be included in the calibration procedure notwithstanding their share in the variance.

In PLS both the matrices of measured values Y and analytical values X are decomposed according to Eqs. (6.89) and (6.90): $Y = TP^T + E_Y$ and $X = TQ^T + E_X$ and thus relations between spectra and concentrations are considered from the outset. The B-matrix of calibration coefficients is estimated by

$$\hat{B} = V(P^T V)^{-1} Q^T \tag{6.96}$$

Because the Y-matrix and X-matrix are interdependently decomposed the B-matrix fits better and more robust than in PCR the calibration. The evaluation is carried out by Eq. (6.88) according to $\hat{X} = Y\hat{B}$. The application of PLS to only one y-variable is denoted as PLS 1. When several y-variables are considered in the form of a matrix the procedure is denoted PLS 2 (Manne [1987]; Høskuldsson [1988]: Martens and Næs [1989]: Faber and Kowalski [1997a, b]).

6.4.3
Validation of Multivariate Calibration

The reliability of multispecies analysis has to be validated according to the usual criteria: selectivity, accuracy (trueness) and precision, confidence and prediction intervals and, calculated from these, multivariate critical values and limits of detection. In multivariate calibration collinearities of variables caused by correlated concentrations in calibration samples should be avoided. Therefore, the composition of the calibration mixtures should not be varied randomly but by principles of experimental design (Deming and Morgan [1993]; Morgan [1991]).

Selectivity. In general, selectivity of analytical multicomponent systems can be expressed qualitatively (Vessman et al. [2001]) and estimated quantitatively according to a statement of Kaiser [1972] and advanced models (Danzer [2001]). In multivariate calibration, selectivity is mostly quantified by the condition number; see Eqs. (6.80)–(6.82). Unfortunately, the condition number does not consider the concentrations of the species and gives therefore only an aid to orientation of maximum expectable analytical errors. Inclusion of the concentrations of calibration standards into selectivity models makes it possible to derive multivariate limits of detection.

Precision. The uncertainty of calibration and prediction of unknown concentrations are expressed by the standard error of calibration (*SEC*), defined as

$$\hat{s}_{cal} = \sqrt{\frac{\sum_{i=1}^{n}\left(y_{i_{calc}} - y_{i_{true}}^{(cs)}\right)^2}{n}} \tag{6.97}$$

and the standard error of prediction (*SEP*), defined as

$$\hat{s}_{pred} = \sqrt{\frac{\sum_{i=1}^{n}\left(y_{i_{calc}} - y_{i_{true}}^{(ts)}\right)^2}{n}} \tag{6.98}$$

where $y_{i_{true}}^{(cs)}$ are the true values of the calibration samples (cs, e.g. standards), $y_{i_{true}}^{(ts)}$ the true values of test samples (ts) with which the prediction power independently is estimated, and $y_{i_{calc}}$ are the respective y-values calculated by the model.

Another measure for the precision of multivariate calibration is the so-called *PRESS*-value (predictive residual sum of squares, see FRANK and TODESCHINI [1994]), defined as

$$s_{res}^2 = \sum_{i=1}^{n} e_i^2 = \sum_{i=1}^{n} (y_{i_{calc}} - y_{i_{true}})^2 \tag{6.99}$$

It can be calculated as usual for *SEP*, see Eq. (6.98) by use of test samples. It is also possible to estimate the *PRESS*-value on the basis of standard samples only applying cross validation by means of the so-called hat matrix H (FABER and KOWALSKI [1997a, b]; FRANK and TODESCHINI [1994]):

$$H = X(X^T X)^{-1} X^T \tag{6.100}$$

The $n \times n$ *hat matrix* transforms the vector of the measured y-values to the vector of the estimated-values. An element h_{ij} of the hat matrix is calculated by

$$h_{ij} = x_i^T (X^T X)^{-1} x_j \tag{6.101}$$

From the elements of the hat matrix some important relations can be derived, e.g. the rank of the X-matrix from the sum of the significant diagonal elements of the hat matrix

$$\text{rank}(X) = \sum_{i=1}^{n} h_{ii} \tag{6.102}$$

(the rank of the hat matrix is equal to its trace) and the residuals

$$\hat{e} = y - \hat{y} = y - X(X^T X)^{-1} X^T y = [I - X(X^T X)^{-1} X^T] y = [I - H] y \tag{6.103}$$

The residuals can be calculated from a given set of calibration samples in a different way. Cross validation is an important procedure to estimate a realistic prediction error like *PRESS*. The data for k samples are removed from the data matrix and then predicted by the model. The residual errors of prediction of cross-validation in this case are given by

$$e_{(cv)} = \frac{\hat{e}_{(k)}}{1 - h_{kk}} \tag{6.104}$$

The *PRESS* value of cross-validation is given by the sum of all the k variations

$$s_{(cv)res}^2 = \sum \hat{e}_{(cv)}^2 \tag{6.105}$$

Prediction limits for the estimation of an unknown concentration x_i can be calculated. The calculation depends on the specific multivariate calibration model

$$\Delta \overline{x}_{i,pred} = \hat{\overline{x}}_i \pm \Delta \hat{x} = \hat{\overline{x}}_i \pm s_x \, t_{1-\alpha,v} \tag{6.106}$$

where $\overline{x}_{i,pred}$ is the mean of the predicted unknown concentration and s_x is the standard deviation of prediction estimated from

$$s_x^2 = \frac{\sum\limits_{i=1}^{n}(y_i - \hat{y}_i)^2}{m - p} \tag{6.107}$$

with $t_{1-\alpha,\nu}$ being the Student-t statistic for ν degrees of freedom at the $1 - \alpha$ confidence level. The variance s_x^2 depends on the number of sensors or wavelengths, m, the number of species, n, the number of parameters p and a factor Λ which takes the form

$$\Lambda = (B^T B)^{-1} \tag{6.108}$$

in the case of classical multivariate calibration. For inverse calibration,

$$\Lambda = 1 + y_0^T (Y^T Y)^{-1} y_0 \tag{6.109}$$

and for cross-validation when the leverage values are applied in calibration,

$$\Lambda = 1 + h_{kk} \tag{6.110}$$

Trueness. Absence of systematic errors can be tested traditionally by means of recovery functions; see Sect. 6.1.2, Fig. 6.3C; BURNS et al. [2002]. For this reason the concentration estimated by the model, $\hat{x}$, is compared with the true concentration value, x_{true}, by a regression model

$$\hat{x} = \alpha + \beta x_{true} \tag{6.111}$$

where the x_{true} can be the known values of an independent set of test samples or reference values estimated on the same samples by means of an independent method which yields true values as is well known. The regression coefficients have to be $\alpha = 0$ and $\beta = 1$ where values outside of the confidence interval $\pm\Delta\alpha$ indicate additive (constant) systematic errors and values exceeding the confidence interval $1\pm\Delta\beta$ upwards or downwards show proportional systematic errors. By means of recovery studies both accuracy can be tested and precision can be estimated.

Multivariate limit of detection. Starting from a model like Eq. (6.74) in which the background vector y_0 is included:

$$y = AX + y_0 \tag{6.112}[8]$$

and its solution for the concentration vector

$$x = A^+ (y - y_0) \tag{6.113}$$

BAUER et al. [1991a] derived the following propagation of uncertainty for x

$$dx = -A^+ (dY - dY_0)X^+ x + dX X^+ x + A^+ (dy - dy_0) \tag{6.114}$$

---

[8]   To avoid confusion, A is used here as symbol for the sensitivity matrix; BAUER et al. [1991a, b] use S for this purpose and B for the background vector (here y_0).

where the d-vectors and -matrixes are errors of the respective quantities. From Eq. (6.114) the variances $s_{x_k}^2$ of the contents x_k are obtained:

$$
\begin{aligned}
s_{x_k}^2 = &\sum_{p=1}^{P}\sum_{n=1}^{N}\sum_{k=1}^{K} (A^+ X^+ x)^2 (s_Y^2 + s_{Y_0}^2) \\
&+ \sum_{n}^{N}\sum_{k}^{K} (X^+ x)^2 s_X^2 + \sum_{p}^{P} (A^+)^2 (s_y^2 + s_{y_0}^2) \quad .
\end{aligned}
\tag{6.115}
$$

By means of these variances the limits of detection $x_{k_{LD}}$ of the k analytes under investigation can be estimated in analogy to the univariate case (see Sect. 7.5), namely on the basis of the critical values $y_{net_{k,c}}$ of the net signals their vector is denoted $y_{k,c}$ in the last term of Eq. (6.116b):

$$
x_{k_{LD}} = u_\alpha \cdot s_{x_k}
\tag{6.116a}
$$

$$
x_{k_{LD}} = u_\alpha \cdot \sqrt{\sum_{p=1}^{P}\sum_{n=1}^{N}\sum_{k=1}^{K} (A^+ X^+ x)^2 (s_Y^2 + s_{Y_0}^2) + \sum_{n}^{N}\sum_{k}^{K} (X^+ x)^2 s_X^2 + \sum_{p}^{P} (A^+)^2 (s_{y_{k,c}}^2 + s_{y_0}^2)}
\tag{6.116b}
$$

A detailed derivation can be found in BAUER et al. [1991b]. The limit of detection according to Eq. (6.116a) corresponds to KAISER's so-called 3σ criterion; see Sect. 7.5., LORBER and KOWALSKI [1988] as well as FABER and KOWALSKI [1997b] take into account errors of the first and second kind. The multivariate detection limits are estimated then in analogy to the univariate limits being twice the 3σ-limit (with $u_\alpha = u_\beta$); see Sect. 7.5 and EHRLICH and DANZER [2006]).

6.5
Calibration by Artificial Neural Networks

Neural networks are systems of information processing that consists of a large number of units which transmit information to another by means of directed connections activating other units. Artificial neural networks (ANN) are simplified and idealised replicas of biological neural networks. An ANN can be considered as a parallel machine that uses a large number of processor units, the *neurons*, with a high degree of connectivity (ZUPAN and GASTEIGER [1991]; FRANK and TODESCHINI [1994]; DANZER et al. [2001]). Neurons are arranged in layers of which modern ANNs contain at least three:

- The *input layer*, their neurons correspond to the prediction variables (the analytical values x_i); input layers are counted to be layer 0
- A *hidden layer* (sometimes more than one), representing the model of calculation and having not connection to the outside world
- The *output layer*, producing the output quantities (the measured values y_j)

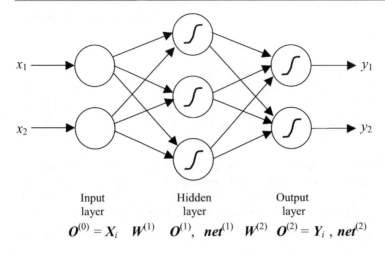

Input layer   Hidden layer   Output layer

$$O^{(0)} = X_i \quad W^{(1)} \quad O^{(1)}, \; net^{(1)} \quad W^{(2)} \quad O^{(2)} = Y_i, \; net^{(2)}$$

Fig. 6.18. Schematic representation of a multilayer perceptron with two input neurons, three hidden neurons (with sigmoid transfer functions), and two output neurons (with sigmoid transfer functions, too)

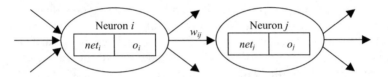

Fig. 6.19. Components and connection of two artificial neurons, according to DANZER et al. [2001]

In general, there is no connection between neurons within the same layer, but each neuron is connected with each neuron of the next layer. The structure of an ANN can be seen from Fig. 6.18.

Neural networks are characterized by their *weights*, w_{ij}, and their respective sums are given by the weight matrixes, $W^{(j)}$, between the diverse layers. The weights represent the strength of the directed connection between neurons i and j; see Fig. 6.19.

Neurons have one or more inputs, an output, o_i, an *activation state*, A_i, an *activation function*, f_{act}, and an *output function*, f_{out}. The *propagation function* (*net function*)

$$net_i(t) = \sum_{i=1}^{k} O_i W_{ij} \tag{6.117}$$

characterizes how the input of the neuron are computed from the outputs O_i of the other neurons and the respective weights W_{ij}.

The activation state A_i of a neuron i at time (or iteration step) $t + 1$ is given by

$$A_i(t+1) = f_{act}\big(A_i(t), net_i(t), \Theta_i\big) \tag{6.118}$$

with f_{act} being the activation function (transfer function), $net_i(t)$ the input and Θ_i the threshold value (bias) of the neuron. This threshold must be exceeded so that the neuron becomes active ("fires").

The output $O_i(t)$ of the neuron is estimated by means of the output function f_{out}

$$O_i(t) = f_{out}\big(A_i(t)\big) \tag{6.119}$$

Often the identity function is chosen as output function. Then the output becomes

$$O_i = f_{act}\big(net_i(t)\big) \tag{6.120}$$

ANNs need supervised learning schemes and can so be applied for both classification and calibration. Because ANNs are nonlinear and model-free approaches, they are of special interest in calibration.

By means of the learning algorithm the parameters of the ANN are altered in such a way that the net inputs produce adequate net outputs. Mostly this is affected only by changing the weights (other procedures like adding or removing of neurons and modification of activation functions are rarely used; see ZELL [1994]).

The most popular techniques of multilayer perceptrons (MLP) are *back-propagation networks* (WYTHOFF [1993]; JAGEMANN [1998]). The weight matrixes W are estimated by minimizing the net error

$$E = \frac{1}{2}\sum_{i=1}^{n}(Y_i - O_i)^2 \tag{6.121}$$

on the basis of the generalized delta rule. According to this algorithm also the weights of inner neurons can be modified. The weights are estimated from the net outputs O_i by means of Eqs. (6.117) and (6.120). Then the error function of the neurons can be calculated according to

$$\delta_{ij} = \begin{cases} f'_{act}\, net_{ij}(Y_{ij} - O_{ij}) & \text{if } j \text{ is an output neuron} \\[2ex] f'_{act}\, net_{ij} \sum_{k=1}^{n_l} \delta_{ik} W_{jk} & \text{if } j \text{ is an hidden neuron} \end{cases} \tag{6.122}$$

where f'_{act} is the first derivation of the activation function, Y_{ij} the jth element of the teaching input and n_l the number of neurons in the respective hidden layer (k is indexing the neurons in the next layer). The correction of the weights in the nth run is carried out by

$$\Delta W_{jk}(n+1) = \eta\, \delta_{ij}\, O_{ik} + \alpha\, \Delta W_{jk}(n) \tag{6.123}$$

where η is the *learning rate* (step size parameter) and α the *momentum term*.

In case of MLP, the activation function has to be monotonous and differentiable (because of Eq. 6.122). Frequently used is the sigmoid function

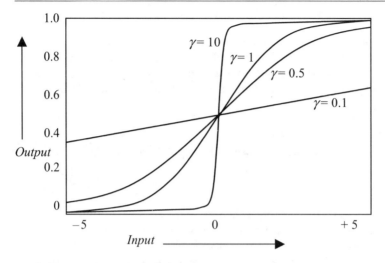

Fig. 6.20. Sigmoid activation function with diverse values of the steepness parameter γ

$$f_{act} = O_j = \frac{1}{1 + \exp(-\gamma \; net_j)} \tag{6.124}$$

with γ being the steepness parameter of the activation function; see Fig. 6.20. The outputs are between 0 and 1.

Although for each neuron a specific activation function can be defined, the most applications use the same function for the neurons of a given layer. When the activation function has a limited range, then X and Y must be scaled correspondingly.

The threshold values Θ of the neuron are considered by using an additional neuron in the hidden layer ("*on neuron*", *bias neuron*) the output of which is always $+1$.

The popular *radial basis function nets* (RBF nets) model nonlinear relationships by linear combinations of basis functions (ZELL [1994]; JAGEMANN [1998]; ZUPAN and GASTEIGER [1993]). Functions are called to be radial when their values, starting from a central point, monotonously ascend or descend such as the CAUCHY function or the modified GAUSS function at Eq. (6.125):

$$h_j(x_i) = \exp\left(-\frac{(X_i - W_j^{(1)})^2}{r^2}\right) \tag{6.125}$$

where $W^{(1)}$ is the vector of the centres of the basis functions and r a scaling parameter. Figure 6.21 shows the dependence of the GAUSSian basis function on the scaling factor. The local character of activation gets lost in case of high values of r.

The determination of the weights $W^{(2)}$ of the output layer is simple in so far as they base on a linear model and can be calculated in one step, in

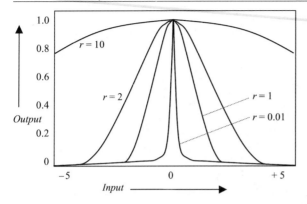

Fig. 6.21. Gaussian radial basis function with diverse values of the scaling parameter r

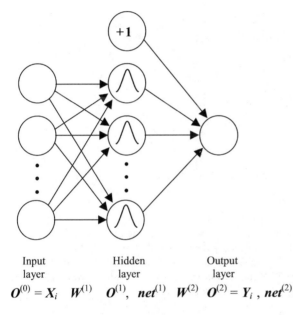

Fig. 6.22. Structure of a RBF net with one hidden layer with bias neuron and one output unit as used for single component calibration (according to FISCHBACHER et al. [1997]; JAGEMANN [1998])

contrast to backpropagation nets in which the weights must be estimated iteratively. The architecture of a RBF net is given in Fig. 6.22.

At first, the input of the hidden layer is calculated from the $n \times m$ net input X and the $k \times m$ weight matrix $W^{(1)}$ according to

$$\mathbf{net}_i^{(1)} = \sqrt{\sum_{j=1}^{k} (X_i - W_j^{(1)})^2} \tag{6.126}$$

With Eq. (6.126) and a GAUSSIAN activation function the output of the hidden neurons (the *RBF design matrix*) becomes

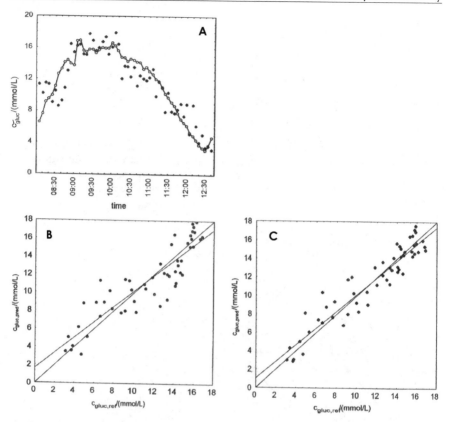

Fig. 6.23. A Glucose profile of a single test person (comparable to that of Fig. 6.13), c_{gluc} reference value of glucose concentration, ○, and the predicted glucose values of NIR measurements, ●, evaluated by PLS. **B** Recovery function and fitted calibration line of the PLS calibration (number of included wavelengths $n = 56$, 6 factors). **C** Recovery function and fitted calibration line of the RBF calibration (number of included wavelengths $n = 56$, 10 hidden layers), according to FISCHBACHER et al. [1995]

$$O^{(1)} = \exp\left(-\frac{net^{(1)}}{r}\right)^2 \tag{6.127}$$

Neural networks are applied in analytical chemistry in many and diverse ways. Used in calibration, ANNs have especially advantages in case of nonlinear relationships, multicomponent systems and single component analysis in case of various disturbances.

The latter applies to NIR spectroscopy used for the non-invasive determination of blood glucose by means of a fibre-optical measuring-head (JAGEMANN et al. [1995]; MÜLLER et al. [1997]; DANZER et al. [1998]). In addition to the weak overtone and combination bands resulting from glucose, strongly disturbing absorption of water, that is the main component

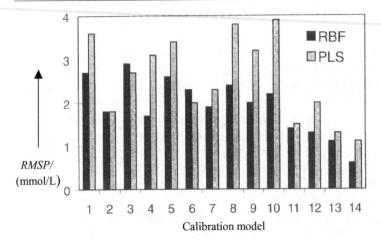

Fig. 6.24. Comparison for cross-validated PLS and RBF calibration model for different test persons (JAGEMANN et al. [1995])

of living tissue, takes place. Furthermore, baseline effects and some time-dependent biological processes like pulse and respiration appear. Therefore, multicomponent calibration by PLS 1 (see Sects. 6.2.6 and 6.4.2) and calibration by a RBF net have been applied.

The results of PLS- and RBF calibration are compared in Fig. 6.23. It can be seen that the calibration points in case of the RBF net deviate slightly lower than that of PLS. This fact can be expressed quantitatively by means of the RMSP value (*root mean standard error of prediction*):

$$RMSP = \sqrt{\frac{\sum\limits_{i=1}^{n} (\hat{x}_i - x_i)^2}{n}} \qquad (6.128)$$

that can be estimated by using leaving-one-out cross validation. Concerning the calibration represented in Fig. 6.23, PLS is characterized by $RMSP = 1.8$ mmol/L and RBF by $RMSP = 1.4$ mmol/L.

More generally, the comparison is represented in Fig. 6.24 where the $RMSP$ values of 14 calibration models ($n = 349$ test persons) are given. In most cases the RBF models greatly surpasses the PLS models.

References

Bauer G, Wegscheider W, Ortner HM (1991a) *Limits of detection in multivariate calibration.* Fresenius J Anal Chem 340:135

Bauer G, Wegscheider W, Ortner HM (1991b) *Selectivity and error estimates in multivariate calibration: application to sequential ICP-OES.* Spectrochim Acta 46B:1185

Burns DT, Danzer K, Townshend A (2002) IUPAC, Analytical Chemistry Division, Commission on General Aspects of Analytical Chemistry: *Use of the terms "recovery" and "apparent recovery" in analytical procedures (IUPAC Recommendations 2002).* Pure Appl Chem 74:2201

Centner V, Massart DL, de Jong S (1998) *Inverse calibration predicts better than classical calibration.* Fresenius J Anal Chem 361:2

Currie LA (1995) IUPAC, Analytical Chemistry Division, Commission on Analytical Nomenclature: *Nomenclature in evaluation of analytical methods including detection and quantification capabilities.* Pure Appl Chem 67:1699

Danzer K (1989) *Robuste Statistik in der analytischen Chemie.* Fresenius Z Anal Chem 335:869

Danzer K (1990) *Problems of calibration in trace-, in situ-micro- and surface analysis.* Fresenius J Anal Chem 337:794

Danzer K (1995) *Calibration – a multidimensional approach.* Fresenius J Anal Chem 351:3

Danzer K (2001) *Selectivity and specificity in analytical chemistry. General considerations and attempt of a definition and quantification.* Fresenius J Anal Chem 369:397

Danzer K, Currie LA (1998) IUPAC, Analytical Chemistry Division, Commission on General Aspects of Analytical Chemistry: *Guidelines for calibration in analytical chemistry. Part 1. Fundamentals and single component calibration* (recommendations 1998). Pure Appl Chem 70:993

Danzer K, Fischbacher C, Jagemann K-U, Reichelt KJ (1998) *Near-infrared diffuse reflection spectroscopy for non-invasive blood-glucose monitoring.* LEOS Newslett 12(2):9

Danzer K, Hobert H, Fischbacher C, Jagemann K-U (2001) *Chemometrik. Grundlagen und Anwendungen.* Springer, Berlin Heidelberg New York

Danzer K, Otto M, Currie LA (2004) IUPAC, Analytical Chemistry Division, Commission on General Aspects of Analytical Chemistry: *Guidelines for calibration in analytical chemistry. Part 2. Multispecies calibration.* (IUPAC Technical Report), Pure Appl Chem 76:1215

Danzer K, Schrön W, Dreßler B, Jagemann K-U (1998) *Matrix sensitivity of solid sampling AAS. Determination of zinc in geological samples.* Fresenius J Anal Chem 361:710

Danzer K, Venth K (1994) *Multisignal calibration in spark- and ICP-OES.* Fresenius J Anal Chem 350:339

Danzer K, Wagner M (1993) *Multisignal calibration in optical emission spectroscopy.* Fresenius J Anal Chem 346:520

Danzer K, Wagner M, Fischbacher C (1995) *Calibration by orthogonal and common least squares – theoretical and practical aspects.* Fresenius J Anal Chem 352:407

Danzer K, Wienke D, Wagner M (1991) *Application of chemometric principles for evaluation of emission spectrographic data.* Acta Cim Hungar 128:623

Deming SN, Morgan SL (1993) *Experimental design: a chemometric approach, 2nd edn.* Elsevier, Amsterdam

Dillon WR, Goldstein M (1984) *Multivariate analysis – methods and applications.* Wiley, New York

Draper NR, Smith H (1981) *Applied regression techniques, 2nd edn.* Wiley, New York

Edgeworth FY (1887) *On observations relating to several quantities.* Hermathena 6:279

Eckschlager K, Danzer K (1994) *Information theory in analytical chemistry.* Wiley, New York

Efron B (1982) *The jackknife, the bootstrap and other resampling techniques.* Society for Industrial and Applied Mathematics, Philadelphia, PA

Efron B, Tibshirani R (1993) *An introduction to the bootstrap.* Chapman and Hall, London

Ehrlich G, Danzer K (2006) *Nachweisvermögen von Analysenverfahren – Objektive Bewertung und Ergebnisinterpretation.* Springer, Berlin Heidelberg New York

Faber K, Kowalski BR (1997a) *Improved prediction error estimates for multivariate calibration by correcting for the measurement error in the reference values.* Appl Spectrosc 51:660

Faber K, Kowalski BR (1997b) *Propagation of measurement errors for the validation of predictions obtained by principal component regression and partial least squares.* J Chemom 11:181

Fischbacher C, Jagemann K-U, Danzer K, Müller UA, Mertes B (1995) *Non-invasive blood glucose monitoring by chemometrical evaluation of NIR-spectra.* Twenty-fourth EAS (Eastern Analytical Symposium), Somerset, NJ, 12-16.11.1995

Fischbacher C, Jagemann K-U, Danzer K, Müller UA, Papenkordt L, Schüler J (1997) *Enhancing calibration models for non-invasive near-infrared spectroscopical blood glucose determination.* Fresenius J Anal Chem 359:78

Flury B (1988) *Common principal components and related multivariate models.* Wiley, New York

Frank IE, Todeschini R (1994) *The data analysis handbook.* Elsevier, Amsterdam

Garden JS, Mitchell DG, Mills WH (1980) *Nonconstant variances regression techniques for calibration-curve-based analysis.* Anal Chem 52:2310

Hampel FR (1980) *Robuste Schätzungen: Ein anwendungsorientierter Überblick.* Biometr J 22:3

Hampel FR, Ronchetti EM, Rousseeuw PJ, Stahel W (1986) *Robust statistics. The approach based on influence functions.* Wiley, New York

Hässelbarth W (1995) *Traceability of measurement and calibration in chemical analysis.* Fresenius J Anal Chem 352:400

Høskuldsson A (1988) *PLS regression methods.* J Chemom 2:211

Huber PJ (1981) *Robust statistics.* Wiley, New York

Hulanicki A (1995) IUPAC, Analytical Chemistry Division, Commission on General Aspects of Analytical Chemistry: *Absolute methods in analytical chemistry*. Pure Appl Chem 67:1905

Inczédy J, Lengyiel JT, Ure AM, Geleneser A, Hulanicki A (eds) (1997) *Compendium of analytical nomenclature, 3rd edn* (IUPAC Orange Book). Blackwell, Oxford

ISO (1993), International Organization for Standardization, *Guide to the expression of uncertainty in measurement*. Geneva

IUPAC Orange Book (1997, 2000)
– printed version: *Compendium of analytical nomenclature (Definitive Rules 1997)*; see Inczédy et al (1997)
– web version (as from 2000): www.iupac.org/publications/analytical_compendium/

Jagemann K-U, Fischbacher C, Danzer K, Müller UA, Mertes B (1995) *Application of near-infrared spectroscopy for non-invasive determination of blood/tissue glucose using neural networks*. Z Physikal Chem 191:179

Jagemann K-U (1998) *Neuronale Netze in der Analytik*, in: Günzler H, Bahadir AM, Danzer K, Engewald W, Fresenius W, Galensa R, Huber W, Linscheid M, Schwedt G, Tölg G, *Analytiker-Taschenbuch* 19:75

Kaiser H (1972) *Zur Definition von Selektivität, Spezifität und Empfindlichkeit von Analysenverfahren*. Fresenius Z Anal Chem 260:252

Lorber A, Kowalski BR (1988) *Estimation of prediction error for multivariate calibration*. J Chemometrics 2:93

Malinowski ER, Howery DG (1980) *Factor analysis in chemistry*. Wiley, New York

Mandel J (1964) *The statistical analysis of experimental data*. Wiley, New York

Mandel J (1984) *Fitting straight lines when both variables are subject to error*. J Quality Technol 16:1

Manne R (1987) *Analysis of two partial-least-squares algorithms for multivariate calibration*. Chemom Intell Lab Syst 2:187

Martens H, Næs T (1989) *Multivariate calibration*. Wiley, New York

Moore EH (1920) *On the reciprocal of the general algebraic matrix*. Bull Amer Math Soc 26:394

Morgan E (1991) *Chemometrics: experimental design*. Wiley, Chichester, UK

Müller UA, Mertes B, Fischbacher C, Jagemann K-U, Danzer K (1997) *Non-invasive blood glucose monitoring by means of near infrared spectroscopy: methods for Improving the reliability of the calibration models*. Int J Artific Organs 20:285

Nimmerfall G, Schrön W (2001) *Direct solid sample analysis of geological samples with SS-GF-AAS and use of 3D calibration*. Fresenius J Anal Chem 370:760

Otto M, Bandemer H (1986) *Calibration with imprecise signals and concentrations based on fuzzy theory*. Chemom Intell Lab Syst 1:71

Penrose R (1955) *A generalized inverse for matrices*. Proc Cambridge Phil Soc 51:406

Rousseeuw PJ, Leroy AM (1987) *Robust regression and outlier detection*. Wiley, New York

Sachs L (1992) *Angewandte Statistik*. Springer, Berlin Heidelberg New York

Sharaf MA, Illman DL, Kowalski BR (1986) *Chemometrics*. Wiley, New York

STATISTICA for Windows 4.5 (1993) Statististical Software Inc

Venth K, Danzer K, Kudermann G, Blaufuß K-H (1996) *Multisignal evaluation in IPC-MS. Determination of trace elements in Mo-Zr-alloys.* Fresenius J Anal Chem 354:811

Vessman J, Stefan RI, van Staden JF, Danzer K, Lindner W, Burns DT, Fajgelj A, Müller H (2001) IUPAC, Analytical Chemistry Division, Commission on General Aspects of Analytical Chemistry: *Selectivity in analytical chemistry (IUPAC recommendations 2001).* Pure Appl Chem 73:1381

Wald A (1940) *The fitting of straight lines if both variables are subject to error.* Ann Math Statist 11:284

Wegscheider W (1996) *Validation of analytical methods.* In: Günzler H (ed) *Accreditation and quality assurance in analytical chemistry.* Springer, Berlin Heidelberg New York

Wold S (1974) *Spline functions in data analysis.* Technometrics 16:1

Wythoff BJ (1993) *Backpropagation neural networks. A tutorial.* Chemom Intell Lab Syst 18:115

Zell A (1994) *Simulation neuronaler Netze.* Addison-Wesley, Bonn

Zupan J, Gasteiger J (1991) *Neural networks: a new method for solving chemical problems or just a passing phase?* Anal Chim Acta 248:1

Zupan J, Gasteiger J (1993) *Neural networks for chemists.* VCH, Weinheim

7 Analytical Performance Characteristics

As a measuring science, analytical chemistry has to guarantee the quality of its results. Each kind of measurement is objectively affected by uncertainties which can be composed of random scattering and systematic deviations. Therefore, the measured results have to be characterized with regard to their quality, namely both the *precision* and *accuracy* and – if relevant – their information content (see Sect. 9.1). Also analytical procedures need characteristics that express their potential power regarding *precision, accuracy, sensitivity, selectivity, specificity, robustness,* and *detection limit.*

Notwithstanding that a large number of definitions exist that have been produced and harmonized by many commissions of international organizations (e.g., IUPAC, IUPAP, ISO, BIPM, IEC, IFCC, OIML), there are still official recommendations and definitions which are not completely satisfying and sometimes confusing (PRICHARD et al. [2001]).

In general, it is well-known that analytical results can be assessed and how it is done in detail. On the other hand, an assessment of analytical operations, carried out within the hierarchy of analytical techniques, methods, procedures, and SOPs, takes place scarcely according to standardized viewpoints.

Figure 7.1 shows how the terms of analytical proceedings (see Chap. 9) can be classified as belonging to the steps of the analytical process which is differently represented here in comparison to Fig. 2.1. The classification is descended from GOTTSCHALK [1975], DANZER et al. [1987], TAYLOR [1983], and PRICHARD et al. [2001]. Table 7.1 illustrates by examples how different the degree of concretization of analytical proceedings is.

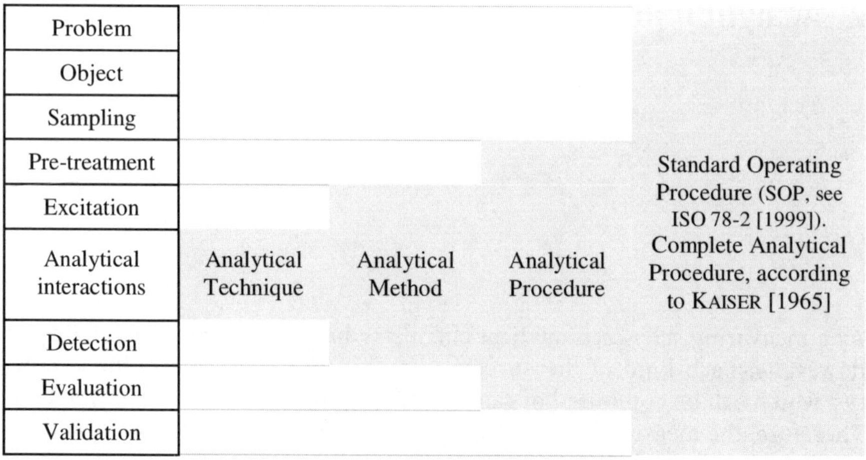

Problem				Standard Operating Procedure (SOP, see ISO 78-2 [1999]). Complete Analytical Procedure, according to KAISER [1965]
Object				
Sampling				
Pre-treatment				
Excitation				
Analytical interactions	Analytical Technique	Analytical Method	Analytical Procedure	
Detection				
Evaluation				
Validation				

Fig. 7.1. The meaning of the terms of analytical proceedings in relation to the steps of the analytical process

Table 7.1. Examples of the various degree of concretization of analytical activities

Ex #	Analytical technique	Analytical method	Analytical procedure	Standard operating procedure
1	OES	ICP-OES	Direct ICP-OES of wines	Complete analytical procedure see THIEL and DANZER [1997]
2	GC	GC-MS	SPME-CGC of wines	Complete analytical procedure as described by DE LA CALLE GARCÍA et al. [1998]
3	IR	NIR	Non-invasive NIR of blood-glucose	Complete analytical procedure see DANZER et al. [1998]

7.1
Reliability of Analytical Measurements

For the characterization of the reliability of analytical measurements the terms *precision*, *accuracy*, and *trueness* have a definite meaning.

7.1.1
Precision

Precision is defined as "*the closeness of agreement between independent test results obtained under stipulated conditions*" (FLEMING et al. [1996b]; PRICHARD et al. [2001]). Precision characterizes the random component of the measurement error and, therefore, it does not relate to the true value.

According to the conditions under which the measurements are carried out, it is to distinguish between

- *Repeatability*, the *"precision under repeatability conditions"* (ISO 3534-1 [1993]) and
- *Reproducibility*, the *"precision under reproducibility conditions"* (ISO 3534-1 [1993]).

Repeatability is *"the closeness of the agreement between the results of successive measurements of the same measurand carried out under the same conditions of measurement"*. Repeatability conditions include: the same measurement procedure, the same observer, the same measuring instrument, used under the same conditions, the same location, and repetition over a short period of time (ISO 3534-1 [1993]).

On the other hand, *reproducibility* is the *"closeness of the agreement between the results of measurements of the same measurand carried out under changed conditions of measurement"*. The changed conditions include: principle of measurement, method of measurement, observer, measuring instrument, reference standards, location, conditions of use, and time. Such variable conditions are typical for *interlaboratory studies* (laboratory intercomparisons).

It should be clearly distinguished whether given terms characterize analytical results or methods. In this respect, there exist some contradictory views, as can be illustrated by the example of repeatability. According to ISO 3534-1 [1993] and PRICHARD et al. [2001] *repeatability* characterizes *"results of measurements"*, on the other hand, according to FLEMING et al. [1996a] repeatability *" is a characteristic of a method not of a result"*. Such different interpretation of terms can occasionally be observed.

Quantification: Precision can be quantified by suitable dispersion characteristics. It is proposed to characterize precision by *standard deviation*, see Eqs. (4.12)–(4.14) and *relative standard deviation*, see Eq. (4.15) (FLEMING et al. [1996b]; PRICHARD et al. [2001]). Because of some uncertainty, the characterization of analytical proceedings in their hierarchy (see Fig. 7.1) and of analytical results, respectively, will be considered in detail.

- *Precision of analytical techniques* should only be described verbally or comparatively (e.g., "the precision of coulometry is high" or "the precision of spectrophotometry is better then that of OES").
- *Precision of analytical methods* is usually characterized semi-quantitatively by giving a rough value of the relative standard deviation (e.g., "the precision of arc-excited OES is about 20%").
- *Precision of an analytical procedure* is commonly expressed by an average relative standard deviation (e.g., "the precision of the determination of Mn in steel by XRF in a given routine control is 1.5%").

- *Precision of a complete analytical procedure*, i.e., a *standard operation procedure (SOP)*, should be characterized by the *uncertainty of measurement* (absolute or relative) as exactly as validation it stipulates.

- *Precision of a result of measurement* (y) is characterized by the *confidence interval* of the result (see Sect. 4.1.2, Fig. 4.5 and Eqs. (4.16) and (4.17)) or the corresponding *uncertainty interval* (see Sect. 4.2 and Eqs. (4.29)–(4.32)).

- *Precision of an analytical result* (x) is expressed by the *prediction interval* which include not only the dispersion of the measured results but additionally the uncertainty of the calibration on which the estimation of x is based; see Sect. 6.2.2.

- *Precision of a measuring system* and with it also of *signals* and *signal functions* obtained by instrumental analytical methods, is characterized by the signal-to-noise ratio.

In analogy to other measuring sciences, also in analytical chemistry the *signal-to-noise ratio*

$$\frac{S}{N} = \frac{\overline{y}}{s_y} = \frac{1}{s_{y_{rel}}} \tag{7.1}$$

is used. The mean of y is mostly applied in the form of the background-corrected *net value* y_{net}

$$\frac{S}{N} = \frac{\overline{y}_{net}}{s_{y_{net}}} = \frac{\overline{y} - \overline{y}_0}{s_{y_{net}}} = \frac{\overline{y - y_0}}{s_{y_{net}}} \tag{7.2}$$

where y_0 is the background signal (blank) and s_{y_0} its standard deviation; see Fig. 7.2.

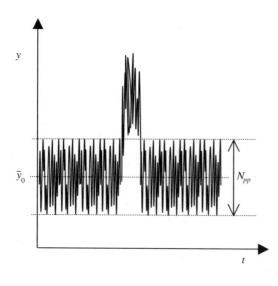

Fig. 7.2. Signal function with noise; N_{pp} noise amplitude (peak-to-peak noise)

The net value $\overline{y} - \overline{y}_0$ may be also estimated by averaging of the individual differences $y - y_0$. In both cases using differences, the belonging standard deviation $s_{y_{net}}$ increases in comparison with s_y

$$s_{y_{net}} = \sqrt{\frac{(n_y - 1) s_y^2 + (n_{y_0} - 1) s_{y_0}^2}{n_y + n_{y_0} - 2} \left(\frac{n_y + n_{y_0}}{n_y \cdot n_{y_0}}\right)} \tag{7.3}$$

provided that the standard deviations s_y and s_{y_0} of the analyte signal and the background signal do not differ significantly from each other. In the case that both are estimated by an equal number of measurements ($n_y = n_{y_0} = n$) and $s_y \approx s_{y_0}$, Eq. (7.3) is simplified as follows:

$$s_{y_{net}} = s_y \sqrt{\frac{2}{n}} \tag{7.4}$$

Noise $N(t)$ is characterized by the following characteristics:

- *Time average* (temporal mean), $\overline{N(t)}$, where ideally is $\overline{N(t)} = 0$, in case that there exist a background, $\overline{N(t)} = \overline{y}_0$
- *Noise amplitude* (peak-to-peak noise), $N_{pp} = N(t)_{max} - N(t)_{min}$; if the difference cannot be estimated directly, an approximation can be obtained by $N_{pp} \approx 2\Delta\overline{y}_0$ (with $P = 0.999$)
- *Noise variance* (squared mean), $\sigma_N^2 = \overline{N(t)^2}$
- *Standard deviation of noise*, σ_N, which be roughly estimated by means of the peak-to-peak distance according to

$$s_N \approx \frac{N_{pp}}{5} \tag{7.5a}$$

(DOERFFEL et al. [1990]) and more exactly by

$$s_N \approx \frac{N_{pp}}{\kappa} \tag{7.5b}$$

where κ can be found in tables (SACHS [1992]; LIECK [1998]).

Signal-to-noise ratio characterizes recorded signals and signal functions with regard of their quality, i.e., their precision. Unfortunately, the signal-to-noise is not uniformly used in analytical chemistry. In addition to the definitions given in Eqs. (7.1) and (7.2), there exist another one, related to the peak-to-peak noise N_{pp}:

$$snr(\overline{y}) = \frac{\overline{y}_{net}}{N_{pp}} = \frac{\overline{y} - \overline{y}_0}{N_{pp}} \approx \frac{\overline{y} - \overline{y}_0}{2(\Delta\overline{y}_0)_{0.999}} \tag{7.6}$$

The signal-to-noise ratio $snr(y)$ defined in this way, can differ from S/N defined by Eq. (7.2) considerably, namely according to $S/N \approx \left(5\sqrt{\frac{\pi}{2}}\right) snr(y) = \left(3,536\sqrt{n}\right) snr(y)$ and $snr(y) \approx \left(\frac{1}{5}\sqrt{\frac{2}{n}}\right) S/N = \left(\frac{0,2828}{\sqrt{n}}\right) S/N$, respectively.

Therefore, it is necessary to inform on the type of definition which is applied, $snr(y)$ or S/N, to avoid confusion.

The signal-to-noise ratio has been used in analytical chemistry since the 1960s. At first, atomic spectroscopy prepared the way for application, and some other spectroscopic disciplines and chromatography are important domains of use.

7.1.2
Precision of Trace Analyses

It is a well-known fact that the precision in trace analysis decreases with diminishing concentration in a similar way as it does with decreasing sample weight (Sect. 2.1). The dependency of the repeatability and reproducibility standard deviation on the concentration of analytes has been investigated systematically at first by HORWITZ et al. [1980] on the basis of thousands of pieces of interlaboratory data (mostly from food analysis). The result of the study has been represented in form of the well-known "HORWITZ trumpet" which is represented in Fig. 7.3.

The HORWITZ relationship agrees with the experience of analysts and has been confirmed in various fields of trace analysis, not only in its qualitative form but also quantitatively. THOMPSON et al. [2004] have estimated the mathematical form of the HORWITZ functiontextscHorwitz function being $s_H = 0.02\,x^{0.85}$, or linearized, $\log s_H = 0.85 \log x$. The agreement of this equation is usually good and, therefore, the HORWITZ functiontextscHorwitz function is sometimes used as a bench-mark for the performance of analytical methods. For this purpose, the so-called "Horrat" (HORWITZ ratio) has been defined, $Horrat = s_{actual}/s_H$, by which the actual standard deviation is compared with the estimate of the HORWITZ function. Serious deviations

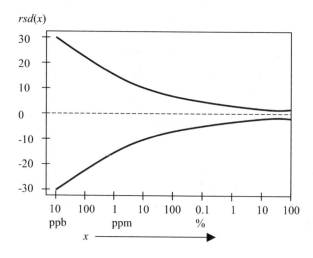

Fig. 7.3. The so-called "HORWITZ trumpet": Dependency of the relative standard deviation, $rsd(x)$ on the concentration (x)

of the *Horrat* from 1 can be used to reject results from interlaboratory comparisons (THOMPSON et al. [2004]).

In general, in trace analysis concessions must be made with regard to precision and accuracy. However, accuracy has strictly to be distinguished from precision.

7.1.3
Accuracy and Trueness

Accuracy is defined in the name of BIPM, IEC, IFCC, ISO, IUPAC, IUPAP, and OIML as "*the closeness of agreement between the result of a measurement and the true value of the measurand*" (ISO 3534-1 [1993]; FLEMING et al. [1996b]). Accuracy characterizes the absence of a relevant bias of a measured value. As can be seen from Fig. 4.1, accuracy is a measure that combines the effects of both random and systematic deviations. Therefore, a bias can only be detected if it exceeds the range of random error.

In contrast to *accuracy* which relates to single test results, *trueness* is defined as "*closeness of agreement between the average value obtained from a large series of test results and an accepted reference value*" (IUPAC ORANGE BOOK [1997, 2000]; FLEMING et al. [1996c]). Both terms, *conventional true value* and *accepted reference value* describe practical approaches to *a true value* which is by nature indeterminable (ISO 3534-1 [1993]). *Conventional true value* is defined as a "*value attributed to a particular quantity and accepted, sometimes by convention, as having an uncertainty appropriate for a given purpose*" (ISO 3534-1 [1993]; PRICHARD et al. [2001]). Accuracy is a qualitative concept (ISO 3534-1 [1993]). However, in analytical practice, accuracy is frequently characterized in a quantitative way by the bias (FLEMING et al. [1996b]); see Eq.(4.1).

> The ISO recommendation [1993] should be followed and accuracy used only as a qualitative term. In case of quantitative characterization (by means of the bias), a problem may appear which is similar to that of precision, namely that a quality criterion is quantified by a measure that has a reverse attribute regarding the property which have to be characterized. If the basic idea of measures can be accepted, which is that a high quality becomes a high value and vice versa, bias is an unsuited measure of accuracy (and trueness). In this sense, accuracy could be defined by means of a measure proposed in the next paragraph.

7.1.4
Remark on the Quantification of Precision, Accuracy and Trueness

Precision, accuracy and trueness are important performance characteristics in analytical chemistry. Each of them is well-defined in a positive sense ("*closeness of agreement ...*"). However, their quantifying is done by means of unfavourable measures, namely by error quantities like, e.g., standard deviation and bias, respectively, which indeed do quantify *imprecision* and *inaccuracy.* This inconsistency has been stated by KAISER and SPECKER [1956] who criticized that "*It is unsightly that the confidence interval is narrow when the confidence is high and vice versa*". Therefore, they proposed an alternative measure to characterize properly the *precision of an analytical measurement*, namely in the form of the reciprocal value of the relative standard deviation

$$\Gamma = \frac{\overline{x}}{s} = \frac{1}{s_{rel}} \tag{7.7}$$

It is difficult to comprehend why this measure has not been applied in analytical chemistry. Instead of this, in the last decades the *signal-to-noise ratio* has increasingly been used. Signal-to-noise ratio, see Eq. (7.1), is the measure that corresponds to Γ in the signal domain. In principle, quantities like S/N (Eq. (7.1)) and Γ (Eq. (7.7)) could represent measures of precision, but they have an unfavourable range of definition, namely $range[\Gamma] = range[S/N] = 0 \ldots \infty$.

Useful measures of **precision** could be derived from relative dispersion measures, namely by their differences from 1, e.g., the *precision of an analytical procedure*

$$prec(x) = 1 - rsd(x) = 1 - \frac{s_x}{x} \tag{7.8}$$

The numerical value of this measure increases with decreasing error. The range is normally $range[prec(x)] = 1 \ldots 0$, i.e. for dispersion-free measurements ($s_x \to 0$) the precision becomes 1. On the other hand, if the dispersion amounts to 100% ($s_x \to x$) the precision becomes 0. Therefore, high precision is characterized by a high value of $prec(x)$. The precision becomes negative if the error exceeds the measured value what corresponds to $rsd(x) > 100\%$. But such cases should appear very rarely in analytical chemistry.

The *precision of an analytical result* can be quantified by means of the relative confidence interval

$$prec(\overline{x}) = 1 - \frac{\Delta \overline{x}}{x} \tag{7.9}$$

The interpretation is similar to that of the precision of analytical procedures.

Accuracy and *trueness* have been defined above and it was mentioned that these terms base on qualitative concepts (ISO 3534-1 [1993]). If it is necessary to have quantitative information, the bias, which is a measure of inaccuracy, should not be used to quantify accuracy and trueness, respectively. Instead of this, the following measures might be applied

$$acc(x) = 1 - \frac{bias(x)}{\Delta \overline{x}} \tag{7.10}$$

and

$$acc(\overline{x}) = 1 - \frac{bias(\overline{x})}{\Delta x} = trn(\overline{x}) \quad . \tag{7.11}$$

Table 7.2. Analytical results of trace elements in blood plasma analyses (STRECK [2004])

Analyte	Reference values/ μmol/L	Analytical results $\bar{x} \pm \Delta\bar{x}/(\mu mol/L)$	Recovery	Precision $prec(\bar{x})$	Trueness $acc(\bar{x})$	Comment: The result is
Ca	2200	2170 ± 170	0.986	0.922	0.824	Correct
Cu	20.0	19.5 ± 0.4	0.975	0.979	−0.250	Incorrect[a]
Zn	23.0	22.8 ± 0.9	0.991	0.961	0.778	Correct

[a] The reference value is outside the (upper) confidence limit and therefore the result is classified to be incorrect. This can also be proved by the t-test

These expressions are characterized by a definition range $range[acc(\bar{x})] = 1 \ldots -\infty$. The following cases can appear:

(i)   If the range of accuracy (trueness) is $1 \geq acc(x) > 0$, then the measured result is conventionally correct (see Fig. 4.1). The more the systematic error decreases the more $acc(x)$ approaches to 1. On the other hand, $acc(x)$ approaches to 0 with increasing bias.

(ii) If the range is $acc(x) < 0$, then the measured result is conventionally *incorrect*. The evidence of that can be seen in Fig. 4.1 and shown by means of the t-test, see Eq. (4.45).

The advantage of the measures as suggested here is that they point into the same direction as it is done by the verbal definitions. High precision and a high degree of accuracy, respectively, are characterized by high numerical values of the measures which approximate to 1 in the ideal case (absence of random and systematic deviations, respectively) and approximate to 0 if the deviations approach to 100%. In the worst cases, the numerical values $\{prec(\bar{x})\}$ and $\{acc(\bar{x})\}$ become negative which indicates that the relative random or systematic error *exceeds 100%*.

The example given in Table 7.2 is taken from a study to verify the trueness of clinical analyses (STRECK [2004]). Recovery rates have been used as the criterion to accept a good agreement between the measured results and the reference values as it is frequently done by analysts.

The evaluation by means of the performance characteristics $prec(\bar{x})$ and $acc(\bar{x})$ shows that they are meaningful measures which can usefully be applied in the daily analytical practice. From the example can also be seen that the recovery rate – solely used – can mislead the analyst if the relation of bias to confidence interval is disregarded.

7.2
Sensitivity

Sensitivity is a significant characteristic in all scientific disciplines which have to do with measurements. *Sensitivity* is defined from the viewpoint of instrumental measuring as "*the change in the response of a measuring instrument divided by the corresponding change in the stimulus*" (ISO

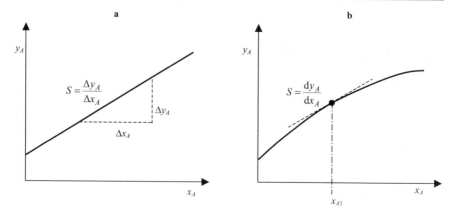

Fig. 7.4. Sensitivity in the case of a calibration straight line (**a**) and a curved calibration function (**b**)

3534-1 [1993]; IUPAC ORANGE BOOK [1997, 2000]; FLEMING et al. [1997b], PRICHARD et al. [2001]).

In analytical chemistry, the sensitivity S_{AA} of an analytical procedure (of the determination of an analyte A) is defined as the change in the measured value divided by the corresponding analytical value (analyte amount or concentration, respectively):

$$S_{AA} = \frac{\mathrm{d}y_A}{\mathrm{d}x_A} \qquad\qquad (7.12)$$

In case of a linear calibration function (see Chap. 6), the sensitivity becomes $S_{AA} = \Delta y_A / \Delta x_A$ and corresponds to the slope b of the calibration straight line; see Fig. 7.4a. If the calibration function is a curved line, then the sensitivity will vary according to the analyte amount or concentration as Fig. 7.4b shows.

It should be noted that the term "sensitivity" sometimes may alternatively be used, namely in analytical chemistry and other disciplines. Frequently the term "sensitivity" is associated with detection limit or detection capability. This and other misuses are not recommended by IUPAC (ORANGE BOOK [1997, 2000]). In clinical chemistry and medicine another matter is denoted by *"sensitivity"*, namely *"the ability of a method to detect truly positive samples as positive"* (O'RANGERS and CONDON [2000], cited according to TRULLOLS et al. [2004]). However, this seems to be more a problem of trueness than of sensitivity.

It is regrettable that a unified use of such an essential term like sensitivity could not be reached until now. All the more so, as the term "sensitivity" is defined in metrology and analytical chemistry in the same sense as it is used in daily life, namely as "reactio per actio" (effect/cause, output/input). A person is called to be sensitive if it strongly reacts on a given impulse (accusation, nudge, or stroke of fate, respectively), and insensitive vice versa. The stock market reacts sensitive (or insensitive) on changes of political and economical facts.

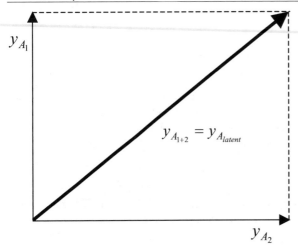

Fig. 7.5. Enhancement of signal intensity by merging two single signals into a latent signal

Some analytical methods open possibilities to evaluate more than only one signal (e.g., OES, MS, XRF). In such cases, the relevant signals $y_{A_1}, y_{A_2}, \ldots$ (the evaluation of each of which corresponds to another analytical *procedure*, see Fig. 7.1), do have different sensitivities $S_{AA_1}, S_{AA_2}, \ldots$, from which the best will be selected, as a rule.

On the other hand, it is possible to evaluate simultaneously more then one signal per analyte (DANZER and WAGNER [1993]) and enhance so the sensitivity of the determination of a given analyte A. The principle can be seen in Fig. 7.5 in which the intensities of two signals y_{A_1} and y_{A_2} are represented towards each other. The intensity of the combined signal, which is in fact a latent signal, is given by

$$y_{A_{latent}} = y_{A_{1+2}} = \sqrt{y_{A_1}^2 + y_{A_2}^2} \tag{7.13}$$

When merging more than two signals, $y_{A_{latent}}$ is obtained either by the n-dimensional principle of PYTHAGORAS or – better – by *principal component analysis (PCA, eigenanalysis*; see Sect. 6.3; DANZER [1990]).

Because all the intensities in Fig. 7.5, y_{A_1}, y_{A_2}, and $y_{A_{latent}}$, are caused by the same analyte concentration, the sensitivity also increases according to

$$S_{A_{latent}} = S_{A_{1+2}} = \sqrt{S_{A_1}^2 + S_{A_2}^2} \tag{7.14}$$

By multisignal evaluation, detection capability can be improved as has been shown for OES and MS by DANZER and VENTH [1994] and VENTH et al. [1996].

In multicomponent systems, the number of signals is (at least) equal in size to the number of components:

$$y_A = y_{A0} + S_{AA}x_A + S_{AB}x_B + \cdots + S_{AN}x_N + e_A$$
$$y_B = y_{B0} + S_{BA}x_A + S_{BB}x_B + \cdots + S_{BN}x_N + e_B$$
$$\vdots$$
$$y_N = y_{N0} + S_{NA}x_A + S_{NB}x_B + \cdots + S_{NN}x_N + e_N$$

(7.15)

The sensitivity of each component is given by $S_{AA}, S_{BB}, \ldots, S_{NN}$, analogous to Eq. (7.12). The interferences which appear between the species can be characterized by *cross sensitivities* as defined in Eq. (3.11):

$$S_{Ai} = \frac{\partial y_A}{\partial x_i}$$

(7.16)

More extensive, multicomponent system are described by the *sensitivity matrix* (*matrix of partial sensitivities* according to KAISER [1972], also called *K*-matrix according to JOCHUM et al. [1981]):

$$S = \begin{pmatrix} S_{AA} & S_{AB} & \cdots & S_{AN} \\ S_{BA} & S_{BB} & \cdots & S_{BN} \\ \vdots & \vdots & & \vdots \\ S_{NA} & S_{NB} & \cdots & S_{NN} \end{pmatrix}$$

(7.17)

In the ideal case only the diagonal elements of the sensitivity matrix are different from zero. Then no component disturbs any other and the analytical procedure works selectively (see Sect. 7.3). The *K*-matrix is defined analogously except that the elements are called k_{IJ}, their definition is the same as the S_{IJ} according to Eq. (7.16).

It may be useful to characterize multicomponent analyses not only by the single sensitivities of each component, but additionally by a *total multicomponent sensitivity* S_{total}. In the case of undisturbed measurements it is given by

$$S_{total} = \prod_{i=A}^{N} S_{ii}$$

(7.18)

If interferences and overlappings appear, the *total (multicomponent) sensitivity* is calculated as the determinant of the (squared) sensitivity matrix (Eq. (7.17)); (KAISER [1972]; SHARAF et al. [1986]; BERGMANN et al. [1987]; MASSART et al. [1988]):

$$S_{total} = |\det(S)|$$

(7.19)

In case of serious overlappings, multivariate techniques (see Sect. 6.4) are used and $p \gg n$ sensors (measuring points z_k) are measured for n components. From this an overdetermined systems of equations results and, therefore, non-squared sensitivity matrixes. Then the *total multicomponent sensitivity* is given by

$$S_{total} = \sqrt{\det(S^T S)}$$

(7.20)

(SHARAF et al. [1986]; BERGMANN et al. [1987]; MASSART et al. [1988]).

Some simple examples should illustrate the problem of sensitivity in multicomponent systems. At first, single component analysis of a system that contains 1 mol/L of the analyte A as well as 0.1 and 0.1 mol/L of the accompanying components B and C, i.e. the concentration vector is $x = (1, 0.1, 0.1)$; let the sensitivity be $S_{AA} = 5$ au/(mol/L)[1] and the cross sensitivities $S_{AB} = 0.1$ au/(mol/L) and $S_{AC} = 0.2$ au/(mol/L)

(i) Analysis of A in presence of the disturbing components yields the following real sensitivity according to Eq. (7.13): $S_{AA,i} = (5 + 0.01 + 0.01)$ au/(mol/L) $= 5.03$ au/(mol/L).

(ii) In case that the concentrations of all the three components are equal, e.g. $x = (1, 1, 1)$ it results for the real sensitivity: $S_{AA,i} = (5 + 1 + 2)$ au/(mol/L) $= 8.00$ au/(mol/L).

(iii) In case of simultaneous multicomponent analysis, at first it should be assumed a full selective analytical procedure corresponding to the sensitivity matrix $S_{(1)}$

$$S_{(1)} = \begin{pmatrix} 5 & 0 & 0 \\ 0 & 3 & 0 \\ 0 & 0 & 6 \end{pmatrix}, \quad S_{(2)} = \begin{pmatrix} 5 & 0.1 & 0.2 \\ 0.1 & 3 & 0.1 \\ 0.1 & 0.2 & 6 \end{pmatrix}, \quad S_{(3)} = \begin{pmatrix} 5 & 1 & 2 \\ 1 & 3 & 1 \\ 1 & 2 & 6 \end{pmatrix}.$$

The three single sensitivities can be found at the diagonal of the matrix $S_{(1)}$. The total multicomponent sensitivity according to Eq. (7.18) is $S_{total} = 90$ au^3/(mol/L)^3.

(iv) Multicomponent analysis in the case of slight disturbances as expressed by the sensitivity matrix $S_{(2)}$. According to Eq. (7.19) the total multicomponent sensitivity is $S_{total} = \det(S_{(2)}) = (5 \times 3 \times 6 + 0.1 \times 0.1 \times 0.1 + 0.2 \times 0.1 \times 0.2) - (0.2 \times 3 \times 0.1 + 5 \times 0.1 \times 0.2 + 0.1 \times 0.1 \times 6) = (90.005 - 0.22) = 89.785$ au^3/(mol/L)^3.

(v) Serious disturbances as given by the sensitivity matrix $S_{(3)}$ results in the following total sensitivity: $S_{total} = \det(S_{(3)}) = (5 \times 3 \times 6 + 1 \times 1 \times 1 + 2 \times 1 \times 2) - (2 \times 3 \times 1 + 5 \times 1 \times 2 + 1 \times 1 \times 6) = (95 - 22) = 73$ au^3/(mol/L)^3.

7.3
Selectivity and Specificity

The concepts *selectivity* and *specificity* are associated with *real analytical multicomponent systems*. Both selectivity and specificity refer to a given state of the analytical system as well as the analytical method, i.e., a completely defined procedure (SOP), as characterized in Fig. 7.1. Each change in operating and physicochemical conditions like pH value, temperature as well as changed composition of the system (accompanying species, masking reagents etc) will alter the procedure and therefore cause a change of sensitivity and specificity.

In such a well-defined analytical system the term *selectivity* is relevant to *multicomponent analysis*. Selectivity of an analytical procedure characterizes the extent to which n given analytes can be measured simultaneously by n sensors (detecting channels) without interferences by other components

---

[1]   au: arbitrary unit

and, therefore, can be detected or determined independently and undisturbedly (KAISER [1972]; DANZER [2001, 2004]).

On the other hand, *specificity* refers to *single component analysis* and means that the one individual analyte can be undisturbedly measured in a real sample by a specific reagent, a particular sensor or a comparable measuring system (e.g., measurement of emitted or absorbed radiation at a fixed wavelength).

Selectivity and specificity are important performance characteristics of analytical procedures, especially in connection with validation processes. Nevertheless, both terms are used mostly verbal and a quantification is avoided, as a rule (IUPAC; see VESSMAN et al. [2001]). Moreover, the concepts of selectivity and specificity are used interchangeably and synonymously. Occasionally, specificity is regarded as an intensification of selectivity, viz. the ultimate of selectivity (DEN BOEF and HULANICKI [1983]; PERSSON and VESSMAN [1998, 2001]; PRICHARD et al. [2001]).

A clear differentiation between selectivity and specificity was firstly given by KAISER [1972] together with a quantification of the both characteristics. According to him an analytical procedure is known as "*selective*" if it can detect and determine simultaneously *several components* independently from each other. On the other hand, a procedure is known as "*specific*" if *only one species* can be detected and determined independently from all the other components which are present in the sample but does not contribute to the signal in this case.

Starting from a relationship like Eq. (6.70a), KAISER [1972] defined *sensitivity, partial sensitivities* (*cross sensitivities*, Eq. (7.16)), and the *sensitivity matrix* Eq. (7.17). From these quantities he derived the following measures:

- *Selectivity*:

$$\Xi = \min_{I=1...N} \left(\frac{S_{IJ}}{\sum\limits_{J=1}^{N} |S_{IJ}| - |S_{II}|} - 1 \right) \tag{7.21}$$

- *Specificity*:

$$\Psi_A = \frac{S_{AA}}{\sum\limits_{J=1}^{N} |S_{JJ}| - |S_{AA}|} - 1 \tag{7.22}$$

These quantifications are inconvenient in two respects:

(i)   The range[2] of both quantities is $R\{\Xi\} = R\{\Psi_A\} = (-1...+\infty)$ and therefore, an interpretation of a given value of Ξ and Ψ_A in the sense that the selectivity/specificity is high or less, may be difficult.

(ii) The definitions at Eqs. (7.21) and (7.22) do not consider the concentration of the disturbing components. However, not only the cross sensitivity but also the concentration of trouble-making components is crucial for the appearance and intensity of disturbing effects; see Eq. (3.10)

---

[2]   The range $R\{x\}$ of values of a quantity x corresponds to its definition range

The last item has been taken into account by FUJIWARA et al. [1980] who introduced a "total selectivity" similar to Kaiser's selectivity (Eq. 7.21) but products $S_{IJ} \cdot c_J$ instead of the partial sensitivities. However, this definition range is $(-1 \ldots + \infty)$. A more acceptable range of "selectivity" results from DOERFFEL's "coefficient of selectivity CS_A" (DOERFFEL et al. [1986]):

$$CS_A = \frac{y_A}{\sum\limits_{I=A}^{N} y_I} = \frac{S_{AA} \cdot c_A}{\sum\limits_{I=A}^{N} S_{AI} \cdot c_I} \tag{7.23}$$

(S_{AA} sensitivity of the detection channel A for the analyte A; S_{IA} partial sensitivities of all components $I = A, B, \ldots, N$ at the detection channel A; c_A concentration of the analyte A; c_I concentrations of all components $I = A, B, \ldots, N$). But unfortunately, DOERFFEL's "selectivity" according to Eq. (7.23) proves to be specificity because it corresponds to KAISER's Eq. (4.54).

Starting from the sensitivity matrix, Eq. (7.17), as it is illustrated in Fig. 7.6a, selectivity can be defined as follows:

$$sel(A, B, \ldots, N) = \frac{\sum\limits_{I=A}^{N} S_{II} \cdot x_I}{\sum\limits_{I=A}^{N} \sum\limits_{J=A}^{N} S_{IJ} \cdot x_J} \tag{7.24}$$

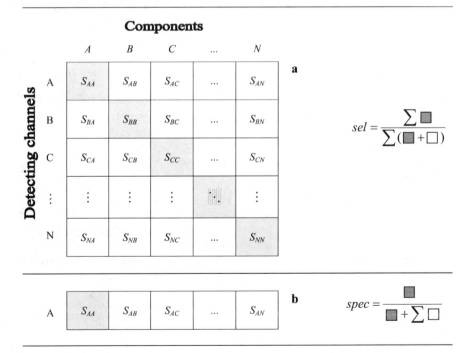

Fig. 7.6. Selectivity matrix (**a**) and specificity vector of the detecting channel A (**b**) with the simplified formulas (all the x_I and $x_J = 1$) for selectivity and specificity (*on the right*)

In the ideal case, only the diagonal elements of the matrix at Eq. (7.17) are different from zero and all the off-diagonal elements $S_{IJ} = 0$ (for $I \neq J$), $sel(A, B, \ldots, N)$ becomes 1. For real analytical systems, Eq. (7.24) expresses a certain degree of selectivity depending on the sensitivities of the analytes I, their contents x_I, and all the cross sensitivities S_{IJ} and contents x_J of potentially disturbing species.

For the characterization of *specificity* only the detecting channel of the analyte, A, is relevant and therefore only the corresponding vector

$$s_A = (S_{AA}, S_{AB}, \ldots, S_{AN}) \tag{7.25}$$

is of interest. It contains the sensitivity S_{AA} and the cross sensitivities S_{AJ} at the channel (sensor, wavelength, etc.) A; see Fig. 7.6b.

Specificity of an analyte A with reference to the accompanying components $B, C, \ldots, N$ can be defined by

$$spec(A/B, \ldots, N) = \frac{S_{AA} \cdot x_A}{\sum\limits_{J=A}^{N} S_{AJ} \cdot x_J} \tag{7.26}$$

The problem is illustrated in a simplified way in Fig. 7.6. The quantities $sel(A, B, \ldots, N)$ and $spec(A, B, \ldots, N)$ not only define absolute characteristics but also a certain degree of selectivity and specificity of an analytical procedure. The measures can take values from 1 for fully selective or totally specific procedures, respectively. The range of selectivity $R\{sel\}$ and the range of specificity $R\{spec\}$ are given by

$$R\{sel\} = R\{spec\} = (0 \ldots + 1) \tag{7.27}$$

For practical purposes, it can be supposed that values of $sel > 0.9$ and $spec > 0.9$ indicate fair selectivity and specificity, respectively, whereas smaller values show more and more insufficient selectivity and specificity.

> To obtain some impression on the characteristics, the same examples as dealt with in Sect. 7.2 (iii)–(v) will be considered in Table 7.3. Some more examples from the application of sensors, photometry, and ion sensitive electrodes have been given by DANZER [2001].

Not only components from the set $A, \ldots, N$ can interfere with the analytes but also additional species which may be partly unknown or will be formed during the measuring process. Such situations occur especially in ICP-MS, where the signal of an isotope A_i may be interfered by isotopes of other elements, B_j, C_k etc., and additionally by molecule ions formed in the plasma (e.g., argon compound ions and ions formed from solvent constituents).

In such cases the sensitivity matrix, Eq. (7.17), will be extended into

$$S = \begin{pmatrix} S_{AA} & S_{AB} & \cdots & S_{AN} & S_{A,N+1} & \cdots & S_{A,N+p} \\ S_{BA} & S_{BB} & \cdots & S_{BN} & S_{B,N+1} & \cdots & S_{B,N+q} \\ \vdots & \vdots & & \vdots & \vdots & & \vdots \\ S_{NA} & S_{NB} & \cdots & S_{NN} & S_{N,N+1} & \cdots & S_{N,N+r} \end{pmatrix} \tag{7.28}$$

Table 7.3. Selectivities and specificities in case of various sensitivity matrixes

Sensitivity matrix (see Sect. 7.2 (iii)–(v))	Selectivity	Specificities
$S_{(1)}$	$sel(A, B, C) = 1.00$	$spec(A/B, C) = 1.00$ $spec(B/A, C) = 1.00$ $spec(C/A, B) = 1.00$
$S_{(2)}$	$sel(A, B, C) = 0.946$	$spec(A/B, C) = 0.943$ $spec(B/A, C) = 0.938$ $spec(C/A, B) = 0.952$
$S_{(3)}$	$sel(A, B, C) = 0.636$	$spec(A/B, C) = 0.625$ $spec(B/A, C) = 0.600$ $spec(C/A, B) = 0.667$

The cross sensitivities $S_{j,N+i}$ can easily be estimated in the case of neighbouring isotopes, viz. from the isotope abundances. On the other hand, estimation in case of molecule ions may be difficult and can only be done experimentally.

The concept of selectivity and specificity has been applied to characterize interferences appearing in two different ICP-MS techniques (HORN [2000]). Classical ICP-MS with pneumatic nebulization and ETV-ICP-MS are compared for the determination of traces of zinc in sea-water. Whereas spectral interferences decrease using the ETV device, nonspectral interferences increase significantly (BJÖRN et al. [1998]). A quantitative comparison of the both analytical procedures, here called *PN* (pneumatic nebulization) and ETV (electrothermal vaporization, STURGEON and LAM [1999]) is possible by means the specificity as a function of the Zn concentration (HORN [2000]). The spectral interferences on the four zinc isotopes are listed in Table 7.4.

Table 7.4. Seawater components and their spectral interferences with zinc isotopes

Ion	64 amu	66 amu	67 amu	68 amu	Concentration in CASS-3
Zn^{2+}	n.a.^{a} 48.6%	n.a. 27.9%	n.a. 4.1%	n.a. 18.8%	Certified 1.24 ng/mL
Cl^-	–	–	$^{35}Cl^{16}O^{16}O^+$	$^{35}Cl^{16}O^{17}O^+$	About 16 mg/L
SO_4^{2-}	$^{32}S^{16}O^{16}O^+$ $^{32}S^{32}S^+$	$^{34}S^{16}O^{16}O^+$ $^{32}S^{34}S^+$ $^{33}S^{33}S^+$ $^{32}S^{17}O^{17}O^+$ $^{32}S^{16}O^{18}O^+$ $^{33}S^{16}O^{17}O^+$	$^{33}S^{34}S^+$ $^{32}S^{17}O^{18}O^+$ $^{33}S^{16}O^{18}O^+$ $^{33}S^{17}O^{17}O^+$ $^{34}S^{16}O^{17}O^+$	$^{36}S^{16}O^{16}O^+$ $^{34}S^{36}S^+$ $^{34}S^{17}O^{17}O^+$ $^{34}S^{16}O^{18}O^+$ $^{34}S^{34}S^+$ $^{33}S^{17}O^{18}O^+$ $^{32}S^{18}O^{18}O^+$	About 2 mg/L
Mg^{2+}	$^{24}Mg^{40}Ar^+$ $^{26}Mg^{38}Ar^+$	$^{26}Mg^{40}Ar^+$	–	–	About 1 mg/L

^{a} Natural abundance

The specificity vectors of the four zinc isotopes are given in matrix form as follows:

$$
S = \begin{pmatrix}
\dfrac{\Delta I_{64}}{\Delta x_{Zn^{2+}}} & \dfrac{\Delta I_{64}}{\Delta x_{Cl^-}} & \dfrac{\Delta I_{64}}{\Delta x_{SO_4^{2-}}} & \dfrac{\Delta I_{64}}{\Delta x_{Mg^{2+}}} \\[3mm]
\dfrac{\Delta I_{66}}{\Delta x_{Zn^{2+}}} & \dfrac{\Delta I_{66}}{\Delta x_{Cl^-}} & \dfrac{\Delta I_{66}}{\Delta x_{SO_4^{2-}}} & \dfrac{\Delta I_{66}}{\Delta x_{Mg^{2+}}} \\[3mm]
\dfrac{\Delta I_{67}}{\Delta x_{Zn^{2+}}} & \dfrac{\Delta I_{67}}{\Delta x_{Cl^-}} & \dfrac{\Delta I_{67}}{\Delta x_{SO_4^{2-}}} & \dfrac{\Delta I_{67}}{\Delta x_{Mg^{2+}}} \\[3mm]
\dfrac{\Delta I_{68}}{\Delta x_{Zn^{2+}}} & \dfrac{\Delta I_{68}}{\Delta x_{Cl^-}} & \dfrac{\Delta I_{68}}{\Delta x_{SO_4^{2-}}} & \dfrac{\Delta I_{68}}{\Delta x_{Mg^{2+}}}
\end{pmatrix}
\tag{7.29}
$$

The results for the PN- and ETV procedure are the following:

$$
S_{PN} = \begin{pmatrix}
 & Zn^{2+} & Cl^- & SO_4^{2-} & Mg^{2+} \\
64\,amu & 616 & -1.22 \cdot 10^{-3} & 1.68 \cdot 10^{-2} & 2.85 \cdot 10^{-2} \\
66\,amu & 378 & -3.48 \cdot 10^{-4} & 6.48 \cdot 10^{-4} & 1.43 \cdot 10^{-2} \\
67\,amu & 63 & 6.03 \cdot 10^{-4} & 1.36 \cdot 10^{-4} & 2.52 \cdot 10^{-3} \\
68\,amu & 265 & -1.99 \cdot 10^{-4} & -2.15 \cdot 10^{-4} & 8.97 \cdot 10^{-2}
\end{pmatrix}
$$

$$
S_{ETV} = \begin{pmatrix}
 & Zn^{2+} & Cl^- & SO_4^{2-} & Mg^{2+} \\
64\,amu & 5669 & -1.02 \cdot 10^{-2} & -1.30 \cdot 10^{-2} & -6.24 \cdot 10^{-2} \\
66\,amu & 3557 & -5.67 \cdot 10^{-3} & -2.30 \cdot 10^{-2} & -5.69 \cdot 10^{-2} \\
67\,amu & 524 & -1.01 \cdot 10^{-3} & -2.53 \cdot 10^{-3} & 2.80 \cdot 10^{-3} \\
68\,amu & 2438 & -4.13 \cdot 10^{-3} & -1.73 \cdot 10^{-2} & -4.40 \cdot 10^{-2}
\end{pmatrix}
$$

where the partial sensitivities are given in counts $\cdot$ mL $\cdot$ s^{-1} $\cdot$ ng^{-1}. The partial sensitivities are the result of both spectral and nonspectral interferences. It can be seen that the partial sensitivities of the four masses due to the interfering ions Cl^-, SO_4^{2-} and Mg^{2+} are low compared to the sensitivities of Zn^{2+}. Furthermore it can be seen that ETV-ICP-MS is about ten times more sensitive than PN-ICP-MS.

It is possible to calculate so-called *specificity functions* which describe the specificity of a procedure in dependency of the concentration of the analyte, here Zn, in the sample:

$$
spec_{PN,64\,amu}\left(Zn^{2+}/Cl^-, SO_4^{2-}, Mg^{2+}\right) = \frac{616 \cdot x_{Zn^{2+}}}{616 \cdot x_{Zn^{2+}} + 93300}
$$

$$
spec_{ETV,64\,amu}\left(Zn^{2+}/Cl^-, SO_4^{2-}, Mg^{2+}\right) = \frac{5669 \cdot x_{Zn^{2+}}}{5669 \cdot x_{Zn^{2+}} + 153000}
$$

The both functions are represented in Fig. 7.7.

By means of the specificity function, concentration-matrix ratios can be estimated for which a reliable determination of analytes may be possible.

For specific applications, measures like *selectivity indices, selectivity factors*, and *selectivity coefficients* have been introduced. Their significance is

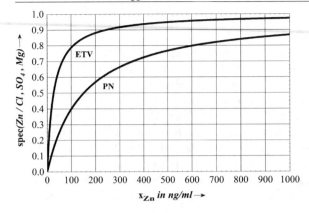

Fig. 7.7. Specificity functions of the determination of Zn in seawater by means of PN-ICP-MS and ETV-ICP-MS

limited to certain fields of analytical chemistry like sensors, ISE, and chromatography as well as concrete applications.

As expected, the concept of selectivity and specificity is closely related to that of sensitivity. The same may be anticipated for the concept of robustness and ruggedness.

7.4
Robustness and Ruggedness

Methods can only usefully applied in analytical practice when they are sufficiently robust and therefore insensitive to small variations in method conditions and equipment (replacement of a part), operator skill, environment (temperature, humidity), aging processes (GC- or LC columns, reagents), and sample composition. This demand makes robustness (ruggedness) to an important validation criterion that has to be proved by experimental studies. The concepts of robustness and ruggedness mostly have been described verbally where it must be stated that their use is frequently interchangeably and synonymously (e.g., HENDRICKS et al. [1996]; KELLNER et al. [1998]; EURACHEM [1998]; ICH [1994, 1996]; WÜNSCH [1994]; WILDNER and WÜNSCH [1997]; VALCARCEL [2000]; KATEMAN and BUYDENS [1993]).

Only in a few cases was there a distinction made between both concepts by their use in *intra laboratory studies* (robustness) and *interlaboratory studies* (ruggedness); see USP 23-NF18 [1995]; USP 24-NF19 [2000]; RODRIGUEZ et al. [1998]; ZEAITER et al. [2004]. WAHLICH and CARR [1990] seem to be the first to use robustness and ruggedness in a hierarchical sense but in a reverse meaning as given above.

ICH [1994] defines *robustness* as follows: "*The robustness of an analytical procedure is a measure of its capacity to remain unaffected by small, but deliberate variations in method parameters and provides an indication of its reliability during normal use*".

USP 23-NF18 [1995] and USP 24-NF19 [2000] accept this definition of robustness and define *ruggedness* so: "*The ruggedness of an analytical method is the degree of reproducibility of test results obtained by the analysis of the same samples under a variety of conditions such as different laboratories, different analysts, different instruments, different lots of reagents, different elapsed assay times, different assay temperatures, different days, etc. Ruggedness is normally expressed as the lack of influence on test results of operational and environmental factors of the analytical method. Ruggedness is a measure of reproducibility of test results under the variation in conditions normally expected from laboratory to laboratory and analyst to analyst.*"

Whereas this definition of ruggedness directly refers to *inter*-laboratory studies, the ICH definition of robustness clearly is related to *intra*-laboratory studies.

Consequently, it was proposed to define (BURNS et al. [2005]): "*Robustness of an analytical procedure is the property that indicates insensitivity against changes of known operational parameters on the results of the method and hence its suitability for its defined purpose*" and "*Ruggedness of an analytical procedure is the property that indicates insensitivity against changes of known operational variables and in addition any variations (not discovered in intra-laboratory experiments) which may be revealed by inter-laboratory studies*" (BURNS et al. [2005]).

The conditions and factors which influence uncertainty and insensitivity of analytical procedures obtained in a given laboratory or in different laboratories can be considered in detail and described quantitatively (DANZER [2004]; BURNS et al. [2005]).

As a starting point the general Eq. (3.15), particularly Eq. (3.15a) will be taken, which describes the gross signal in dependence from all signal effects from the analyte as well as from blank, interferences, and influencing factors:

$$
\begin{aligned}
y_A &= f(x_A; x_B, x_C, \ldots, x_N; f_1, \ldots, f_m) \\
&= y_{A0} + S_{AA}x_A + \sum_{i=B}^{N} S_{Ai}x_i + \sum_{j=1}^{m} I_{Aj}x_j + e_A
\end{aligned}
\tag{7.30}
$$

The *robustness* of an analytical procedure for the determination of the analyte A in presence of some accompanying species $i = B, \ldots, N$ under influence of various factors $f_j(j = 1, \ldots, m)$ according to Eq. (4.30) is in reciprocal proportion to the sum of all their cross sensitivities, S_{Ai}, multiplied by the actual amounts, x_i, and the specific influencing strengths, I_{Aj}, of the factors multiplied by their actual values (in relation to x_A); see DANZER (2004). Because of the way measurements are obtained, the range of their values is $range = (0 \ldots \infty)$, so it makes sense to calculate the *relative robustness* which includes the analyte sensitivity and amount itself, $S_{AA} \cdot x_A$, as follows:

$$
rob(A/B, \ldots, N; f_1, \ldots, f_m) = \frac{S_{AA}x_A}{S_{AA}x_A + \sum_{i=B}^{N} |S_{Ai}| x_i + \sum_{j=1}^{m} |I_{Aj}| x_j}
\tag{7.31}
$$

Relative robustness will have values between 0 (no robustness) and 1 (ideal robustness). BURNS et al. [2005] defines the relative robustness of an analytical procedure as "*the ratio of the ideal signal for an uninfluenced method compared to the signal for a method subject to known operational parameters determined in an intra-laboratory experiment*" (BURNS et al. [2005]).

For the definition of ruggedness, which refers to inter-laboratory situations, some further (unknown) effects u_{Ak} must be added to Eq. (7.30):

$$y_A = f(x_A; x_B, x_C, \ldots, x_N; f_1, \ldots, f_m; u_1 \ldots, u_p)$$

$$= y_{A0} + S_A x_A + \sum_{i=B}^{N} S_{Ai} x_i + \sum_{j=a}^{m} I_{Aj} x_j + \sum_{k=1}^{p} u_k + e_A \tag{7.32}$$

and therefore, relative ruggedness is obtained according to

$$rug(A/B, \ldots, N; f_1, \ldots, f_m; u_1, \ldots, u_p)$$

$$= \frac{S_{AA} x_A}{S_{AA} x_A + \sum_{i=B}^{N} |S_{Ai}| x_i + \sum_{j=1}^{m} |I_{Aj}| x_j + \sum_{k=1}^{p} |u_k|} \tag{7.33}$$

The range of relative ruggedness is $range\{rug\} = (0 \ldots 1)$, too. According to BURNS et al. [2005], the relative ruggedness of an analytical procedure is given "*by the ratio of the ideal signal for an uninfluenced method compared to the signal for a method subject to known and unknown operational parameters as determined in an inter-laboratory experiment*".

The problem, particularly in analytical practice is that most of the parameters needed for the calculation of relative robustness and ruggedness may frequently be unknown.

In such cases it may be possible to check on robustness and ruggedness by means of statistical tests (see Sect. 4.3). All the variations to the measured signal, apart from that of the analyte, can be considered in form of error terms; see Eq. (3.6):

$$\sum_{i=B}^{N} S_{Ai} x_i = e_i \tag{7.34a}$$

$$\sum_{j=1}^{m} I_{Aj} x_j = e_j \tag{7.34b}$$

$$\sum_{k=1}^{p} u_k = e_k \tag{7.34c}$$

and Eq. (7.30) turns into

$$y_A = y_{A0} + S_A x_A + e_i + e_j + e_A = y_{A0} + S_A x_A + e_{ij} + e_A \tag{7.35}$$

(see Eqs. (3.16a)–(3.16c)) and Eq. (7.32) turns into

$$y_A = y_{A0} + S_A x_A + e_i + e_j + e_k + e_A = y_{A0} + S_A x_A + e_{ijk} + e_A \quad (7.36)$$

where the error contributions are additive: $e_i + e_j = e_{ij}$ are the intra-laboratory variations and $e_i + e_j + e_k = e_{ijk}$ the intra-laboratory variations.

- *Test of robustness*: On the basis of Eq. (7.35), robustness as an *intra*-laboratory property can be tested in three ways:

  (i) Influence on precision can be tested as usual by FISHER's F-test with the null hypothesis H_0: $\sigma_{total} = \sigma_A$ and therefore H_0': $\sigma_{ij} = 0$; see Sect. 4.3.3:

  $$\hat{F} = \frac{\sigma_{total}^2}{\sigma_A^2} = \frac{\sigma_{ijA}^2}{\sigma_A^2} = \frac{\sigma_{ij}^2 + \sigma_A^2}{\sigma_A^2} \quad (7.37)$$

  If $\hat{F} \le F_{1-\alpha,\nu_1,\nu_2}$, then the null hypothesis cannot be rejected and the procedure may be considered as to be robust (strictly speaking robust with regard to method precision).

  (ii) Influence on sensitivity can be tested by STUDENT's t-test. The null hypothesis is H_0: $S_{AA}^{real} = S_{AA}^{ideal}$ where S_{AA}^{ideal} is the sensitivity under ideal, i.e. robust conditions and S_{AA}^{real} the sensitivity under the influence of i interferents and j factors:

  $$\hat{t} = \frac{\left| S_{AA}^{ideal} - S_{AA}^{real} \right|}{\sigma_{S_{AA}} t_{\alpha,\nu}} \quad (7.38)$$

  If $\hat{t} > t_{1-\alpha,\nu}$ then the real sensitivity differs significantly from the ideal one and a nonlinear error is proved.

  (iii) More in detail, each factor (interferents i and influence factors j) can individually be tested by means of multifactorial experiments where each factor is usually varied at two levels. The planning and evaluation of such a multifactorial design is given in Sect. 5.1.2. The significance of each cross sensitivity and factor can be tested by an individual t-test:

  $$\hat{t} = \frac{|S_{AI}|}{\sigma_A \, t_{1-\alpha,\nu}} \quad , \quad \hat{t} = \frac{|I_{AJ}|}{\sigma_A \, t_{1-\alpha,\nu}} \quad (7.39)$$

  If $|S_{AI}|$ exceeds the confidence interval $\sigma_A t_{1-\alpha,\nu}$ of the experimental error e_A, then the influence of the factor concerned is significant and robustness against this factor is missing. On the other hand, $|S_{AI}| < \sigma_A \, t_{1-\alpha,\nu}$ shows robustness against the particular interferent or factor, respectively.

- *Test of ruggedness*: Robustness is regarded to be an *inter*-laboratory property. In this case, all the terms in Eq. (7.36) are relevant and ruggedness can be tested similar to robustness in the same three ways:

(iv) Influence on precision:

$$\hat{F} = \frac{\sigma_{total}^2}{\sigma_A^2} = \frac{\sigma_{ijkA}^2}{\sigma_A^2} = \frac{\sigma_{ij}^2 + \sigma_k^2 + \sigma_A^2}{\sigma_A^2} \tag{7.40}$$

The total error has to be calculated here in different way compared with robustness. Whereas in (i) σ_{ijA}^2 is the variance within a laboratory, σ_{ijkA}^2 is the variance between laboratories plus that within the laboratories, $\sigma_{ijkA}^2 = \sigma_k^2 + \sigma_{ijA}^2$. The interpretation is analogous to (i), if $\hat{F} \le F_{1-\alpha,\nu_1,\nu_2}$, then the null hypothesis cannot be rejected and the procedure can be considered as to be rugged. Advantageously, the test can be carried out by schemes of ANOVA (analysis of variance) as given in Sect. 5.1.1.

In most cases, tests analogous to (ii) and (iii) which exceed the situation of a single laboratory do not make much sense because there do not exist an overall sensitivity for all the laboratories. Also the cross sensitivities are individual for each lab, as a rule. Therefore, such individual measures will not be merged.

In the following example, robustness and ruggedness of a procedure used in two laboratories, L1 and L2, will be considered. It is supposed that they use the same procedure to determine the analyte A in the presence of the interferents B and C under the influence of the factors a, b, and c. The input data for the computation are given in the following table.

Lab	Sensitivity S_{AA}	Cross sensitivities S_{AB}	S_{AC}	Influence strengths I_{Aa}	I_{Ab}	I_{Ac}	Experimental error s_A	Degrees of freedom ν
L1	10	1.2	1.5	0.25	−0.13	0.08	1.2	10
L2	9	1.4	1.9	0.22	−0.14	0.10	1.5	10

(a) Robustness of the procedure in lab L1 where the following conditions exist: $x_B = 0.11$, $x_C = 0.08$, $x_a = 1.5$, $x_b = 0.7$, $x_c = 1.3$

$$rob_{L1}(A/B, C; a, b, c)$$

$$= \frac{10}{10 + 0.132 + 0.120 + 0.375 + 0.091 + 0.104}$$

$$= \frac{10}{10 + 0.822} = 0.924$$

Thus the relative robustness of the lab L1 is 92.4%.

Using the F-test,

$$\hat{F} = \sigma_{ijA}^2 / \sigma_A^2$$

$$= \frac{0.132^2 + 0.12^2 + 0.375^2 + 0.091^2 + 0.104^2 + 1.2^2}{1.2^2}$$

$$= 1.123 < F(0.05; 10; 10) = 2.79.$$

Hence the procedure is robust in lab L1.

(b) Robustness in lab L2 where the following conditions exist:
$x_B = 0.14$, $x_C = 0.13$, $x_a = 1.8$, $x_b = 1.7$, $x_c = 1.6$

$$rob_{L2}(A/B, C; a, b, c)$$

$$= \frac{10}{10 + 0.196 + 0.247 + 0.396 + 0.238 + 0.160}$$

$$= \frac{10}{10 + 1.237} = 0.808$$

Thus the relative robustness of the lab L2 is 80.8%.
Using the F-test,

$$\hat{F} = \sigma_{ijA}^2 / \sigma_A^2$$

$$= \frac{0.196^2 + 0.247^2 + 0.396^2 + 0.238^2 + 0.160^2 + 1.5^2}{1.5^2}$$

$$= 1.15 < F(0.05; 10; 10) = 2.79.$$

Thus the procedure is robust in lab L2.

(c) Ruggedness of the procedure to determine analyte A in the two laboratories, L1 and L2: The analysis of the same sample produced the following results:

L1: 1.47, 1.32, 1.11, 1.37, 1.21, 1.40, 1.51, 1.27, 1.72, 1.13, 1.05 ($v=10$),

L2: 1.38, 1.23, 1.71, 1.60, 1.58, 1.45, 1.67, 1.21, 1.48, 1.51, 1.42 ($v=10$), from which have been estimated:

L1: $\bar{x}_1 = 1.3236$, $s_1 = 0.1988$, $s_1^2 = 0.0395$ ($= {}_1\sum e_{ijA}^2/v$), ${}_1\sum e_{ij} = 0.822$, ${}_1\sum e_{ij}^2 = 0.192$ (from(a)),

L2: $\bar{x}_2 = 1.4764$, $s_2 = 0.1625$, $s_2^2 = 0.0264$ ($= {}_2\sum e_{ijA}^2/v$), ${}_2\sum e_{ij} = 1.237$, ${}_2\sum e_{ij}^2 = 0.338$ (from(b))

The variance between the labs is 0.1284 and the variance within the labs is 0.03295; therefore the F-test gives:

$$\hat{F} = \frac{s_{between}^2}{s_{within}^2} = \frac{0.1284}{0.03295} = 3.897 > F(0.05; 10; 10) = 2.98$$

and *ruggedness* cannot be confirmed. The unknown influences, $\sum e_k$ can be estimated from the total variance $s_{between}^2 = \sum e_{ijkA}^2/v$ and the averaged sums of e_{ijA}. The sums of e_{ij} are given only for completeness. From this the robustness of the separate procedures has been proven as shown in (a) and (b).

7.5
Limit Values

Limits characterize the detection capability of analytical methods and can be related to both analytical domains, *sample domain* as well as *signal domain*. Although there are several limits, namely *lower* and *upper limits*[3] as well as *thresholds*, the most important problem in analytical chemistry is the distinction between real measurement values and "zero" values or blanks, respectively.

Notwithstanding different mathematical derivations and experimental determination, the following three measures are necessary as well as sufficient for the characterization of analytical procedures:

(1) The *critical measurement value* (*critical value, CV*), y_c, being the lowest signal which can significantly be distinguished from that of a blank sample

(2) The *limit of detection* (*detection limit, LD*), x_{LD}, being the lowest content (concentration) which produces a signal $y \geq y_c$ with high statistical probability

(3) The *limit of quantification* (*LQ, limit of determination, quantitation limit*), x_{LQ}, being the lowest content which can quantitatively be determined with a given precision (mostly expressed by the relative uncertainty of measurement)

These limits are the fundamental measures to characterize the detection and quantification abilities of analytical procedures and make decisions on measured values and analytical results. Whereas *CV* and *LD* can be determined in a general way on the basis of objective statistical conditions, *LQ* can only be estimated on the basis of subjective demands resulting from a given analytical problem.

In addition, there exist a lot of different terms, symbols, definitions, and meaning for limits within the scientific community. Some of them seem to be confusing in definition and application and contradicting amongst themselves, too; see CURRIE [1992, 1995, 1997]; EHRLICH and DANZER [2006]. In general the limits can be derived in a simple and understandable way but care has to be taken in interpreting them.

Before dealing with the limits *CV*, *LD*, and *LQ*, their meaningful position within the system of analytical domains will be shown (Table 7.5)

Whereas the limits *ACV* (KAISER's "3σ" limit, KAISER and SPECKER [1956]), *MLD*, and *MLQ* are only of special interest within the corresponding domains and, therefore, of indirect importance for analytical problems, *CV*, *LD*, and *LQ* are the relevant quantities to characterize the performance of a complete analytical procedure according to KAISER [1965] and a SOP according to ISO 78-2 [1999].

---

3   Because the distinction from upper limits (100%) represents only the reverse situation to that from lower limits and is relatively infrequent, too, it will not be treated here. The mathematical models are equivalent to that used for evaluation of lower limits (DOERFFEL [1990], pp 97; DOERFFEL et al. [1990], pp 123)

Table 7.5. Analytical limits within the signal and sample domain the most meaningful of which are emphasized

Signal domain (measured values)	Sample domain (analytical values)
Critical measuring value, CV	*Analytical value at CV (ACV)*
Measured value at LD (MLD)	Limit of detection, LD
Measured value at LQ (MLQ) (also minimum quantifiable value, IUPAC [1995])	Limit of quantification, LQ

The statistical fundamentals of the definition of *CV* and *LD* are illustrated by Fig. 7.8 showing a quasi-three-dimensional representation of the relationship between measured values and analytical values which is characterized by a calibration straight line $y = a + bx$ and their two-sided confidence limits and, in addition (in z-direction) the probability density function the measured values.

The following assumptions are made for the statistical treatment of the detection problem:

(i) *Feasibility of repeated analyses,* namely of the complete analytical procedure from sampling to data evaluation, not only repetitions of instrumental measurements

(ii) *Normal distributions* of the measured values

(iii) *Homogeneity of variances* (homoscedasticity)

(iv) *Known calibration* (absolutely known or estimated by calibration experiments); in the last-mentioned case the uncertainty intervals of the calibration function must be known

(v) *Purpose of evaluation:* it must be clear if the limits are estimated to characterize the *performance of an analytical procedure* before of its application (a priori) or *to interpret an analytical result* obtained by this procedure (a posteriori)

The *critical value* y_c represents the smallest measurement value that can be distinguished from the blank y_{BL} with a given level of significance $P = 1 - \alpha$. In the most general case, the critical value is estimated from the average blank and its uncertainty interval $U(\overline{y}_{BL})$

$$y_c = \overline{y}_{BL} + U(\overline{y}_{BL}) \quad . \tag{7.41a}$$

As a rule, the average blank is estimated from repetition measurements of a – not too small – number of blank samples as arithmetic mean $\overline{y}_{BL}$. If there is information that another than normal distribution applies, then the mean of this other distribution should be estimated (see textbook of applied statistics; see ARNOLD [1990]; DAVIES and GOLDSMITH [1984]; GRAF et al. [1987]; HUBER [1981]; SACHS [1992]).

The uncertainty of the blank (or the background noise) is estimated from the blank scattering which is characterized by corresponding error quanti-

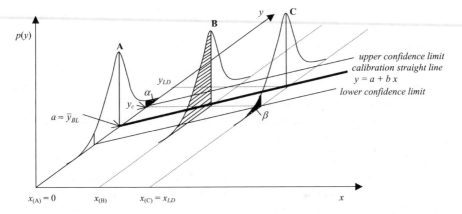

Fig. 7.8. Schematic three-dimensional representation of a calibration straight line of the form $y = a + bx$ with the limits of its two-sided confidence interval and three probability density function (pdf) $p(y)$ of measured values y belonging to the analytical values (contents, concentrations) $x_{(A)} = 0$ (A), $x = x_{(B)}$ (B) and $x_{(C)} = x_{LD}$ (C); y_c is the critical value of the measurement quantity; a the intercept of the calibration function; $\overline{y}_{BL}$ the blank; $x_{(B)}$ the analytical value belonging to the critical value y_c (which corresponds approximately to KAISER's "3σ-limit"); x_{LD} limit of detection

ties, viz statistical or combined one, such as standard deviation s_{BL} or combined standard uncertainty $u(y_{BL})$; see Sect. 4.2; AMC [1995]; EURACHEM [1995]; THOMPSON [1995]. The extended uncertainty will be obtained by means of a coverage factor k according to Eq. (4.29). With this the critical value is given by

$$y_c = \overline{y}_{BL} + k \cdot u(\overline{y}_{BL}) \tag{7.41b}$$

and in the special case of statistical estimation

$$y_c = \overline{y}_{BL} + k \cdot s_{BL} \quad . \tag{7.41c}$$

Applying Eq. (7.41b) or Eq. (7.42c), the factor k can be chosen in various ways, namely

- $k = 3$: This value was proposed by KAISER [1965] and it is high enough to tolerate differences from the normal distribution; the levels of significance are about $P \approx 0.998$ in case of normally distributed values, $P \approx 0.95$ for non-normal unimodal distributions, and $P \approx 0.90$ for any one distribution.
- $k = 2$: If the blanks are normally distributed with high probability, then a significance level $P = 0.95$ is already given for a sample size of $n = 6$.
- $k = f_{1-\alpha,\nu}$: The quantile of STUDENT's t-distribution represents – also assuming a normal distribution – an individual factor which considers explicitly both the experimental expense ($\nu = n - 1$) and the significance level $P = 1 - \alpha$.

- k may also be estimated by means of special knowledge of the concrete distribution or on the basis of distribution-free estimations, e.g., *resampling techniques* (see EHRLICH and DANZER [2006]).

When the critical value is exceeded, $y \geq y_c$, then the analyte under examination is proved to be present in the sample. A quantification of the critical value by means of a calibration factor $S = b$ or function $y = a + bx$ is not very meaningful, but when it is done notwithstanding, then two consequences must be taken into account:

(1) The analytical value ACV (see Table 7.5) which corresponds to $x_{ACV} = x_{(B)}$ in Fig. 7.8 have a relative error of about 33% if the calculation is carried out according to Eq. (7.41c) with $k = 3$ (from $x_{ACV} = 3s_{BL}/b \approx 3s_{ACV}$ it results in $s_{ACV}/x_{ACV} \approx 1/3$ and, therefore, $s_{ACV,rel} = 0.33$. The relative uncertainty of x_{ACV} is $x_{ACV}/\Delta x_{ACV} \approx x_{ACV}/3s_{ACV} = 1$ and amounts therefore to 100%

(2) As can be seen from the distribution function B in Fig. 7.8, an analytical value x_{ACV} produces only in 50% of all cases signals $y \geq y_c$. Whereas the error of the first kind (classifying a blank erroneously as real measurement value) by the choice of $k = 2 \ldots 3$ can be aimed at $\alpha \approx 0.05$, the error of the second kind (classifying a real measured value erroneously as blank) amounts $\beta \approx 0.5$. Therefore, this analytical value – which sometimes, promoted by the early publications of KAISER [1965, 1966], plays a certain role in analytical detection – do not have any significance as a reporting limit in case of $y < y_c$, when no relevant signal have been found. For this purpose, the limit of detection, x_{LD}, has to be used.

The formal transformation of the critical value into the sample domain is necessary to estimate correctly the limit of detection. If the sensitivity is known and without of any uncertainty (e.g. in case of error-free calibration constants b), then the analytical value at CV is calculated by

$$x_{ACV} = U(\overline{y}_{BL})/b \quad . \tag{7.42}$$

In case of experimental calibration, the uncertainty of both the blank and the calibration coefficient, $U(\overline{y}_{BL}, b)$, have to be consider, e.g. according to

$$x_{ACV} = \frac{U(\overline{y}_{BL}, b)}{b} = \frac{s_{BL} \, t_{1-\alpha,v}}{b} \sqrt{\frac{1}{N} + \frac{1}{n}} \tag{7.43}$$

with n being the number of blanks ($v = n - 1$ are the corresponding degrees of freedom) and N being the number of repetition measurements at the analysis. Because the relation at Eq. (7.43) is unsuited as a limit in the concentration domain because of the large risk of error $\beta = 0.5$, a useful measure must be derived.

In order to obtain a comparable risk for the error of second kind ($\alpha = \beta \approx 0.05$), a definition of the limit of detection has to consider confidence intervals of both blanks *and* measurement values (KAISER [1965, 1966]; EHRLICH [1969]; CURRIE [1992, 1995, 1997]):

$$x_{LD} = x_{ACV} + \frac{s_{CV}\, t_{1-\beta,v}}{b} \sqrt{\frac{1}{N} + \frac{1}{n}} \qquad (7.44)$$

Assuming that the standard deviation of the critical value is approximately equal to the standard deviation of the blank, $s_{CV} \approx s_{BL}$, it results for $\alpha = \beta$ in

$$x_{LD} = 2\, \frac{s_{BL}\, t_{1-\alpha,v}}{b} \sqrt{\frac{1}{N} + \frac{1}{n}} \qquad (7.45)$$

With this the *LD* is the double of the *ACV* (Eq. 7.43).

The limit of detection can also be estimated by means of data of the calibration function, namely the intercept a which is taken as an estimate of the blank, $a \approx \overline{y}_{BL}$, and the confidence interval of the calibration straight line:

$$x_{LD_{cal}} = x_{ACV_{cal}} + s_{x0}\, t_{1-\beta,v} \sqrt{\frac{1}{N} + \frac{1}{n} + \frac{\overline{x}_{cal}^2}{S_{xx}}} \qquad (7.46)$$

with $s_{x0} = s_{y.x}/b$ being the standard deviation of the calibration procedure at the position $x_i = 0$, $\overline{x}_{cal}$ is the average of all the calibration contents (centre of calibration data) and $S_{xx} = \sum_{i=1}^{n}(x_i - \overline{x})^2$ is the sum of squares of the contents of the calibration samples. The number of the degrees of freedom is $v = n - 2$ in this case. With $s_{x0}(x_i = 0) \approx s_{x0}(x_i = x_{ACV})$ Eq. (7.46) becomes

$$x_{LD_{cal}} = s_{x0}\left(t_{1-\alpha,v} + t_{1-\beta,v}\right) \sqrt{\frac{1}{N} + \frac{1}{n} + \frac{\overline{x}_{cal}^2}{S_{xx}}}$$

$$x_{LD_{cal}} = 2\, s_{x0}\, t_{1-\alpha,v} \sqrt{\frac{1}{N} + \frac{1}{n} + \frac{\overline{x}_{cal}^2}{S_{xx}}} \approx 2\, x_{ACV_{(cal)}} \qquad (7.47)$$

The *limit of detection* is *that* analytical value that always will be found, apart from the small risk of error $\alpha = \beta$. The detection limit characterizes analytical procedures with regard to minimal value which can be detected with high significance. The limit of detection can, therefore, be reported as limit content when no signal is found.

The corresponding measured value at *LD* (see Table 7.5) is not of crucial importance in analytical chemistry. It characterizes that signal which can significantly be distinguished from the blank considering both types of error (α and β).

Because of its relatively high uncertainty of measurement of 50%, the limit of detection is mainly used as a detection criterion that the analyte is present in the sample. Taking as the basis a coverage factor of $k = 6$, contents at the LD are characterized by a relative standard deviation of about 17% (from $x_{LD} = 6s_{BL}/b \approx 6s_{LD}$ it results $s_{LD}/x_{LD} \approx \frac{1}{6}$ and, therefore, $s_{LD,rel} \approx 0.17$). The relative uncertainty of x_{LD} is $x_{LD}/\Delta x_{LD} \approx x_{LD}/3s_{LD} = 0.5$ and amounts therefore to 50%.

For this reason, a number of analysts uses a further limit quantity, namely the *limit of quantification, x_{LQ}, (limit of determination)*, from which on the analyte can be determined quantitatively with a certain given precision (KAISER [1965, 1966]; LONG and WINEFORDNER [1983]; CURRIE [1992, 1995, 1997]; IUPAC [1995]; EHRLICH and DANZER [2006]). This limit is not a general one like the critical value and the detection limit which are defined on an objective basis. In contrast, the limit of quantification is a subjective measure depending on the precision, expressed by the reciprocal uncertainty $x_{LQ}/\Delta x_{LQ} = k$, which is needed and set in advance. The limit of quantification can be estimated from blank measurements according to

$$x_{LQ} = k \; \frac{s_{BL} \; t_{1-\alpha,\nu}}{b} \sqrt{\frac{1}{N} + \frac{1}{n}} \qquad\qquad (7.48)$$

or from calibration data

$$x_{LQ_{cal}} = k \; s_{x0} \; t_{1-\alpha,\nu} \sqrt{\frac{1}{N} + \frac{1}{n} + \frac{(x_{LQ_{cal}} - \overline{x}_{cal})^2}{S_{xx}}} \qquad\qquad (7.49)$$

In the analytical practice, LQ bases frequently on a precision that is characterized by a relative uncertainty of 10% and, therefore, $x_{LQ} = 10u(x_{LQ})$ is applied. A relative uncertainty of 5% would require $x_{LQ} = 20u(x_{LQ})$, 1% requires $x_{LQ} = 100u(x_{LQ})$ and so on.

The quantification limit is relevant to concentration domain. Reporting LQs should always be completed by the conditions of precision.

The limits CV, LD, and LQ are widely applied in analytical chemistry but, unfortunately, not in a standardized manner. Various concepts are in use as well as different terms, symbols, definitions and meanings. Overviews are given and consequences are shown among others by CURRIE [1995, 1997]; LONG and WINEFORDNER [1983]; FLEMING et al. [1997a]; GEISS and EINAX [2001]; EHRLICH and DANZER [2006].

The term *detection capability (detection power)* represents a generic term of the performance of analytical methods at the lower limit of applicability. Mostly it is used descriptively (detection capability of a method is "high", "good" or "sufficient") or for giving order of magnitudes (detection capability is in the "ppm-range" or "ppb-range").

A quantification of the detection power is possible, in principle, namely in analogy to diverse definitions of resolving power like spectral resolving power, e.g., $R_\lambda = \lambda/\Delta\lambda$, or geometric resolving power, e.g. of microprobes, expressed by $R_a = a/\Delta a$ (a being the area under investigation and Δa the resolved area element) or by the number of pixels as known from digital cameras, too.

For this purpose, KAISER [1966] proposed the measure $\Pi = 1/x_{ACV}$ being the reciprocal value of the analytical value which corresponds to the critical value (KAISER's

"3σ"-limit). From today's experience, detection power, DP, should be expressed – when quantified – by the reciprocal value of detection limit

$$DP = 1/x_{LD} \tag{7.50}$$

For a detection limit of 0.5 µg/g, the detection power would amount 2 g/µg (2×10^6) and of 1 ng/g, correspondingly 1 g/ng (10^9) and so on[4].

As shown above, limits can be estimated on two ways: from blanks and from calibration data. Modern analytical methods such as spectrometry and chromatography use a third way to an increasing extent: the estimation of detection limit by means of the signal-to-noise ratio SNR.

As shown in Sect. 7.1, signal-to-noise ratio S/N can be used to characterize the precision of analytical methods. Noise is a measure of the uncertainty of dynamic blank measurements (of the "background").

Starting from the standard deviation of a net signal intensity $s_{y_{net}}$, see Eq. (7.3), and applying some simplifying conditions (pairwise measurement of signal y and neighboured background y_0, and, therefore, $n_y = n_{y_0} = n$), WINEFORDNER and VICKERS [1964]; WINEFORDNER et al. [1967] and ST. JOHN et al. [1966, 1967] derived the following expression for the *critical signal-to-noise ratio*:

$$(S/N)_c = t_{1-\alpha,v} = \frac{y_c - \overline{y}_0}{s_{y_0}\sqrt{2/n}} \quad . \tag{7.51}$$

In ST. JOHN et al. [1966], tables of the critical SNR have been given and, in addition, the comparability of the SNR concept to that of KAISER has been shown. From Eq. (7.51):

$$y_c = \overline{y}_0 + t_{1-\alpha,v}\, s_{y_0} \sqrt{2/n} \,\hat{=}\, y_{BL} + k\, s_{BL} \tag{7.52}$$

can be directly derived. Similar ideas, called the *SBR-RSDB* concept, were developed by BOUMANS [1991, 1994]. In analogy to the groups of WINEFORDNER and Vickers [1964], WINEFORDNER et al. [1967], and ST. JOHN et al. [1966, 1967], detection limits, mostly on the basis of $(S/N)_c = 3$, have been derived for a discrete measurement of background.

But the main advantage of the SNR concept in modern analytical chemistry is the fact that the signal function is recorded continuously and, therefore, a large number of both background and signal values is available. As shown in Fig. 7.9, the principles of the evaluation of discrete and continuous measurement values are somewhat different. The basic measure for the estimation of the limit of detection is the confidence interval of the blank. It can be calculated from Eq. (7.52). For $n = 10$ measurements of both blank and signal values and a risk of error of $\alpha = 0.05$ one obtains a critical signal-to-noise ratio $(S/N)_c = t_{0.95,9} = 1.83$ and $\alpha = 0.01$: $(S/N)_c = t_{0.99,9} = 2.82$. The common value $(S/N)_c = 3$ corresponds to a risk of error $\alpha = 0.05 \ldots 0.02$ in case of a small number of measurements ($n = 2 \ldots 5$). When $n \geq 6$, a

---

[4] This may be interpreted verbally that in 2 g, 1 µg of the analyte can be detected resp. 1 µg in 1 g, or mass proportions of 2×10^6 resp. 10^9 can be detected.

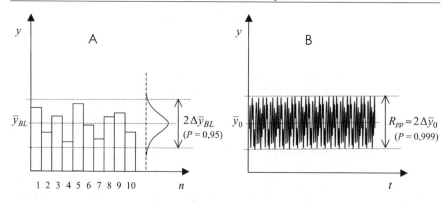

Fig. 7.9. Evaluation of blank variations in case of discrete blank y_{BL} measurements (**A**) and continuous recording of the baseline (background) y_0 (**B**)

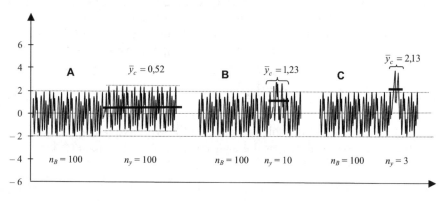

Fig. 7.10. Critical values $\bar{y}_c$ in case that the background noise is estimated from $n_B = 100$ values and the signal value from $n_y = 100$ (**A**), $n_y = 10$ (**B**) and $n_y = 3$ (**C**) values

signal-to-noise ratio of $(S/N)_c = 2$ is sufficient to make significant differentiations to the background (LIECK [1998]).

Modern analytical methods like spectrometry and chromatography produce a large amount of background data, sometime this is the case for signal data, too. As a rule, it can be assumed that $n_B = 100 \ldots 1000$ baseline values are available. Taking as a basis for computation $n_B = 100$, in Fig. 7.10 the signal-to-noise ratios for several numbers of signal values are given (EHRLICH and DANZER [2006]).

Figure 7.10 shows that in case of a large number of baseline values relative small differences between the baseline average and the critical value results.

It must be considered, though, that limits derived from the *SNR* characterize mainly instrumental noise and do not, as a rule, include "chemical" noise, viz such variations of measurement values which come from sample inhomogeneities, sample preparations in the course of the entire analytical process. This should be taken into account by carrying out real repetition measurements at several samples, if it is possible, or fixing a higher critical value $(S/N)_c = 2$ or 3 as be done in official regulations, e.g. ICH [1996].

7.6
Resolving Power

In modern analytical chemistry such methods play an increasing role which have a high resolution with regard to:

- Signal range and signal separation
- Spatial resolution of microprobes
- Temporal resolution of time-resolved methods

Each type of resolved measurement increases the amount of information obtainable by an analytical method, namely with regard to its capability of multielemental, spatial or temporal differentiation.

The *analytical resolving power* (*signal resolving power* R_z) is determined by the size of the recording range of the signals $z_{max} - z_{max}$ and the signal resolution Δz. If the signal range is a linear one, as illustrated in Fig. 7.11a, the analytical resolving power is given by

$$R_z = \frac{z_{max} - z_{min}}{\Delta z} \tag{7.53}$$

In case of variable signal half-width Δz as shown in Fig. 7.11b the following general expression applies (DANZER [1975]):

$$R_z = \int_{z_{min}}^{z_{max}} f(z)\, dz \tag{7.54}$$

In the case of constant signal half-width it results in Eq. (7.53) whereas, in the case of constant resolution $R(z)$, it follows that (KAISER [1970]; DANZER [1975])

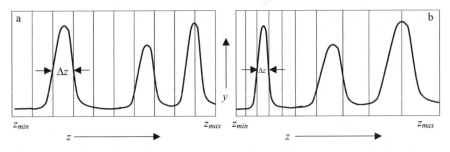

Fig. 7.11. Analytical resolving power for $\Delta z = $ const. (a) and $(\Delta z)^{-1} = f(z)$ (b)

$$R_z = R(z) \, \ln \frac{z_{max}}{z_{min}} \qquad\qquad (7.55)$$

The analytical resolving power is applied in several analytical fields in form of well-known expressions such as, e.g., *spectral resolving power* $R_\lambda = \lambda/\Delta\lambda$ or *mass resolving power* $R_M = M/\Delta M$.

Estimated values of the analytical resolving power of several analytical methods as given in ECKSCHLAGER and DANZER [1994] are shown in Table 7.6.

Table 7.6. Analytical resolving power R_z of several analytical methods

Method	R_z
Spot tests	1
Titration	10
Optical emission spectroscopy	
by grating instruments	200,000
by prism instruments	10,000
by Quantometers	60
Mass spectroscopy	500
High-resolution MS	200,000
UV-Vis spectrophotometry	50
X-ray spectrometry	
Wavelength-dispersive	5,000
Energy-dispersive	500
Infrared spectrometry	1,000
Gas chromatography	1,000
Capillary GC	10,000

The analytical resolving power can be interpreted as being the maximum number of signals which can find place within a given registration range. Therefore, it is evident that R_z is a measure of the multielement efficiency of analytical methods and influences strongly *selectivity*.

In the same way, *spatial resolving power* is a measure of the efficiency of distribution-analytical methods in micro- and surface analysis as well as scanning methods. From all the systematic representations of distribution-analytical problems given in DANZER et al. [1991], the mostly relevant are represented in Fig. 7.12.

The spatial resolving power will be described for three concrete cases:

(i) *Lateral resolving power* characterizing line scans as shown in Fig. 7.12b

$$R_{lateral} = \frac{l}{\Delta l} \qquad\qquad (7.56a)$$

where Δl is the lateral resolution (e.g., spot diameter) and l the line distance under inspection. $R_{lateral}$ is also relevant to point analysis

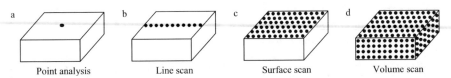

Fig. 7.12. Relevant types of distribution analysis

(Fig. 7.12a). In the special case where l is fixed perpendicular to the surface of the sample the depth resolving power $R_{depth} = d/\Delta d$ is relevant which becomes important for depth profile and thin-film analysis.

(ii) *Surface resolving power* (*area resolving power*) characterizing surface scans (Fig. 7.12c)

$$R_{surface} = \frac{a}{\Delta a} \tag{7.56b}$$

with Δa being the area resolution (spot area) and a the area under inspection. Important methods for micro-, lateral- and surface analysis are microprobe techniques like EPMA and SIMS, laser vaporization and excitation techniques and analytical microscopes.

(iii) *Volume resolving power* characterizing "volume scans" practically realized mostly by combination of lateral analytical methods and sputtering techniques or by 3D-SIMS

$$R_{volume} = \frac{v}{\Delta v} \tag{7.56c}$$

Estimated values of the lateral resolving power of some distribution-analytical methods are shown in Table 7.7 (ECKSCHLAGER and DANZER [1994])

In analogy to the spatial resolution a *temporal resolution power* R_t can be defined:

$$R_t = \frac{t}{\Delta t} \tag{7.57}$$

with t being the duration of the process under control and Δt the time resolution of the analytical method. The order of magnitude of the processes may be characterized by years, months and days for environmental processes, quality control of production or laboratory processes on the one hand and seconds up to femtoseconds in molecular-dynamic studies by means of ultra-short time spectroscopy.

Table 7.7. Spatial resolving power $R_{surface}$ of selected analytical methods. For reasons of comparability an area of 1 mm² has been taken; in the case of 3-D SIMS a volume of $v = 0.1$ mm³ is taken as the basis

Method		$R_{spatial}$
Microspark OES		$1 \ldots 100$
Spark source mass spectroscopy		$100 \ldots 10^4$
Laser OES, laser MS		$100 \ldots 10^4$
Induced autoradiography		$10^4 \ldots 10^6$
Electron probe microanalysis (EPMA)		$10^4 \ldots 10^6$
Ion probe microanalysis (SIMS)	$R_{surface}$	$10^4 \ldots 10^6$
3-D SIMS	R_{volume}	$10^8 \ldots 10^{10}$
Electron microscopy		$10^8 \ldots 10^{11}$
Field ion microscopy (atom probe)		$10^{12} \ldots 10^{14}$

References

AMC (1995) The Analytical Methods Committee: *Uncertainty of measurement – implications for its use in analytical science.* Analyst 120:2303

Arnold SF (1990) *Mathematical statistics.* Prentice-Hall, Englewood Cliffs, NJ

Bergmann G, von Oepen B, Zinn P (1987) *Improvement in the definitions of sensitivity and selectivity.* Anal Chem 59:2522

Björn E, Frech W, Hoffmann E, Lüdke C (1998) *Investigation and quantification of spectroscopic interferences from polyatomic species in inductively coupled plasma mass spectrometry using electrothermal vaporization or pneumatic nebulization for sample introduction.* Spectrochim Acta 53B:1766

Boumans PWJM (1991) *Measuring detection limits in inductively coupled plasma emission spectrometry using the "SBR-RSDB approach" – I. A tutorial discussion of the theory.* Spectrochim Acta 46B:431

Boumans PWJM (1994) *Detection limits and spectral interferences in atomic emission spectrometry.* Anal Chem 66:459A

Burns DT, Danzer K, Townshend A (2005) IUPAC, Analytical Chemistry Division: *Use of the terms "Robust" and "Rugged" and the associated characteristics of "Robustness" and "Ruggedness" in descriptions of analytical procedures.* Draft 2005

Currie LA (1992) *In pursuit of accuracy: nomenclature, assumtions, and standards.* Pure Appl Chem 64:455

Currie LA (1995) IUPAC, Analytical Chemistry Division, Commission on Analytical Nomenclature: *Nomenclature in evaluation of analytical methods including detection and quantification capabilities.* Pure Appl Chem 67:1699

Currie LA (1997) *Detection: International update, and some emerging dilemmas involving calibration, the blank, and multiple detection decisions.* Chemometrics Intell Lab Syst 37:151

Danzer K (1975) *Zur Ermittlung der Informationsmenge bei spektrochemischen Analysenverfahren*. Z Chem 15:158

Danzer K (1990) *Problems of calibration in trace-, in-situ-micro-, and surface analysis.* Fresenius J Anal Chem 337:794

Danzer K (2001) *Selectivity and specificity in analytical chemistry. General considerations and attempt of a definition and quantification.* Fresenius J Anal Chem 369:397

Danzer K (2004) *A closer look at analytical signals.* Anal Bioanal Chem 380:376

Danzer K, Fischbacher C, Jagemann K-U, Reichelt KJ (1998) *Near-infrared diffuse reflection spectroscopy for non-invasive blood-glucose monitoring.* LEOS Newsletter 12:2-9

Danzer K, Schubert M, Liebich V (1991) *Information theory in analytical chemistry. III. Distribution-analytical investigations.* Fresenius J Anal Chem 341:511

Danzer K, Than E, Molch D, Küchler L (1987) *Analytik – Systematischer Überblick, 2nd edn.* Akademische Verlagsgesellschaft Geest & Portig, Leipzig/Wissenschaftliche Verlagsgesellschaft, Stuttgart

Danzer K, Wagner M (1993) *Multisignal calibration in optical emission spectroscopy.* Fresenius J Anal Chem 346:520

Danzer K, Venth K (1994) *Multisignal calibration in spark- and ICP-OES.* Fresenius J Anal Chem 350:339

Davies OL, Goldsmith PL (1984) *Statistical methods in research and production.* Longman, London

De la Calle García D, Reichenbächer M, Danzer K, Hurlbeck C, Bartzsch C, Feller K-H (1998) *Use of solid-phase microextraction capillary gas chromatography (SPME-CGC) for the varietal characterization of wines by means of chemometrical methods.* Fresenius J Anal Chem 360:784

den Boef G, Hulanicki A (1983) IUPAC, Analytical Chemistry Division, Commission on Analytical Reactions and Reagents: *Recommendations for the usage of selective, selectivity and related terms in analytical chemistry.* Pure Appl Chem 55:553

Doerffel K (1990) *Statistik in der analytischen Chemie.* Deutscher Verlag für Grundstoffindustrie, Leipzig

Doerffel K, Müller H, Ullmann M (1986) *Prozessanalytik.* Deutscher Verlag für Grundstoffindustrie, Leipzig

Doerffel K, Eckschlager K, Henrion G (1990) *Chemometrische Strategien in der Analytik.* Deutscher Verlag für Grundstoffindustrie, Leipzig

Eckschlager K, Danzer K (1994) *Information theory in analytical chemistry.* Wiley, New York

Ehrlich G (1969) *Entwicklung und gegenwärtiger Stand der Bemühungen um eine objektive Charakterisierung des Nachweisvermögens analytischer Verfahren.* Wiss Z TH Leuna-Merseburg 11:22

Ehrlich G, Danzer K (2006) *Nachweisvermögen von Analysenverfahren. Objektive Bewertung und Ergebnisinterpretation.* Springer, Berlin Heidelberg New York

EURACHEM (1995) *Quantifying uncertainty in analytical measurement.* Teddington

EURACHEM (1998) *The fitness for purpose of analytical methods.* Teddington

Fleming J, Albus H, Neidhart B, Wegscheider W (1997a) *Glossary of analytical terms (VII)*. Accr Qual Assur 2:51

Fleming J, Albus H, Neidhart B, Wegscheider W (1997b) *Glossary of analytical terms (VIII)*. Accr Qual Assur 2:160

Fleming J, Neidhart B, Tausch C, Wegscheider W (1996a) *Glossary of analytical terms (I)*. Accr Qual Assur 1:41

Fleming J, Albus H, Neidhart B, Wegscheider W (1996b) *Glossary of analytical terms (II)*. Accr Qual Assur 1:87

Fleming J, Neidhart B, Albus H, Wegscheider W (1996c) *Glossary of analytical terms (III)*. Accr Qual Assur 1:135

Fujiwara K, McHard JA, Foulk SJ, Bayer S, Winefordner JD (1980) *Evaluation of selectivity in atomic absorption and atomic emission spectrometry*. Canadian J Spectrosc 25:18

Geiß S, Einax JW (2001) *Comparison of detection limits in environmental analysis – is it possible? An approach on quality assurance in the lower working range by verification*. Fresenius J Anal Chem 370:673

Gottschalk G (1975) *Standardisierung quantitativer Analysenverfahren – I. Allgemeine Grundlagen*. Fresenius Z Anal Chem 275:1

Graf U, Henning H-U, Stange K, Wilrich P-Th (1987) *Formeln und Tabellen der angewandten mathematischen Statistik*. Springer, Berlin Heidelberg New York (3. Aufl)

Hendricks MMWB, de Boer AH, Smilde AK (1996) *Robustness of analytical chemical methods and pharmaceutical technological products*. Elsevier, Amsterdam

Horn R (2000) *Kopplung einer elektrothermischen Verdampfungseinheit mit einem ICP-Massenspektrometer zur Verbesserung der Nachweisgrenze und der Spezifität*. Diploma Thesis, Friedrich Schiller University of Jena

Horwitz W, Kamps LR, Boyer KW (1980) *Quality assurance in the analysis of foods for trace constituents*. J Assoc Off Anal Chem 63:1344

Huber PJ (1981) *Robust statistics*. Wiley, New York

ICH (1994) ICH Topic Q2A (ICH Harmonised Tripartite Guideline): *Validation of analytical methods: Definitions and terminology* (CPMP/ICH/381/95), ICH, London

ICH (1996) ICH Topic Q2B (ICH Harmonised Tripartite Guideline): *Validation of analytical methods: Methodology* (CPMP/ICH/281/95), ICH, London

Inczédy J, Lengyiel JT, Ure AM, Geleneser A, Hulanicki A (Eds) (1997) *Compendium of analytical nomenclature, 3rd edn* (IUPAC Orange Book), Blackwell, Oxford

ISO 3534-1 (1993), International Organization for Standardization (BIPM, IEC, IFCC, ISO, IUPAC, IUPAP, OIML), *International vocabulary of basis and general terms in metrology*. Geneva

ISO 78-2 (1999) International Organisation for Standarization: *Layout for standards – 2. Methods of chemical analysis*. Geneva

IUPAC Orange Book (1997, 2000)
– printed version: *Compendium of analytical nomenclature (Definitive Rules 1997)*; see Inczédy et al. (1997)
– web version (from 2000 on): www.iupac.org/publications/analytical_compendium/

IUPAC (1995) Analytical Chemistry Division, Commission on Analytical Nomenclature: *Nomenclature in evaluation of analytical methods including detection and quantification capabilities,* prepared for publication by LA Currie, Pure Appl Chem 67:1699

Jochum C, Jochum P, Kowalski BR (1981) *Error propagation and optimal performance in multicomponent analysis.* Anal Chem 53:85

Kaiser H (1965) *Zum Problem Nachweisgrenze.* Fresenius Z Anal Chem 209:1

Kaiser H (1966) *Zur Definition der Nachweisgrenze, der Garantiegrenze und der dabei benutzten Begriffe.* Fresenius Z Anal Chem 216:80

Kaiser H (1970) *Quantitation in elemental analysis. Part I.* Anal Chem 42(2):24A, *Part II.* 42(4):26A

Kaiser H (1972) *Zur Definition von Selektivität, Spezifität und Empfindlichkeit von Analysenverfahren.* Fresenius Z Anal Chem 260:252

Kaiser H, Specker H (1956) *Bewertung und Vergleich von Analysenverfahren.* Fresenius Z Anal Chem 149:46

Kateman G, Buydens L (1993) *Quality control in analytical chemistry, 2nd edn.* Wiley, New York

Kellner R, Mermet J-M, Otto M, Widmer HM (eds) (1998) *Analytical chemistry.* Wiley-VCH, Weinheim

Lieck G (1998) *Nachweisgrenze und Rauschen.* LaborPraxis 22/June:62

Long GL, Winefordner JD (1983) *Limit of detection. A closer look at the IUPAC Definition.* Anal Chem 55:712A

Massart DL, Vandeginste BGM, Deming SN, Michotte Y, Kaufman L (1988) *Chemometrics: a textbook.* Elsevier, Amsterdam

O'Rangers JJ, Condon RJ (2000) In: Kay JP, MacNeil JD, O'Rangers JJ (eds), *Current issues in regulatory chemistry.* (AOAC Int., Gaithersburg, MD, p 207

Persson B-A, Vessman J (1998) *Generating selectivity in analytical chemistry to reach the ultimate – specifity.* Trends Anal Chem 17:117

Persson B-A, Vessman J (2001) *The use of selectivity in analytical chemistry – some considerations.* Trends Anal Chem 20:526

Prichard E, Green J, Houlgate P, Miller J, Newman E, Phillips G, Rowley A (2001) *Analytical measurement terminology – handbook of terms used in quality assurance of analytical measurement.* LGC, Teddington, Royal Society of Chemistry, Cambridge

Rodriguez LC, Garcia RB, Garcia Campana AM, Bosque Sendra JM (1998) *A new approach to a complete robustness test of experimental nominal conditions of chemical testing procedures for internal analytical quality assessment.* Chemom Intell Lab Syst 41:57

Sachs L (1992) *Angewandte Statistik, 7th edn.* Springer, Berlin Heidelberg New York [1383]

Sharaf MA, Illman DL, Kowalski BR (1986) *Chemometrics.* Wiley, New York

St John PA, McCarthy WJ, Winefordner JD (1966) *Application of signal-to-noise theory in molecular luminescence spectrometry.* Anal Chem 38:1828

St John PA, McCarthy WJ, Winefordner JD (1967) *A statistical method for evaluation of limiting detectable sample concentrations.* Anal Chem 39:1495

Streck S (2004) *Die schnelle und effiziente analytische Bestimmung von Mengen-, Spuren- und Ultraspurenelementen im Blutplasma von Patienten mit unterschiedlichen Erkrankungen und der statistische Vergleich mit einer Kontrollgruppe.* Doctoral thesis, Friedrich Schiller University of Jena

Sturgeon RE, Lam JW (1999) *The ETV as a thermochemical reactor for ICP-MS sample introduction.* J Anal Atomic Spectrom 14:785

Taylor JK [1983] *Validation of analytical methods.* Anal Chem 55:600A

Thiel G, Danzer K (1997) *Direct analysis of mineral components in wine by inductively coupled plasma optical emission spectrometry (ICP-OES).* Fresenius J Anal Chem 357:553

Thompson M (1995) *Uncertainty in an uncertain world.* Analyst 120:117N

Thompson M, Statistical Subcommittee of the Analytical Method Committee (2004) *The amazing Horwitz function.* AMC Techn Brief 17

Trullols E, Ruisanchez I, Rius FX (2004) *Validation of qualitative analytical methods.* Trends Anal Chem 23:137

USP 23-NF18 (1995) *The US pharmacopoeia & the national formulary.* US Pharmacopeial Convention Inc, Rockville, MD

USP 24-NF19 (2000) *The US pharmacopoeia & the national formulary.* US Pharmacopeial Convention Inc, Rockville, MD

Valcarcel M (2000) *Principles of analytical chemistry.* Springer, Berlin Heidelberg New York

Venth K, Danzer K, Kundermann G, Blaufuß K-H (1996) *Multisignal evaluation in ICP-MS. Determination of trace elements in Mo-Zr-alloys.* Fresenius J Anal Chem 354:811

Vessman J, Stefan RI, van Staden JF, Danzer K, Lindner W, Burns DT, Fajgelj A, Müller H (2001) IUPAC, Analytical Chemistry Division, Commission on General Aspects of Analytical Chemistry: *Selectivity in analytical chemistry.* Pure Appl Chem 73:1381

Wahlich JC, Carr GP (1990) *Chromatographic system suitability tests – what should we be using?* J Pharmac Biomed Anal 8:619

Wildner H, Wünsch G (1997) *Neue Ansätze zur Quantifizierung der Robustheit als Gütekennzahl analytischer Systeme im Hinblick auf Bewertbarkeit und Vergleichbarkeit.* J prakt Chem 339:107

Winefordner JD, Vickers TJ (1964) *Calculation of limit of detectability in atomic absorption flame spectrometry.* Anal Chem 36:1947

Winefordner JD, Parsons ML, Mansfield JM, McCarthy WJ (1967) *Derivation of expressions for calculation of limiting detectable atomic concentration in atomic fluorescence flame spectrometry.* Anal Chem 39:436

Wünsch G (1994) *Robustheit und Anfälligkeit als Kenngrößen zur Bewertung analytischer Systeme.* J Prakt Chem 339:107

Zeaiter M, Roger J-M, Bellon-Maurel V, Rutledge DN (2004) *Robustness of models developed by multivariate calibration. Part I. The assessment of robustness.* Trends Anal Chem 23:157

8 Presentation, Interpretation and Validation of Analytical Results

Modern analytical chemistry has a large number of hightech methods at its disposal which are able to carry out measurements with both high reliability and high throughput. In this way, an enormous amount of data is produced daily all over the world. Apart from the fact that a large part of these data remain uninterpreted and land up in "data cemeteries", a certain part (hopefully a small part) of analytical results is inadequately evaluated and presented. This is the more regrettable as the measured results obtained from the analytical instruments are not only highly reliable but also expensive.

8.1
Presentation of Analytical Results

If it can be supposed that the measured values, y, obtained in the course of the analytical process, are both precise and accurate, then the analytical results, x, do not have automatically the same quality. For this, some preconditions must be fulfilled, namely

(1) The analytical results have to be estimated from a correct evaluation model; see Sects. 6.2.5 and 6.2.6.

(2) It is desirable that the sensitivity of calibration model $S = \Delta y / \Delta x$ is not lower than 1, otherwise the uncertainty of x will increase compared with that of y, $\Delta x > \Delta y$; see Sect. 6.2.2.

(3) A realistic uncertainty interval has to be estimated, namely by considering the statistical deviations as well as the non-statistical uncertainties appearing in all steps of the analytical process. All the significant deviations have to be summarized by means of the law of error propagation; see Sect. 4.2.

It is in the nature of people to present results – whatever kind – with the highest possible precision. From this point of view it seems to be unnatural and absurd to collect facts which may decrease the precision and, therefore, the quality of the measured data. However, that is asked of the analyst, not only for the sake of truthfulness but also for responsible comparisons of analytical results with reference values, as will be shown in Sect. 8.2.

Analytical results should be given always in the form

$$(\overline{x} \pm U(\overline{x})) \cdot [x] \tag{8.1}$$

provided it concerns an arithmetic mean $\overline{x}$. According to Eq. (8.1), together with the mean of the analytical value, $\overline{x}$, its extended uncertainty, $U(\overline{x})$, has to be reported. Naturally, the unit of the measurand, $[x]$, must be given. The extended uncertainty $U(x)$ is obtained by multiplying the combined uncertainty $u(x)$ by a reasonable covering factor k; see Sect. 4.2. It is recommended, that the type of uncertainty interval should be specified as being, e.g., a statistical confidence interval, $cnf(\overline{x})$, prediction interval, $prd(\overline{x})$, or extended uncertainty interval, $u(\overline{x})$. Useful additional information may be the number n of replicate analyses from which the mean and the uncertainty has been estimated, the belonging level of significance, $P = 1 - \alpha$, and – if relevant – the sort of non-statistical uncertainties that have been considered.

For example, copper in sewage sludge is determined by ICP-MS. The reading (output) of the instrument may be 2.58685 for the average of duplicate analyses and 0.09513 for its confidence limit; see Sect. 4.1.2, Eq. (4.16), each given in μg/g. As usual, in the case of routine analyses, the confidence limit is calculated by means of a standard deviation that has been estimated independently from a large number of measurements. On the basis of the readings the analytical result should be given as follows:

(2.59 ± 0.10) μg/g Cu $(n = 2, \ P = 0.95)$

or, in the case that nonstatistical uncertainties have been considered:

(2.59 ± 0.13) μg/g Cu $(n = 2, \ k = 2,$ uncertainty considers sampling and sample preparation)

In each case, the number of digits should be rounded in such a way that no insignificant precision is feigned.

Results of ultra trace analyses are sometimes characterized by relatively high uncertainties up to more than 100%. In such cases it is not allowed that the lower uncertainty limit falls below zero. Results like, e.g., (0.07 ± 0.10) must be replaced by such as $(0.07 + 0.10 / -0.07)$ or $\left(0.07^{+0.10}_{-0.07}\right)$, respectively. That means, the total uncertainty interval (confidence interval, prediction interval is $0 \ldots 0.17$). In general, when the confidence interval includes a negative content (concentration), the result has to be given in the form

$$\left(\overline{x}^{+U(\overline{x})}_{-\overline{x}}\right) \cdot [x] \tag{8.2}$$

Other means like the *median* (see Eq. (4.22)) or the *geometrical mean* (see Eq. (4.18)) etc. have to be reported in a similar way together with the belonging uncertainty interval, e.g.,

$$(med\{x_i\} \pm U \ (med\{x_i\})) \cdot [x] \tag{8.3}$$

Together with a median it has to be said which type of median has been computed (see the footnote in Sect. 4.1.2) and also which kind of uncertainty (derived from *median absolute deviation, mad{x_i}*, or *quantiles*; see DANZER [1989]; HUBER [1981]; HAMPEL et al. [1986]; ROUSSEEUW and LEROY [1987]).

Geometrical means have unsymmetrical uncertainty intervals which are characterized by a dispersion factor v (see Sect. 4.1.2, Eq. (4.20)) and a covering factor k (see Sect. 4.2). Corresponding results should be given in the form

$$\left(\overline{x}_{geom}\cdot U(\overline{x}_{geom})^{\pm 1}\right)[x] = \left(\overline{x}_{geom}\cdot 10^{\Delta x_{geom}}\right)[x] = \left(\overline{x}_{geom}\cdot 10^{k\cdot s_{\lg x}}\right)[x] \qquad (8.4a)$$

That means the uncertainty interval covers the range $\overline{x}/\left(U(\overline{x}_{geom})\right)\ldots$ $\overline{x}\cdot U(\overline{x}_{geom}) = \overline{x}/(kv)\ldots\overline{x}\cdot kv$. Negative values of concentration at the lower limit of uncertainty do not appear in this case.

> Example: Uranium has been found in wine in a concentration of 2.0 ng/L. The dispersion factor v has been estimated $v = 1.2$ and the coverage factor has been chosen $k = 2$. Then the uncertainty factor amounts $v_A = 1.45$ and the analytical result has to be presented in one of the following ways:
>
> - $\left(2.0\cdot(2.4)^{\pm 1}\right)\frac{ng}{L}$ U (according to Eq. (8.4)) or
> - $(0.83\ldots 4.8)\frac{ng}{L}$ U or
> - $\left(2.0^{+2.80}_{-1.17}\right)\frac{ng}{L}$ U.

8.2
Factual Interpretation of Analytical Results

Analytical investigations are always carried out to serve a definite purpose. In this respect analytical results have to be evaluated and interpreted. In modern fields of applications like environmental monitoring, foodstuff control, medical laboratory diagnostics etc., conclusions have to be drawn about the presence of given species and their amount as well as the exceeding of limit values or falling significantly below them.

In many cases, when no extreme results (near a given limit) have been obtained, the analytical results may be reported as represented in Sect. 8.1. Attention has to be paid when results are situated nearby of critical values as it is the case with limit values of all type, especially *detection limits* and *reference values*.

8.2.1
Presentation of Results Near the Limit of Detection

The situation with detection limit has been discussed in Sect. 7.5. According to KAISER and SPECKER [1956]), KAISER [1965, 1966], LONG and WINEFORDNER [1983], CURRIE [1992, 1995, 1997], and EHRLICH and DANZER [2006] etc., the following decisions and reports are recommended:

(1) In the case that the measured value y_i is larger than the critical measuring value y_c:

> $y_i \geq y_c$: "*(the analyte)* **A has been detected** *(in sample X)*".

As shown in Sect. 7.5, the analyte under examination is definitely present in the sample (apart from a remaining risk of error α) when $y_i \geq y_c$.

Formally, an analytical result x_i can be calculated from y_i by means of the corresponding calibration function. When this result (from repeated measurements) should be reported, it must be taken into account that the relative uncertainty amounts minimally 100% (see Sect. 7.5, item (1) p. 201) and, therefore, it holds that $(\overline{x} \pm \overline{x})$. That means, that the uncertainty interval of analytical results calculated from measured values nearby the critical value covers a range of about $0 \dots 2\overline{x}$. As additional information, the limit of quantification, x_{LQ}, should be given.

(2) No signal being significantly larger than blank signals could be found:

> $y_i < y_c$: "*(the analyte)* **A could not be detected** *(in sample X)*. **The limit of detection** *(on the basis of the 6σ criterion)* **is x_{LD}**".

(3) The measured value y_i exceeds the measured value at the limit of detection, y_{LD} (MLD, see Table 7.5):

> $y_i \geq y_c$: "*(the analyte)* **A has been detected** *(in sample X)*. **The limit limit of quantification is x_{LQ}**".

As in case (1), the analyte is definitely detected, apart from the remaining risk of error α. Also in this case, an analytical result x_i can be calculated *formally* from y_i by means of the corresponding calibration function. But it must be noted that the relative uncertainty of results nearby of the detection limit amounts about 50% (see Sect. 7.5, item (2), p 201).

(4) The uncertainty of the results between the detection limit and the limit of quantification decreases continuously up to the precision set in advance by the precision factor k. In reaching and exceeding the quantification limit, analytical results can be reported as usual; see Sect. 8.1.

The problematic nature of the presentation of analytical results will be illustrated by examples of results of the Mo determination in several wine samples (fictional, but on the basis of own studies; see THIEL et al. [2004]; GEISLER [1999]) by means of ICP-MS (nuclide Mo 95):

Sample 1:   Mo: not detected (detection limit $DL = 47$ ng/L) or Mo: < 47 ng/L

Sample 2:   Mo: detected (25 ± 25) ng/L (i.e. the concentration is slightly above of that at the critical value; $CV = 23.5$ ng/L)

Sample 3:   Mo: detected (48 ± 24) ng/L (i.e., slightly above the detection limit)

Sample 4[1]:   Mo: (3580 ± 370) ng/L

---

[1]   Real sample #162, see GEISLER [1999]

8.2.2
Missing Data

There are situations in analytical chemistry where it is not advisable or not allowed to present data sets that contain zero values, "less than" values, "*LTV*"s, and "non detectables", "*ND*"s). The first one (zero data) normally come from missing data because of the breakdown of instruments or lacking samples – for whatever reason. In many cases, authorities and institutions do not accept data protocols with missing data. On the other hand, chemometrical data analyses cannot be carried out in such cases. Therefore, a lot of ideas has been suggested, how missing data can be substituted (see, e.g., LITTLE and RUBIN [1987]; HELSEL [1990]; NEITZEL [1996]; ALLISON [2002]; HOWELL [2002]). The mostly applied procedures are substitutions of "*LTV*"s and "*ND*"s, respectively, by concrete numerical values estimated in a meaningful way and completions of missing data by such substitutes which are the most probable data in the given situation (e.g., data lists, time series, or surface measurement nets).

Substitutions of "LTV"s and "ND" results. A meaningful value to substitute an *LTV* or *ND* is

$$x_{substit} = \frac{x_{LD}}{4} = \frac{x_{CV}}{2} \tag{8.5}$$

where x_{LD} corresponds to the 6σ-limit and x_{CV} to the 3σ-limit; see Sect. 7.5. The reason for this substitution can be derived from Fig. 8.1 that shows an unsymmetrical distribution of analytical values, e.g., a lognormal distribution. This is the most probable distribution because the lower limit of x is zero and cannot be crossed (as it would be done in case of a normal distribution). Also the assumption of a uniform distribution, as sometimes be done, is scarcely justified because the probability $p(x)$ approximates to zero as x it does. The substitute $x_{substit}$ is characterized by a relative uncertainty of 200% (in comparison: x_{CV} by 100%, and x_{LD} by 50%; see Sects. 7.5 and 8.2.1, (1) and (3)).

$p(x)$

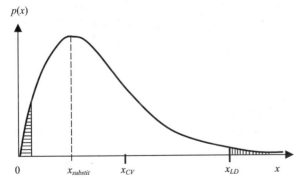

Fig. 8.1. Most probable distribution of analytical values, x, being situated below the detection limit x_{LD} (6σ-limit); x_{CV} is the analytical value at the critical measurement value (3σ-limit)

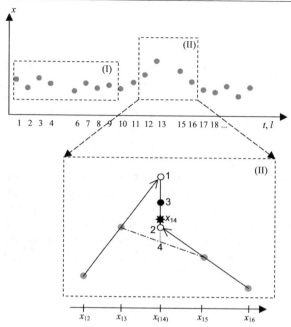

Fig. 8.2. Substitution of values in temporal or spatial successions of analytical values:
(I) random series
(II) maximum expected; estimation of the substitute of x_{14} in enlargement: 1: extrapolated x_{12} and x_{13}; 2: extrapolated x_{15} and x_{16}; 3: averaged 1 and 2; x_{14}: averaged 3 and 4 (point at the connection line between x_{13} and x_{15}) (deadening)

Completion of missing data. The proceeding of completion depends on the character of the incomplete data set. In *conventional data lists of random character* such as a $n \times m$ data list (data matrix) which contain results of several analytes (m) in diverse samples (n), missing data are frequently substituted by the mean of the respective analyte over all the samples.

Another proceeding is recommended in the case of *time series* or *lateral line scans*. When the values randomly scatter and – possibly – at the same time continuously slope up or down, then the mean between the two (or three) preceding and following should be taken; see Fig. 8.2a (I). On the other hand, if an extreme value, e.g., a maximum has to be expected, then it should be interpolated after extrapolation, e.g., as illustrated in Figs. 8.2a (II) and 8.2b. Such a situation is indicated by successively increasing and then decreasing values (or vice versa), to be precise at least three each of them.

In case (I), the missing values x_5 may be calculated as an average of the preceded and subsequent values x_4 and x_6, a weighted average of four or six neighbours, or may be generated by a random number out of a given random interval.

In cases of extremes (II), the missing value (here x_{14}) may be approximated by means of a spline function. If x_{14} is generated by extrapolation, a deadening of the extrapolated value (as can be seen from Fig. 8.2b) should be applied. The extrapolation is then, e.g., carried out according to $x_{m_{(14)}} = \frac{3x_{m-1}+3x_{m+1}-x_{m-2}-x_{m+2}}{8}$. Using more than the four neighbouring values, corresponding estimations by averaging or linear or nonlinear regression have to be carried out.

Completion of missing data in two-dimensional data fields. The goal of investigations carrying out in form of measurement points of a network – whether it be in micro- (surface analysis) or macro dimensions (deposits) – is to detect inhomogeneities, structures, gradients, "hot spots", etc. Therefore, missing data should not be substituted by averages as in the case of random data. Total averages could simulate non-evident sub-extremes when they are situated in the surrounding of real minima or maxima. Figure 3.12(b) in Sect. 3.4 may illustrate the situation. A value missed in a position where a gradient exists pointing to the maximum in the foreground at the left side should be substituted by the mean of the closely neighbouring measured points, e.g., in the data matrix

$$
\begin{array}{ccccc}
\vdots & \vdots & \vdots & & \\
\cdots & x_{i-1,j-1} & x_{i,j-1} & x_{i+1,j-1} & \cdots \\
\cdots & x_{i-1,j} & - & x_{i+1,j} & \cdots \\
\cdots & x_{i-1,j+1} & x_{i,j+1} & x_{i+1,j+1} & \cdots \\
& \vdots & \vdots & \vdots & \\
\end{array}
$$

this can be calculated as the average of the four nearest neighbours

$$
x_{(ij)_4} = \frac{1}{4} \sum \left(x_{i,j-1} + x_{i-1,j} + x_{i+1,j} + x_{i,j+1} \right)
$$

or of the eight neighbours

$$
x_{(ij)_8} = \frac{1}{8} \sum \left(x_{i,j-1} + x_{i-1,j} + x_{i+1,j} + x_{i,j+1} + x_{i-1,j-1} + x_{i-1,j+1} + x_{i+1,j-1} + x_{i+1,j+1} \right)
$$

or weighted, e.g., in the case of equidistant measuring points as follows:

$$
x_{(ij)_{8,w}} = \sum \frac{x_{i,j-1} + x_{i-1,j} + x_{i+1,j} + x_{i,j+1} + \frac{1}{\sqrt{2}} \left(x_{i-1,j-1} + x_{i-1,j+1} + x_{i+1,j-1} + x_{i+1,j+1} \right)}{4 + 4/\sqrt{2}}
$$

The decision on weighting and its type should be made according to the experimental facts. Supposed the data matrix given above is as follows then the following substitutes can be calculated:

$$
\begin{array}{cccc}
\vdots & \vdots & \vdots & \\
\cdots \ 1.25 & 1.22 & 1.53 \ \cdots & \quad x_{(ij)_4} = 1.265 \\
\cdots \ 1.17 & - & 1.44 \ \cdots & \quad x_{(ij)_8} = 1.293 \\
\cdots \ 1.09 & 1.23 & 1.41 \ \cdots & \quad x_{(ij)_{8,w}} = 1.288. \\
\vdots & \vdots & \vdots & \\
\end{array}
$$

In more complicated situations as they can be found in case of geochemical studies, sophisticated methods of spatial interpolation have been developed. In principle, *global* and *local* methods can be distinguised (BURROUGH [1986]). In global models (e.g., *trend surface analysis*, FOURIER *series*), each point is related to all the other points in the field of study. On the other hand, local models (e.g., *spline interpolation, moving averages*) estimate missing values only from neighbouring points and can, therefore, manifest local anomalies. Frequently used are methods like Inverse Distance Weighting (IDW, see BURROUGH and McDONNELL [1998] and Kriging (AKIN and SIEMES [1988], EINAX et al [1997]).

8.2.3
Analytical Results in Relation to Fixed Values

The relation of measured results to given values, e.g., *critical levels, legally fixed values, regulatory limits, maximum acceptable values,* is of continual relevance in analytical chemistry. In the analytical reality, the problematic nature of detection leads to the test statistics, strictly speaking to the t-test (CURRIE [1995, 1997]; EHRLICH and DANZER [2006]). By means of that, it is tested, if the determined analytical result is significantly different from the average blank of the critical value, respectively.

In the same way, the relation of measured results to limits of various kind (critical levels, x_{CL}) can be treated in general. The situation is shown in Fig. 8.3.

Frequently, limit values are given as a numerical value without any exceeding level. Then, according to the t-statistics, a significant exceeding has to be stated if $\overline{x} - \Delta\overline{x} > x_{CL}$, as illustrated in Fig. 8.3a. On the other hand,

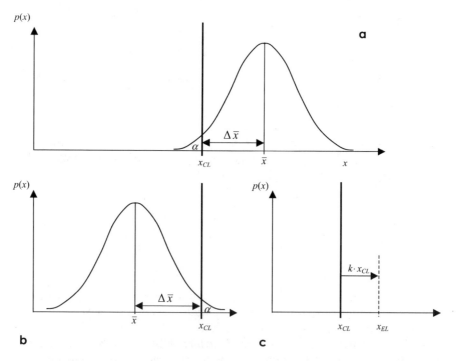

Fig. 8.3. Relationships between critical limits x_{CL} and analytical results $\overline{x}$: **a** the critical limit is significantly exceeded; **b** the result falls significantly below the limit value; **c** given critical exceeding limit $x_{EL} = k \cdot x_{CL}$

if $\bar{x}$ has to be (significantly) smaller than x_{CL}, the condition $\bar{x} + \Delta\bar{x} < x_{CL}$ must be fulfilled; see Fig. 8.3b. When an exceeding level x_{EL} is given, as shown in Fig. 8.3c, the simple relation $\bar{x} > x_{CL}$ indicates the exceeding of the critical level.

Comparison of test values with a conventional true value ("reference value") of a (certified) reference material (RM, CRM). In method development and validation of analytical procedures, the comparison of experimental results with standards of diverse kind (laboratory standards, certified reference materials, primary standards) plays an essential role. The decision as to whether an experimental result hits the reference value depends not only from the result itself but also from its uncertainty interval.

When working with standards, there are two cases regarding their precision:

(i) The reference value is given without any uncertainty or with a very small uncertainty interval in comparison with typical experimental results of laboratories ($U(\bar{x}_{RM}) \ll U(\bar{x})_{exp}$, *high quality standards* like primary standards, pure elements or compounds).

(ii) As a rule, the reference value is given with an uncertainty which is significant and can, therefore, not be disregarded in comparison with that of experimental results.

In case (i), the comparison of an experimental result, $\bar{x}_{exp}$, with the reference value, $\bar{x}_{RM}$, of a high quality standard is simply carried out by a specified t-test according to Eq. (4.45):

$$\hat{t} = \frac{\left|\bar{x}_{exp} - \bar{x}_{RM}\right|}{s_{x_{exp}}}\sqrt{n_{exp}} = \frac{\left|\bar{x}_{exp} - \bar{x}_{RM}\right|}{s_{\bar{x}_{exp}}} \tag{8.6}$$

where n_{exp} is the number of replicates which results in $\bar{x}_{exp}$. If the uncertainty of the reference value cannot be disregarded, the generalized t-test (WELCH's T_W-test; see Sect. 4.3.4, Eq. (4.43))

$$\hat{T}_W = \frac{\left|\bar{x}_{exp} - \bar{x}_{RM}\right|}{\sqrt{\frac{s^2_{x_{exp}}}{n_{exp}} + s^2_{\bar{x}_{RM}}}} = \frac{\left|\bar{x}_{exp} - \bar{x}_{RM}\right|}{\sqrt{s^2_{\bar{x}_{exp}} + \frac{U^2(\bar{x}_{RM})}{k^2}}} \tag{8.7}$$

has to be applied. The variance of the reference value can be estimated from its uncertainty interval divided by the squared coverage factor k (frequently $k = 2$).

In case (ii), the comparison of the result experimentally found and the reference value is carried out by the common t-test; see Sect. 4.3.4, Eq. (4.41):

$$\hat{t} = \frac{\left|\bar{x}_{exp} - \bar{x}_{RM}\right|}{s_{av}}\sqrt{\frac{n_{exp}\, n_{RM}}{n_{exp} + n_{RM}}} \tag{8.8}$$

Normally, the averaged standard deviation s_{av} is estimated according to Eq. (4.42). If no information is given on n_{RM}, it can be taken from the assumption that mostly $n_{RM} \geq 20$.

Both in cases (i) and (ii) the experimental uncertainty has to be estimated realistically as is done for the reference value, too. All the sources of variations and deviations have to be included in the calculation of the uncertainty. Efforts of analysts to shine with excellent analytical performance characteristic can have a detrimental effect as the examples in Table 8.1 demonstrate.

Table 8.1. Comparison of experimentally obtained results of copper content ($\mu g/g$) and the reference value of the material BCR277 (Institute for Reference Materials and Measurement, Belgium); number of repetitions $n_{exp}=10$

Case	Uncertainty includes	$\overline{x}_{Cu_{exp}}/(\mu g/g)$ experimental result	$\overline{x}_{Cu_{RM}}/(\mu g/g)$ certified value	Assessment on the basis of t-test (8.6)
A	Measurement	95.1 ± 3.2		The result is false because the uncertainty interval does not include the certified value
B	+ Calibration	95.1 ± 5.1	**101.7**	
C	+ Preparation	95.1 ± 6.9		The uncertainty interval includes the certified value; therefore, it cannot be considered to be false

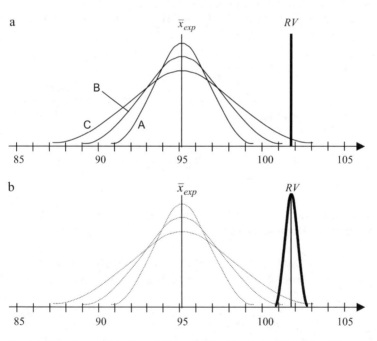

Fig. 8.4. Found result $\overline{x}_{exp}$ in relation to the reference value RV: **a** illustrates the location of RV without the uncertainty intervals A and B on the one hand and within the interval C on the other; **b** represents in addition the uncertainty interval of the reference value

In Table 8.1 three different analytical results are listed, the uncertainties of which are estimated in several ways: (A) measurement uncertainty only, as sometimes can be done in analytical practice, (B) additionally uncertainty of calibration considered, and (C) uncertainty of sample preparation included (partially nonstatistically estimated). Whereas in cases (A) and (B) the results are judged to be significantly false, in case (C) the difference is statistically not significant. The situation is illustrated in Fig. 8.4a when a comparison is carried out on the basis of the t-test (Eq. 8.6).

Another assessment will be found if the uncertainty of the reference value is considered likewise and, therefore, the t-test according to Eqs. (8.7) or (8.8) is applied. The corresponding $\hat{t}$-values ($\hat{t}_A = 5.96$, $\hat{t}_B = 4.00$, $\hat{t}_C = 3.02$) are larger in each case than the critical t-value 2.05 (on the basis $\alpha = 0.05$, $n_{exp} = 10$, and $n_{RM} = 20$). That means that the comparison will become sharper if the relatively small uncertainty interval of RV is included into the test and, therefore, the result C is assessed to be false, too.

8.2.4
Interlaboratory Studies

Analytical laboratories need to check their performance with regard to the production of accurate results with satisfactory precision. The most desirable way to ensure the reliability of analytical results is the participation of laboratories into regular interlaboratory tests. An interlaboratory study has to be understood as "*a study in which several laboratories measure a quantity in one or more 'identical' portions of homogeneous, stable materials under documented conditions, the result of which are compiled into a single document*" (IUPAC [1994]; Prichard et al. [2001]).

There are several types of interlaboratory studies (Horwitz [1994]):

(1) *Method performance study*: All laboratories follow the same written protocol and use the same test method to measure a quantity (usually concentration of an analyte) in sets of identical test samples. The results are used to estimate the *performance characteristics of the method*, which are usually *within-laboratory-* and *between-laboratory precision* and – if relevant – additional parameters such as sensitivity, limit of detection, recovery, and internal quality control parameters (IUPAC ORANGE BOOK [1997, 2000]).

(2) *Laboratory performance study*: Laboratories use the method of their choice to measure one or more quantities on one or more homogeneous and stable test samples in order to assess the *performance of the laboratory or analyst*. The reported results are compared among themselves, with those of other laboratories, or with the known or assigned reference value, usually with the objective of evaluating or improving laboratory performances (IUPAC ORANGE BOOK [1997, 2000]).

(3) *Material certification study*: Study that assigns a *reference value* ("*true value*") to a quantity (concentration or property) in the test mate-

rial. As a rule, the participating laboratories are selected according to their competence regarding the candidate reference material (IUPAC ORANGE BOOK [1997, 2000]).

Proficiency testing is a special type of laboratory performance studies defined as "*study of laboratory performance by means of ongoing interlaboratory test comparisons*" (ISO GUIDE 33 [1989]; ISO/REMCO N 280 [1993]; IUPAC [1993]; PRICHARD et al. [2001]). Proficiency testing is an essential part of external quality assessment schemes and performance checks.

The principle of proficiency testing schemes consists in analyzing one or more samples sent to the laboratories by an external body. The analytical results returned to the organizer are evaluated in comparison to the assigned value(s) of the sample(s).

Proficiency testing can be assessed in different ways. One of the most used evaluation system is that of so-called *z-scores* which are defined as follows (LAWN et al. [1993]; THOMPSON and WOOD [1993]):

$$z = \frac{|\bar{x}_{lab} - x_{ctv}|}{s_{x_{ctv}}} \tag{8.9}$$

where $\bar{x}_{lab}$ is the mean of the laboratory, x_{ctv} the assigned value of the quantity (mostly the conventional true value) and $s_{x_{ctv}}$ the target standard deviation. The following convention is used:

- $z \leq 2$ is considered to be satisfactory.
- $2 < z < 3$ is a cause for concern.
- $z \geq 3$ is unsatisfactory. As a rule, proficiency testing has to be repeated in such a case.

In general, the evaluation of interlaboratory studies can be carried out in various ways (Danzer et al. [1991]). Apart from z-scores, multivariate data analysis (nonlinear mapping, principal component analysis) and information theory (see Sect. 9.2) have been applied.

8.3
Chemometrical Interpretation of Analytical Data

The ultimate purpose of analytical studies is rarely a set of analytical results (data set) but frequently chemical information about the subject under investigation. Chemical information may concern properties like:

- *Purity, genuineness,* and *activity* of products
- Typical pattern of particular constituents in samples characterizing their *authenticity* and *provenance*
- Typical substance groups being indicators for *pollution* in environmental media

– Characteristic spacial and temporal structures in objects under study, etc.

Data interpretation always becomes difficult if several properties of various objects (samples) have to be considered simultaneously.

8.3.1
Principles of Data Analysis

Usually multivariate analytical information is represented in form of a data matrix:

$$X = (x_{ij}) = \begin{pmatrix} x_{11} & x_{12} & \cdots & x_{1m} \\ x_{21} & x_{22} & \cdots & x_{2m} \\ \vdots & \vdots & & \vdots \\ x_{n1} & x_{n2} & \cdots & x_{nm} \end{pmatrix} \qquad (8.10)$$

where the n rows represent the n objects under investigation which are characterized each by m features (properties, e.g. contents of various chemical constituents). Different from the principles of analytical dimensionality (see Sect. 3.4), the structure of a data set X according to Eq. (8.10) is represented in an m-dimensional space. The goals of the various methods of data analysis are:

(i)   Recognition of structures in multivariate data
(ii)   Identification, reproduction and quantitative separation of such data structures
(iii) Exploration of the causes for this structures
(iv)   Graphical visualization of the structures in two- or three-dimensional diagrams

From the viewpoint of data analysis, these objectives are achieved by means of the following fundamental methods:

(i)   *Cluster analysis*
(ii)   *Classification* methods, viz *multivariate variance and discriminant analysis* (MVDA)
(iii) *Factor analysis (Principal component analysis*, PCA)
(iv)   *Display methods*

In addition, methods of artificial intelligence (artificial neural networks and genetic algorithms) are applied.

The proceeding of common methods of data analysis can be traced back to a few fundamental principles the most essential of which are *dimensionality reduction, transformation of coordinates*, and *eigenanalysis*.

The principle of reduction of dimensionality will be illustrated schematically. In case that the property $(+/-)$ of an object depends mainly on one

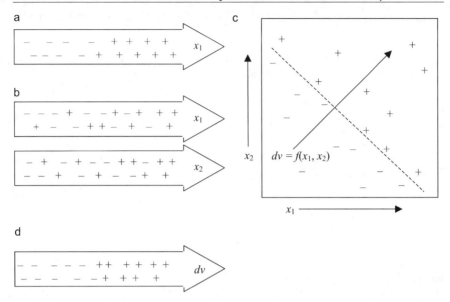

Fig. 8.5. Discrimination of object quality $(+/-)$ by means of one variable x_1 (**a**), of two variables x_1 and x_2 (**b**) which are represented as biplot in (**c**) from which the discriminant variable dv (discriminant function df) (**d**) can be derived

parameter, x_1, a classification of the quality can simply be done by means of common test statistics (e.g., univariate t-test). The proceeding is shown in Fig. 8.5a. On the other hand, if two parameters, x_1 and x_2, influence significantly the object property, then a situation as shown in Fig. 8.5b may arise. Whereas the classification of the object quality can clearly be seen in Fig. 8.5a, the classification by means of both variables, x_1 and x_2, seems to be chaotic. However, by using a biplot diagram (Fig. 8.5c), the situation concerning the object properties becomes clear. In addition, it can be seen that a new discriminant variable, $dv = f(x_1, x_2)$, can be formed by means of which simple discriminations between the object quality $(+/-)$ can be made; see Fig. 8.5d.

In this way, the discrimination problem by means of two variables which are represented in two dimensions is reduced to one dimension by means of a new variable $dv = f(x_1, x_2)$. All the reductions of dimensionality, from m to graphically presentable three- or two dimensions, happen according to this principle where uncorrelated variables are generated.

The basis of all data-analytical procedures is the data matrix (Eq. 8.10). In many cases the original data x_{ij} have to be transformed, either into standardized data:

$$z_{ij} = \frac{x_{ij} - \bar{x}_j}{s_j} \tag{8.11}$$

or suitable measures of multivariate distances, e.g., Euclidean distances:

$$d_{ij} = \sqrt{\sum_{k=1}^{m} (x_{ik} - x_{jk})^2} \qquad (8.12)$$

The use of standardized data (*variable standardization* or *column autoscaling*, see FRANK and TODESCHINI [1994]) results in data which are independent of the unit of measurement. Other types of standardization like object standardization, row autoscaling, or global standardization (global autoscaling, $(x_{ij} - \overline{\overline{x}})/s$) do not play a large role in data analysis.

Important methods of data analysis base on evaluation of the *covariance matrix* (*variance-covariance matrix*)

$$S = (s_{ij}) = \begin{pmatrix} s_{11} & s_{12} & \cdots & s_{1m} \\ s_{21} & s_{22} & \cdots & s_{2m} \\ \vdots & \vdots & & \vdots \\ s_{m1} & s_{m2} & \cdots & s_{mm} \end{pmatrix} \qquad (8.13)$$

which represents a symmetric matrix of pairwise *covariances* and *variances* at the diagonal; see Sect. 6.1.3. If standardized data are used the covariance matrix equals the *correlation matrix*

$$R = (r_{ij}) = \begin{pmatrix} 1 & r_{12} & \cdots & r_{1m} \\ r_{21} & 1 & \cdots & r_{2m} \\ \vdots & \vdots & & \vdots \\ r_{m1} & r_{m2} & \cdots & 1 \end{pmatrix} \qquad (8.14)$$

The essential information on the structure of data can be extracted by means of the fundamental equation of data analysis

$$(R - \lambda I)v = 0 \qquad (8.15)$$

where λ is the eigenvalue, v the eigenvector (latent vector) and I the identity matrix. Eigenvalues and eigenvectors are essential quantities for factor analysis (FA), principle component analysis (PCA) and multivariate variance and discriminant analysis (MVDA). They can be estimated by eigenanalysis (DILLON and GOLDSTEIN [1984]; FRANK and TODESCHINI [1994]).

8.3.2
Cluster Analysis: Recognition of Inherent Data Structures

Inhomogeneities in data can be studied by cluster analysis. By means of cluster analysis both structures of objects and variables can be found without any pre-information on type and number of groupings (*unsupervised learning, unsupervised pattern recognition*).

Geometrically illustrated, clusters are continuous regions of a high-dimensional space, each of them containing a relatively high density of points (e.g., objects), separated from each other by regions that are relatively empty (low density of points). The belonging of points (objects) to

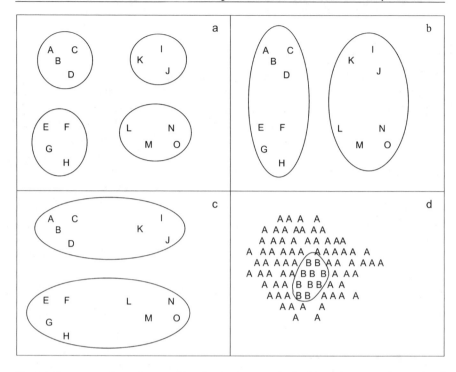

Fig. 8.6. Representation of 14 multivariately characterized objects in a two-dimensional space of variables where the clusters are connected into four groups (**a**) and classified into two differently chosen groups (**b, c**), respectively; **d** shows a nested clustering of B within A

certain clusters is not unchangingly fixed but may alter in dependence of the method of clustering used, the measure of similarity chosen, the number of clusters supposed and factual reasons as well. An example about that is shown in Fig. 8.6.

Methods of cluster analysis may be distinguished into two groups:

(i)   Hierarchical clustering
(ii) Non-hierarchical clusteringclustering!non-hierarchical

Hierarchical clustering can be carried out in an agglomerativeclustering!agglomerative and a divisiveclustering!divisive way. The results are mostly represented in form of tree-like diagrams, so-called *dendrograms*. The relation of the objects shown in Fig. 8.6a–c is represented in Fig. 8.7 in the form of a dendrogram. It can be seen that the objects are united at first pairwise and then step by step groupwise according to their multivariate similarity (agglomerative clustering). On the other hand, the entire population is divided stepwise into two, three, four clusters, node by node up to the single objects.

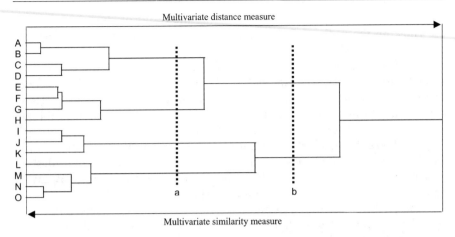

Fig. 8.7. Schematic representation of hierarchical clustering of the 14 objects shown in Fig. 8.6; the separation lines a and b corresponds to the clusters in 8.6a,b

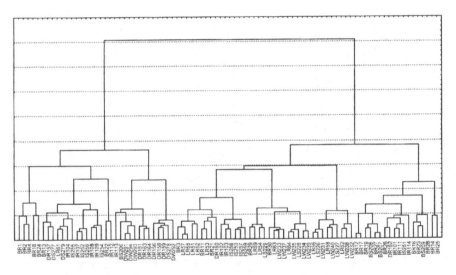

Fig. 8.8. Result of cluster analysis of 88 German wines according to WARD's method (THIEL et al. [2004])

There exist several methods of hierarchical clustering which use diverse measures of distance or similarity, respectively, e.g., *single linkage, complete linkage, average linkage, centroid linkage,* and *Ward's method* (SHARAF et al. [1986], MASSART et al. [1988], OTTO [1998]; DANZER et al. [2001]).

It is possible to treat dendrograms according to graph-theoretical principles (FRANK and TODESCHINI [1994]). However, in general, the results of clustering are evaluated qualitatively and taken as a basis of extensive studies on data structure.

In Fig. 8.8 an example is given showing the result of cluster analysis of 88 wines on the basis of the concentrations of 13 elements (As, B, Be, Cs, Li, Mg, Pb, Si, Sn, Sr, Ti, W, and Y).

The provenance of the wines was from four areas, Dienheim and Ingelheim in Rhine-Hesse and Bad Dürkheim and Landau in Rhinelande-Palatinate. At first glance, two clusters can be seen; looking at them a second time, four groups of may be supposed and taken as the basis of discriminant analysis the result of which is shown in Fig. 8.11.

Non-hierarchical cluster methods have in common with classification methods that pre-information on the number of classes is needed or desired to start an iteration process. In the course of the clustering a rearrangement of objects between several clusters is possible.

One of the most used techniques of non-hierarchical cluster analysis is the *density method* (*potential method*). The high density of objects in the *m*-dimension that characterizes clusters is estimated by means of a density function (potential function) P. For this, the objects are modelled by Gaus-

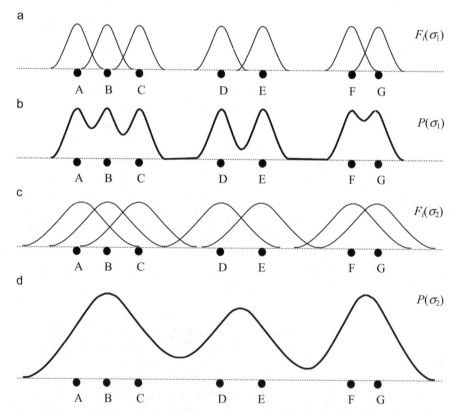

Fig. 8.9. Gaussian functions with different smoothing parameters σ_1 and σ_2 (**a,c**) and the corresponding potential functions (**b,d**)

sian functions $F(\sigma)$ with a parameter σ ("*standard deviation*", "*smoothing parameter*") set in advance as shown in Fig. 8.9a,c. The potential function $P(\sigma)$ results as superimposing curve of the Gaussian functions; see Fig. 8.9b,d.

Whereas the superposition of the GAUSSian functions $F_i(\sigma_1)$ ($i = A, B$, ..., G) do not produce a satisfactory separation, the $F_i(\sigma_2)$ do superimpose to three meaningful clusters.

Cluster analysis is important in all situations where homogeneity of data on the one hand and latent structures on the other hand play a significant role in evaluation and interpretation of analytical results. This applies in particular for single objects with extreme properties like outliers, hot spots etc that can easily be recognized being *singletons* among clusters.

8.3.3
Classification: Modelling of Data Structures

Classification methods consist generally of two steps, the learning phase (training step) and the working phase (classification phase). The assignment of objects to several classes is carried out on the basis of classification rules that have to be learned by means of a set of objects – their belonging to various classes is well-known. Such classes may concern quality, sorts, origins in the widest sense (provenance, producers, counterfeits etc) where sometimes it is a matter of alternative categories like good/poor, healthy/ill, genuine/imitated etc.

The basis of classification is supervised learning where a set of known objects that belong unambiguously to certain classes are analyzed. From their features (analytical data) classification rules are obtained by means of relevant properties of the data like dispersion and correlation.

One of the powerful classification methods is *multivariate variance and discriminant analysis* (MVDA) (DILLON and GOLDSTEIN [1984]; AHRENS and LÄUTER [1974]; DANZER et al. [1984]).

$$\boldsymbol{df}_j = \boldsymbol{v}_j \boldsymbol{x}_i^{\mathrm{T}} \tag{8.16}$$

By means of eigenanalysis multivariate discriminant functions, df_i, can be derived with eigenvectors $\boldsymbol{v}_j$ ($j = 1, 2, \ldots, p$) where $p \leq m$ is the rank of the matrix $\boldsymbol{R}$; see Eq. (8.14), m is the number of original variables ($i = 1, 2, \ldots, m$). With Eq. (8.16) the discriminant functions are linear combinations of the original variables

$$\boldsymbol{df}_j = v_{1j}x_1 + v_{2j}x_2 + \ldots + v_{mj}x_m \tag{8.17}$$

in which they play an especially important role – their variance within the classes is small and between the classes large. Besides, such variables are weighted slightly and are correlated with others. In this way, discriminant functions are obtained; their information content decreases in the order $\boldsymbol{df}_1 > \boldsymbol{df}_2 > \boldsymbol{df}_3 > \ldots > \boldsymbol{df}_p$. At best, the measures $\boldsymbol{df}_1$ and $\boldsymbol{df}_2$ can embody

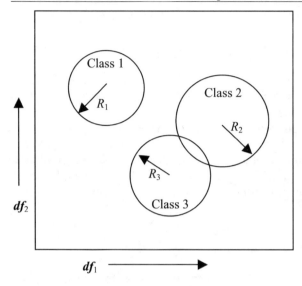

Fig. 8.10. Schematic representation of three classes in a two-dimensional discriminant space df_1 vs df_2; R_1, R_2, R_3 are the confidence radii of the respective classes

80–95% of the total information and, therefore, a two-dimensional plot of them may represent a realistic illustration of the respective classification as is schematically shown in Fig. 8.10. The boundaries between the classes may be estimated according to

$$R_i^2 = \frac{2(n - k)(n_k + 1)\ F_{1-\alpha, f_1=2, f_2=n-k-1}}{n_k(n - k - 1)} \tag{8.18}$$

In a corresponding way, measures of other class figures, e.g., ellipses, can also be calculated.

Pattern recognition has been successfully applied early in the fields of criminalistics (Duewer and Kowalski [1975]; Saxberg et al. [1978]), archaeology (Danzer et al. [1984, 1987]) as well as characterization of food (Forina and Armanino [1982]; Borszeki et al. [1986a]), and wine (Borszeki et al. [1986b]).

Wine data of German growing areas, the clustering of which has been shown in Fig. 8.8, have been classified afterwards by MVDA (Thiel et al. [2004]). The result can be seen in Fig. 8.11.

Whereas the separation of wines from Bad Dürkheim, Landau and Ingelheim/Dienheim succeeds satisfactorily, an overlap has to be recognised between Ingelheim and Dienheim. The total prediction rate, estimated by cross validation, is 88%.

Classification of wines according to the grape variety succeeds better, in general, because there are many more typical bouquet components (several hundreds) than mineral and trace elements being typical for the origin of wine. The organic compounds can be analyzed easily and reliably by Headspace Solid-Phase Microextraction Capillary Gas Chromatography and afterwards used for classification (De la Calle et al. [1998]). An example

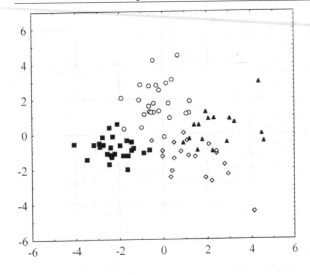

Fig. 8.11. Representation of classification of 88 German wines by discriminant analysis (1st vs 2nd discriminant function)
○ Bad Dürkheim (Rhinelande-Palatinate)
● Landau (Rhinelande-Palatinate)
◇ Ingelheim (Rhine-Hesse)
▲ Dienheim (Rhine-Hesse)

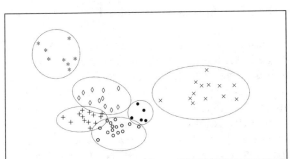

Fig. 8.12. Representation of the both first discriminant functions df_1 vs df_2 obtained by discriminant analysis (MVDA) of 65 German wines from five wine-growing regions according to six different grape varieties by means of 58 features (bouquet components)

● Müller-Thurgau
○ Riesling
◇ Silvaner
× Scheurebe
+ Weissburgunder
∗ Gewürztraminer
◆ Portugieser

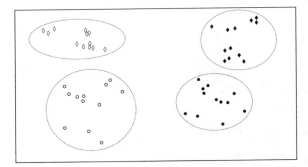

Fig. 8.13. Discriminant analysis of 36 wines from the wine-growing region of Alzey according to four grape varieties (see above) by means of 58 features

showing the classification of 65 German wines from five wine-growing regions according to six grape varieties is given in Fig. 8.12 (DANZER et al. [1999]), whereas in Fig. 8.13 the discrimination of four grape varieties from only one single region (Alzey) can be seen.

As expected, the separation improves if the sources of variation are reduced as in the situation represented in Fig. 8.13 in comparison with Fig. 8.12 where various wine-growing areas are considered.

There are many classification methods apart from linear discriminant analysis (DERDE et al. [1987]; FRANK and FRIEDMAN [1989]; HUBERTY [1994]). Particularly worth mentioning are the SIMCA method (Soft independent modelling of class analogies) (WOLD [1976]; FRANK [1989]), ALLOC (COOMANS et al. [1981]), UNEQ (DERDE and MASSART [1986]), PRIMA (JURICSKAY and VERESS [1985]; DERDE and MASSART [1988]), DASCO (FRANK [1988]), etc.

In general, the quality of classifications depends on several factors such as:

- All the essential features should be contained in the data set.
- The total number of objects (n) should be large enough (as a rule n should be about $3m$).
- The number of objects in every class (n_k) should be of comparable size.
- The variables (features) should be normally distributed.

If normal distribution cannot be assumed, nonparametric techniques of classification should be applied. Widely used is the *k-nearest neighbours method* (*k*NN) (see COVER and HART [1967]; SHARAF et al. [1986]; FRANK and TODESCHINI [1994]; DANZER et al. [2001]). The *k*NN technique is both a very simple and powerful method of data classification. An unknown object is assigned according to the majority votes of its k nearest neighbours in the learning set in the m-dimensional space. The respective votes are given on the basis of multivariate distances, e.g., according to Eq. (8.12). In Fig. 8.14 the principle of object assignment is illustrated. To avoid ambiguities, an odd number of nearest neighbours should be chosen.

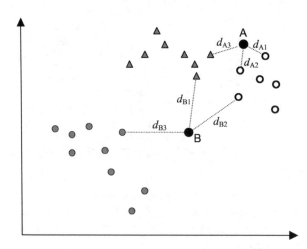

Fig. 8.14. Result of 3NN for two unknown objects A and B: A is assigned to class o, B cannot be classified significantly to any of the given classes

Table 8.2. Euclidian distances d_E and distance's ranks of an unknown wine sample to 16 samples with known growing region (GR), according to DANZER et al. [2001]

No.	1	2	3	4	5	6	7	8	9	10	11	12	13	14	15	16
GR	S-U	S-U	S-U	R	R	R	B	B	B	B	B	R-H	R-H	R-H	R-H	R-H
d_E	8.73	7.85	7.74	6.11	7.13	7.27	5.18	7.37	8.18	6.01	4.86	6.55	7.63	5.95	7.26	7.39
Rank	16	14	13	5	7	9	2	10	15	4	1	6	12	3	8	11

In the context of wine classification, an unknown wine sample had to be identified as being from Saale-Unstrut (S-U), Rhineland (R), Baden (B), or Rhine-Hesse (R-H). By means of the distances and ranks given in Table 8.2, the unknown was assigned to B (by 1NN, 2NN [2B], 3NN [2B/R-H], 4NN [3B/R-H], and 5NN [3B/R-H,R].

With the development of methods of artificial intelligence, neural networks are used increasingly for classification; see Sect. 8.3.6.

8.3.4
Factor Analysis: Causes of Data Structures

As a rule, the properties of objects are determined neither by only one feature nor by all the features equally. Instead the measured parameters influence variously the properties in a complex way because the variables are usually not independent from each other but are correlated to a certain degree.

The goal of factor analysis (FA) and their essential variant *principal component analysis* (PCA) is to describe the structure of a data set by means of new uncorrelated variables, so-called common *factors* or *principal components*. These factors characterize frequently underlying real effects which can be interpreted in a meaningful way.

The principle of FA and PCA consists in an orthogonal decomposition of the original $n \times m$ data matrix X into a product of two matrixes, F ($n \times k$ matrix of *factor scores, common factors*) and L ($k \times m$ matrix of *factor loadings*)

$$X = L\,F \tag{8.19a}$$

that means

$$\begin{pmatrix} x_{11} & \cdots & x_{1m} \\ \vdots & \ddots & \vdots \\ x_{n1} & \cdots & x_{nm} \end{pmatrix} = \begin{pmatrix} l_{11} & \cdots & l_{1m} \\ \vdots & \ddots & \vdots \\ l_{k1} & \cdots & l_{km} \end{pmatrix} \cdot \begin{pmatrix} f_{11} & \cdots & f_{1k} \\ \vdots & \ddots & \vdots \\ f_{n1} & \cdots & f_{nk} \end{pmatrix} \tag{8.19b}$$

and the linear model of any x_{ij} becomes

$$x_{ij} = l_{1j}f_{i1} + l_{2j}f_{i2} + \ldots + l_{kj}f_{ik} \tag{8.19c}$$

This is the *complete factor solution* which admittedly contains uncorrelated variables but all the k factors are extracted completely and no reduction in dimensionality occurs. However, it is usually the aim of data analysis to reduce the variables up to the essential one on one hand and the unimportant on the other. Therefore, the *reduced factor solution* is more important which decomposes X into L, F and an additional $n \times m$ matrix, the residual matrix (error matrix) E:

$$X = L\,F + E \tag{8.20a}$$

$$
\begin{pmatrix} x_{11} & \cdots & x_{1m} \\ \vdots & \ddots & \vdots \\ x_{n1} & \cdots & x_{nm} \end{pmatrix} = \begin{pmatrix} l_{11} & \cdots & l_{1m} \\ \vdots & \ddots & \vdots \\ l_{q1} & \cdots & l_{qm} \end{pmatrix} \cdot \begin{pmatrix} f_{11} & \cdots & f_{1q} \\ \vdots & \ddots & \vdots \\ f_{n1} & \cdots & f_{nq} \end{pmatrix}
$$

$$
+ \begin{pmatrix} e_{11} & \cdots & e_{1m} \\ \vdots & \ddots & \vdots \\ e_{n1} & \cdots & e_{nm} \end{pmatrix} \tag{8.20b}
$$

$$x_{ij} = l_{1j}f_{i1} + l_{2j}f_{i2} + \ldots + l_{qj}f_{iq} + e_{ij} \tag{8.20c}$$

where $q = k - p$ is the number of essential factors and p the number of insignificant factors which can be transferred to the error term.

By means of this reduction of dimensions the information in the form of variance is subdivided into essential contributions (common and specific variance) on one hand and residual variance on the other:

$$var(x_{ij})_{total} = var(x_{ij})_{comm} + var(x_{ij})_{spec} + var(x_{ij})_{res} \tag{8.21}$$

The main difference between factor analysis and principal component analysis is the way in which the variances of Eq. (8.20) are handled. Whereas the interest of FA is directed on the *common variance* $var(x_{ij})_{comm}$ and both the other terms are summarized as *unique variance*

$$var(x_{ij})_{unique} = var(x_{ij})_{spec} + var(x_{ij})_{res}$$

and therefore

$$var(x_{ij})_{total} = var(x_{ij})_{comm} + var(x_{ij})_{unique}$$

PCA separates a so-called *true variance*,

$$var(x_{ij})_{true} = var(x_{ij})_{comm} + var(x_{ij})_{spec}$$

from the residual variance

$$var(x_{ij})_{total} = var(x_{ij})_{true} + var(x_{ij})_{res}$$

Apart from this varied handling of variances, the proceedings of FA and PCA are similar corresponding to Eq. (8.20).

Generally, factor analysis consists of two steps:

(1) The *factor extraction* according to Eq. (8.20) in the course of which the number of common factors are estimated by *rank analysis* and coefficients of the factors (factor loadings) are calculated.

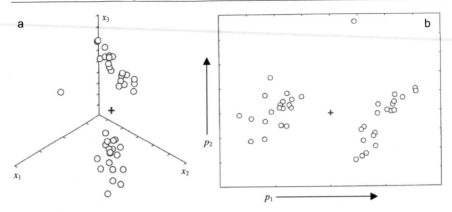

Fig. 8.15. Plot of original data (**a**) and two-dimensional PC plot (**b**); p_1 (91.4%) and p_2 (5.6%) declare 97% of the total variance and, therefore, information of the data

(2) The *factor rotation* by which the factors are transformed into more interpretable variables and can be tested concerning hypothetical data structures, respectively. There are various techniques of factor rotation with specific advantages in several fields of application (FRANK and TODESCHINI [1994]).

The PCA can be interpreted geometrically by rotation of the m-dimensional coordinate system of the original variables into a new co-ordinate system of principal components. The new axes are stretched in such a way that the first principal component p_1 is extended in direction of the maximum variance of the data, p_2 orthogonal to p_1 in direction of the remaining maximum variance etc. In Fig. 8.15 a schematic example is presented that shows the reduction of the three dimensions of the original data into two principal components.

Figure 8.16 shows a principal component plot of that data the classification of which by MVDA was given in Fig. 8.11. It can be seen that a certain structure can be imagined which becomes clearer by the discrimination algorithm.

An important application field of factor and principal component analysis is environmental analysis. EINAX and DANZER [1989] used FA to characterize the emission sources of airborne particulates which have been sampled in urban screening networks in two cities and one single place. The result of factor analysis basing on the contents of 16 elements (Al, B, Ba, Cr, Cu, Fe, Mg, Mn, Mo, Ni, Pb, Si, Sn, Ti, V, Zn) determined by Optical Atomic Emission Spectrography can be seen in Fig. 8.17. In Table 8.3 the common factors, their essential loadings, and the sources derived from them are given.

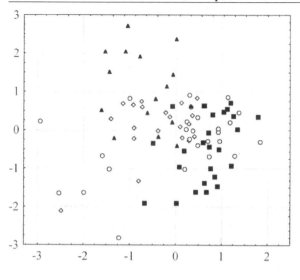

Fig. 8.16. Representation of a principal component plot (p_1 vs p_2) of 88 German wines, see Fig. 8.11 and THIEL et al. [2004]
○ Bad Dürkheim (Rhinelande-Palatinate)
● Landau (Rhinelande-Palatinate)
◇ Ingelheim (Rhine-Hesse)
▲ Dienheim (Rhine-Hesse)

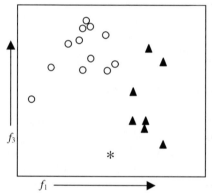

Fig. 8.17. Representation of the factors f_1 vs f_3 of two cities and one single measuring point (∗); according to EINAX and DANZER [1989]

Table 8.3. Common factors of immissions of airborne particulates during the heating period (bold: key elements); according to EINAX and DANZER [1989]

Factor	Variance in %	Essential factor loadings by	Genesis
1	34.3	Si, Mg, Pb, Ti, Mn, Fe	Raised geogene material (secondary dust)
2	17.2	Al, V, Ti, Ni	Industrial and urban heating systems and steam-raising plants
3	16.5	Cu, Mn, Cr, Sn, Mo, Mg	Industrial immission of metallurgy and metal-processing industries
4	15.9	Zn, Ni, Ti	Ubiquitous elements, sedimented by precipitates

8.3.5
Exploratory Data Analysis and Display Methods: Visualization of Data Structures

Sometimes the interpretation of analytical data does not need the deepest mathematical analysis but it is sufficient to get an impression on the structure of the data. Although the basic idea of graphical data interpretation is ancient (e.g., BRINTON [1914]), the fundamentals of modern *explorative data analysis* (EDA) has been developed in the 1960s (TUKEY [1962, 1977]).

The goal of EDA is to reveal structures, peculiarities and relationships in data. So, EDA can be seen as a kind of detective work of the data analyst. As a result, methods of data preprocessing, outlier selection and statistical data analysis can be chosen. EDA is especially suitable for interactive proceeding with computers (BUJA et al. [1996]). Although graphical methods cannot substitute statistical methods, they can play an essential role in the recognition of relationships. An informative example has been shown by ANSCOMBE [1973] (see also DANZER et al. [2001], p 99) regarding bivariate relationships.

The most important methods of explorative data analysis concern the study of the distribution of the data and the recognition of outliers by *boxplots* (Fig. 8.18), *histograms* (Fig. 8.19), *scatterplot matrices* (Fig. 8.20), and various *schematic plots*.

- **Boxplots** *(Box-Whisker plots)*

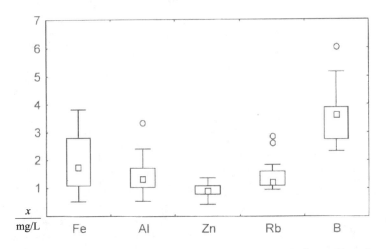

Fig. 8.18. Representation of some trace elements in wines in form of boxplots, constructed as follows: box: lower quartile, median □, and upper quartile; whiskers: minima and maxima within box ±1.5 of the quartiles' difference; outliers ○: outside of box ±1.5 of quartiles' difference (according to DANZER et al. [2001])

- **Histograms**

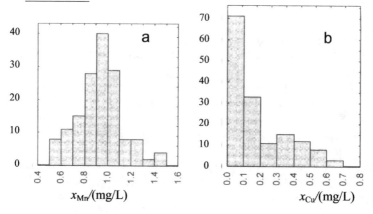

Fig. 8.19. Histograms of trace elements in wine for manganese (**a**) and copper (**b**) (according to DANZER et al. [2001])

- **Scatterplot matrixes**

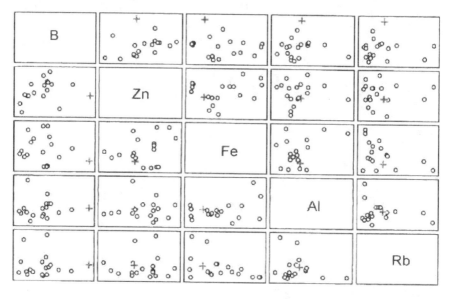

Fig. 8.20. Scatterplot matrix of five trace elements in 19 wine samples; one of the samples is marked (+) in all the special plots to have the opportunity for comparison (according to DANZER et al. [2001])

- **Icon plots** (*schematic plots*). Under this term various schematic graphs are put together (KLEINER and HARTIGAN [1981]; DU TOIT et al. [1986]; NAGEL et al. [1996]) such as plots of *suns, stars, glyphes, diamants, faces, flowers, trees, castles*, etc. From these, such graphs are privileged which easily can be differentiated visually. This applies particularly to faces and trees. In Fig. 8.21 some icon plots are presented to illustrate the high information and discrimination power of graphs.

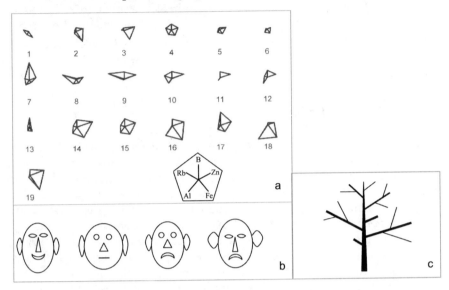

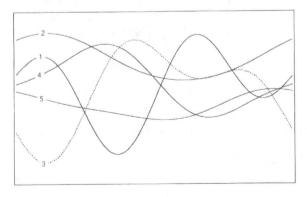

Fig. 8.21. Starplots of five trace elements (directed as shown below) in 19 wine samples (a) CHERNOFF-type faces symbolizing five elements coded in the forms of the face, eyes, mouth, nose, and ears in four wine samples (b) (see CHERNOFF [1973]), and schematic representation of a tree plot of one sample characterized by 15 variables (c)

• *Andrews plots.* In Andrews plots (ANDREWS [1972]) the features (variables set) are used as coefficients of a linear combination of trigonometric functions

$$F_i(t) = \frac{x_{i1}}{\sqrt{2}} + x_{i2}\sin(t) + x_{i3}\cos(t) + x_{i4}\sin(2t) + x_{i5}\cos(2t) + \cdots \quad (8.22)$$

where $-\pi \le t \le \pi$. Plots with similar coefficients results in similar curves; see Fig. 8.22.

Fig. 8.22. Andrews plot of five wines, according to DANZER et al. [2001]

- **Minimal spanning trees** and **dendrograms.** Both methods represent the transition to *cluster analysis* which belongs – in a wider sense – also to EDA. Dendrograms are well-known graphical representations of clustering which have been shown in Figs. 8.7 and 8.8. Minimal spanning tree (KRUSKAL [1956]; PRIM [1957]; CHERITION and TARJAN [1976]) is a method of graph-theoretical clustering by which a tree is build stepwise in such a way that, step by step, the link with the respective smallest distance is added.

8.3.6
Methods of Artificial Intelligence

From the large repertoire of methods of artificial intelligence (AI), viz visual perception learning, inference, problem solving, speech recognition, language understanding and translating, chess-playing, hypothesis testing and theorem proving, only a limited number of fields are significant in analytical chemistry. The most important fields for data interpretation are *expert systems, neural networks* and, partly, *genetic algorithms.*

Expert systems (ES) are computer programs which contain heuristic knowledge basing on the experience of experts in the respective field of application. An ES consists of *data bases* being a collection of facts and figures, on the one hand, and a system of rules of *mathematical logic* (e.g. in the form of BOOLEAN algebra) on the other hand. Using principles such as combining, linking, inferring, and deciding, ES are schematically constructed as represented in Fig. 8.23.

The proceeding of an ES by decision sequences may be illustrated by a (non-computerized) example from undergraduate training in analytical

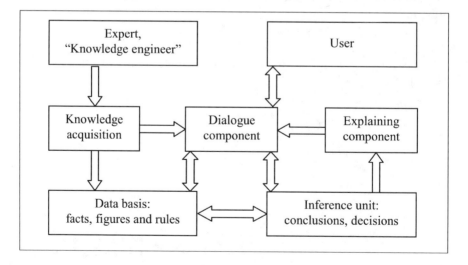

Fig. 8.23. Structure of an expert system

chemistry: the so-called "systematic separation scheme of cations" going back to FRESENIUS [1841]. Having a sample in aqueous solution, the following subproceeding is applied to test if Pb^{2+}, (Hg_2^{2+}), and Ag^+, respectively, are present in the sample.

IF HCl is added to the solution
AND white precipitation appears
THEN (AgCl and/or Hg_2Cl_2 and/or $PbCl_2$)
    IF   precipitation is extracted by hot H_2O, afterwards solution/filtrate is cooled + HCl
    AND white precipitation ($PbCl_2$)
    THEN *Pb* is detected
    ELSE no Pb^{2+} is contained in the sample solution
       IF NH_4OH is added to the precipitation
       AND blackening appears (Hg + $HgNH_2Cl$)
       THEN *Hg* is detected
       ELSE no Hg_2^{2+} is contained in the sample solution
          IF NH_4OH is added to the precipitation, then HNO_3 is added to the filtrate
          AND white precipitation appears (AgCl)
          THEN *Ag* is detected
          ELSE no Ag^+ is contained in the sample solution
ELSE neither Pb^{2+} nor Hg_2^{2+} nor Ag^+ are contained in the sample solution

---

ES is based on comparable decision trees in computerized form. The development and arrangement is carried out by means of

- *Procedural computer languages* (e.g., FORTRAN, PASCAL, BASIC, C/C++) by algorithmic links
- *Logical computer languages* (e.g., PROLOG, LISP) by logical links
- *Shell systems* in which specific factual knowledge can be implemented

Today, shell systems are most widely used in analytical applications because of their simplicity, flexibility, and universality. The main advantage of ESs is the fact that implemented and learned knowledge easily and faultlessly can be reproduced and copied, so that the knowledge may be available for a wide community of users or customers.

An early field of application in analytical chemistry is structure elucidation. DENDRAL was one of the first ES in general, designed to the identification of organic compounds from mass spectrometric data (BUCHANAN and FEIGENBAUM [1978]). In the 1980s and 1990s a flood of expert systems has been developed in analytical chemistry for different types of application, viz:

- *Structure elucidation* by various molecule spectroscopic methods (X-PERT, ELYASHBERG et al. [1997]; MOLGEN, BENECKE et al. [1997], SPECINFO, BREMSER and GRZONKA [1991]; CANZLER and HELLENBRANDT [1992]; BARTH [1993]; NEUDERT and PENK [1996]; SCHUUR [1997]).

- *Interpretation of measurements of methods*: X-ray fluorescence spectrometry (JANSSEN and VAN ESPEN [1986]; ARNOLD et al. [1994]), X-ray diffraction spectra (ADLER et al. [1993]), NMR spectra (HIPS, WEHRENS et al. [1993a]), HPLC retention indices (RIPS, WEHRENS [1994]), KARL FISCHER titration (HELGA, WÜNSCH and GANSEN [1989]).

- *Evaluation of data and validation*: multivariate data analysis (MULTIVAR, WIENKE et al. [1991]), evaluation of interlaboratory studies (INTERLAB, WIENKE et al. [1991]), ruggedness expert system (RES, VAN LEEUWEN et al. [1991]).

- *Sampling strategies* (BIAS, WEHRENS et al. [1993b]).

Laboratory Information Management Systems (LIMS) have been developed exclusively for practical application, viz, the organization and control of working course and information flow in the laboratory (NILSEN [1996]).

LIMS comprises:

- Acceptance, registration, and appointment of samples, generation of working sheets
- Detailing the course of sample treatment and analytical procedures
- Analyzing test samples (if necessary by several methods)
- Evaluation and validation of analytical results and their uncertainty
- Reporting or exporting of the results and data archiving

By means of LIMS the reliability of the tests, the traceability of the results, and the security of the data can be guaranteed. In addition, the effectivity of the costs may be estimated and supply, service and maintenance of instruments managed. LIMS always has to be designed in an individual way according to its specific profile. Information on LIMS can be found mainly on the internet (see LIMSOURCE, LAPITAJS and KLINKNER, etc.).

Another form of artificial intelligence is realized in *artificial neural networks* (ANN). The principle of ANNs has been presented in Sect. 6.5. Apart from calibration, data analysis and interpretation is one of the most important fields of application of ANNs in analytical chemistry (TUSAR et al. [1991]; ZUPAN and GASTEIGER [1993]) where two branches claim particular interest:

(i)   *Classification* of objects on the basis of multicomponent analytical data
(ii) *Spectra interpretation* and spectrum-structure correlation

Both cases can be dealt with both by supervised and unsupervised variants of networks. The architecture and the training of supervised networks for spectra interpretation is similar to that used for calibration. The input vector consists in a set of spectral features $y_i(z_i)$ (e.g., intensities y_i at selected wavelengths z_i). The output vector contains information on the presence and absence of certain structure elements and groups fixed by learning rules (Fig. 8.24). Various types of ANN models may be used for spectra interpretation, viz mainly such as Adaptive Bidirectional Associative Memory (BAM) and Backpropagation Networks (BPN). The correlation

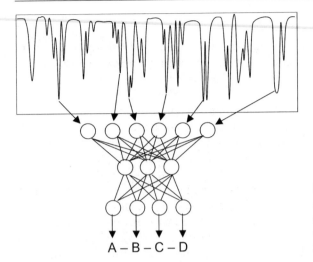

Fig. 8.24. Derivation of element composition or structural features from a measured spectrum by means of a trained ANN

between spectra and structure improves by the extent of training sets. However, it has to be considered that trained nets have excellent abilities in interpolation but worse in extrapolation.

Applications of classical ANNs for spectra interpretation can be found in STONHAM et al. [1975], WYTHOFF et al. [1990], ROBB and MUNK (1990), ALLANIC et al. [1992], ANKER and JURS (1992), BALL and JURS [1993], and BOGER and KARPAS [1994].

In contrast to common ANNs, Kohonen networks produce self-organized topological feature maps (KOHONEN [1982, 1984]). The basic idea of Kohonen mapping is that information in data usually contains not only an algebraic but also a topological aspect. These double aspect is shown schematically in Fig. 8.25 where the data and the structure of them are composed.

(a) 1111111111111122222222222222233332333333333333333344333334333334334433444335555555355

(b)

```
1 1 1 1 1 1 1 1 1
1 1 1 1 1 2 2 2 2
2 2 2 2 2 2 2 2 2
3 3 3 3 2 3 3 3 3
3 3 3 3 3 3 3 3 3
3 3 4 4 3 3 3 3 3
4 3 3 3 3 3 4 3 3
4 4 3 3 4 4 4 3 3
5 5 5 5 5 5 3 5 5
```

(c)

```
1
111
11111
1111122
222222222
22333323333
3333333333344
333334333334334
43344433555555355
```

(d)

```
11111111111
1122222
222222
223333
233333
333333
333344
333334
333334
33  44
33   4
33    44
33
555555  3
55
```

Fig. 8.25. Various topological arrangement of data which are arithmetically identical

Kohonen nets are constructed in such a way that neighbourhoods in the objects generate neighbouring points in the data projection.

Kohonen nets are one-layered networks which treat the inputs in such a way that similar signals excite closely neighbouring neurons which on their part generate neighbouring points in the data projection map. A spectacular example for the efficiency of Kohonen nets is the classification of 572 Italian oils from nine various regions according to their contents of eight different fatty acids (FORINA and ARMANINO [1982]; ZUPAN and GASTEIGER [1993]).

Neural networks are helpful tools for chemists, with a high classification and interpretation capacity. ANNs can improve and supplement data arrangements obtained by common multivariate methods of data analysis as shown by an example of classification of wine (LI-XIAN SUN et al. [1997]).

Genetic algorithms (see Sect. 5.3), representing the third group of methods of artificial intelligence, have their advantages in optimization, simulation and modelling (LUCASIUS and KATEMAN [1993]; HIBBERT [1993]). Applications to data analysis are mostly focused on feature selection (variables reduction); see LI et al. [1992]; FISCHBACHER et al. [1995].

8.4
Analytical Images

Modern methods of surface microanalysis and analytical microscopy produce analytical information mainly in the form of images. Furthermore, analytical results of distribution analysis, obtained in various ways, by direct sample scanning, discontinuous sampling or even by remote sensing, may be presented by images.

Analytical images represent such geometrical information that can be transferred to real *sample positions* given by the field of vision of the respective analytical instrument. On the other hand, with representations showing intensities or related variables in dependence of *signal positions* (analytical z-variables) such as wavelengths, retention values etc. (e.g., hyphenated techniques), no analytical images exist but higher-dimensional analytical information; see Sect. 3.4.

The analytical information obtained as a function of sample coordinates, l_x, l_y and/or l_z, can characterize

(a) Chemical species, viz

- In the form of two-dimensional images (which realize in fact three-dimensional information; see Sect. 3.4), $z = f(l_x, l_y)$, representing so-called *species images*; see, e.g., Figs. 3.12c and 8.26 (O, Al, Fe, and Cu by EPMA)

- Quasi-three-dimensional images, $y_z = f(l_x, l_y)$, representing *concentration profiles*; see, e.g., Figs. 3.12a,b and 8.26 (Al, Fe and Cu by SIMS)

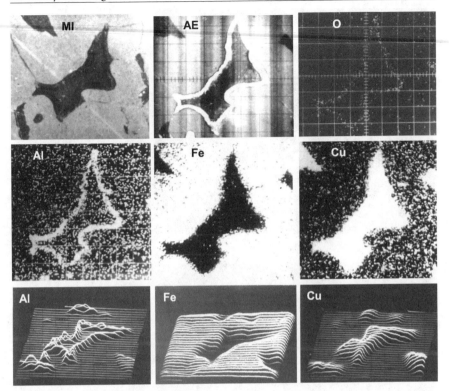

Fig. 8.26. Segregation of copper in an iron copper soak alloy: MI metallographic image, AE absorbed electrons measured by EPMA, and four element-specific X-ray scanning images by EPMA, below: three elemental-specific relief plots by SIMS; according to EHRLICH et al. [1979]

- Three-dimensional images, $z = f(l_x, l_y, l_z)$, representing the presence of species in a given volume element, e.g., by SIMS, or $y_z = f(l_x, l_y, l_z)$, representing the spatial concentration distribution of a certain component in a given volume element; see, e.g., Fig. 8.27

(b) Chemical structure, mostly presented in form of two-dimensional images, $S = f(l_x, l_y)$, as can be seen exemplary in Fig. 8.28

(c) Topological structure in form of metallographic images (see Fig. 8.26, on the left of the first row), images by *electron microscopy* (TEM, SEM, FEM), *ion microscopy* (FIM, FIM), *scanning tunnelling microscopy* (STM), and *atomic force microscopy* (AFM, AFP). By some of these methods, apart from topological structures, chemical species can also be recognized by use of special techniques, additional equipment, and advanced evaluation. For example, from metallographic (light-microscopic) images phases and, therefore, chemical compounds can

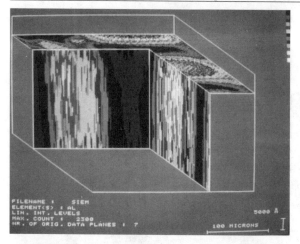

Fig. 8.27. Quantitative 3D analysis of Al on an integrated circuit ("pie section"); different concentrations are represented by various coded colours (with kind permission from RÜDENAUER and STEIGER [1986])

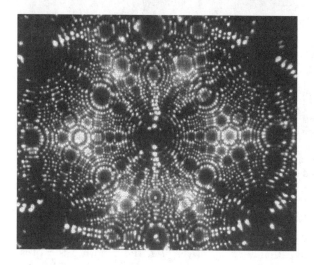

Fig. 8.28. Tungsten crystal tip of 32 nm radius, obtained with FIM by EW MÜLLER (Pennsylvania State University), according to WEISSMANTEL et al. [1968]

be identified by polarized radiation, etching procedures etc. On the other hand, by coupling a field ion microscope with a time-of-flight mass spectrometer with single ion sensitivity, the *field ion atom probe* (FIM-AP) is produced that can recognize atomic structures at solid state tips (as can be seen in Fig. 8.28) and also identify ions at certain lattice points by selective high voltage desorption.

The dimensions of analytical images can cover a range of many orders of magnitude. The one extreme is given by country areas, mountain ranges, and ore deposits imaged by satellites as well as ranges in the atmosphere studied by remote sensing. The other one is characterized by the micro-, nano- and atomic structure of materials, microelectronic devices (integrated circuits) and crystals (see Figs. 8.27 and 8.28). Between these extremes, common analytical images are situated, e.g., concerning species distribution on environmental compartments as well as materials having ordinary dimensions and being controlled by sampling and common analytical methods. Images play an important role not only in chemistry but also in medicine and biology, geology and agriculture.

Image techniques can be classified into three categories (GELADI et al. [1992a]: GELADI and GRAHN [1996]):

(i) *Direct imaging* using area detectors like diode arrays, CCD cameras, and video tubes

(ii) *Scanning* by moving a pointlike detector in l_x and l_y direction over an object (instead of the detector alternatively the radiation source or the object may be moved)

(iii) *Reconstructing* of latent images from data as in tomography and SIMS

The most important properties of analytical images can be summarized as follows:

– Images are two- or three-dimensional and, therefore, each image point (pixel or voxel) is characterized by two or three indices.

– Images are limited in space, not infinite.

– There must be some spatial autocorrelation in the image, that means that the local frequency have to be low or medium according to Fig. 2.6b,c (in contrast, in cases of zero- and high local frequencies, see Fig. 2.6a,d, there is no autocorrelation).

– Spatial distances between object points, each of them characterized by given indices, can be measured and calculated.

– Images have a certain resolution both in spatial distances and intensity.

Image processing needs digitization where two critical aspects have to be considered. On the one hand the *sampling theorem* (STANLEY [1975]; RABINER and GOLD [1975]), according to which at least twice of the limiting frequency (f_{lim}, highest frequency in the given image), should be used for digitization, and on the other hand the *signal-to-noise ratio* (see Sect. 7.1.1), to be precise the noise digitization, should be lower than the intrinsic noise (GELADI et al. [1992a]).

The relationship between data, image and related techniques is shown in Fig. 8.29.

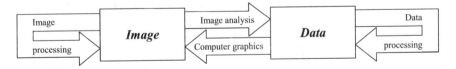

Fig. 8.29. The relationship between image and data, expressed by image processing, image analysis, data processing, and scientific visualization (computer graphics), according to GELADI et al. [1992a]

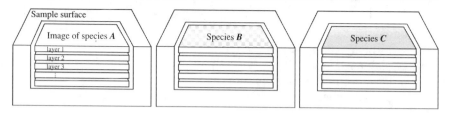

Fig. 8.30. Schematic dismantling of the $m + 3$ dimensions of multidimensional SIMS images; each original data layer contains five dimensions $A, B, C = f(l_x, l_y)$ in the case represented here

Image acquisition as well as *image analysis* and *image processing* concern data of all fields of analytical chemistry regardless of the spatial dimensions, in principle. Nevertheless, the most essential applications can be found in multidimensional micro- and nanoanalysis. This is because some of the so-called 3D techniques like SIMS (RÜDENAUER [1989]) and AUGER microprobe (FERGUSON [1989]) are characterized in fact by more than three dimensions (see Sect. 3.4). In the 3D mode SIMS, the secondary ions of various species are recorded, each of them in three spatial dimensions, as schematically represented in Fig. 8.30.

As a result of 3D SIMS analysis, a stack of multivariate images is stored which can be described in case of qualitative species information by

$$Q = f(l_x, l_y, l_z) \tag{8.23a}$$

and in case of quantitative information by

$$y_Q = f(l_x, l_y, l_z). \tag{8.23b}$$

Consequently, a multivariate image can be considered being a set of congruent images of the same object for various variables and, therefore, being an image the pixels (or voxels) of which are vectors instead of scalars (GELADI et al. [1992a,b]).

The concrete situation as illustrated in Fig. 8.30 is characterized by $A, B, C = f(l_x, l_y, l_z)$ and $y_A, y_B, y_C = f(l_x, l_y, l_z)$. From the multivariate stack of data there can be generated various one-, two- and three-dimensional (two-, three- and four-dimensional information) as, for example, shown in Fig. 8.31.

Multidimensional image information can be processed in the same way as signal functions in general. In many cases, the basis of image processing is the two-dimensional FOURIER analysis

$$F(u, v) = \iint f(l_x, l_y) \exp\left(-j2\pi(u \cdot l_x + v \cdot l_y)\right) \, dl_x dl_y \tag{8.24}$$

an example of which is shown in Fig. 8.32.

FOURIER- and other transformations (e.g., LAPLACE-, HADAMARD-, and Wavelet transformation) are the bases to transfer information complete and

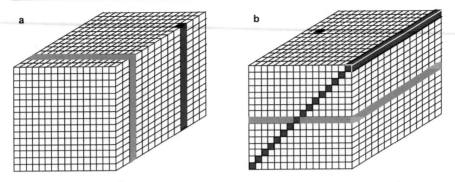

Fig. 8.31. Image processing in 3D SIMS: generation of a *local depth profile* and a *transaxial layer image* (**a**) and of a *coaxial layer image*, a *spatial diagonal image*, and a *point analytical information* (**b**); the representation is inspired from RÜDENAUER [1989]

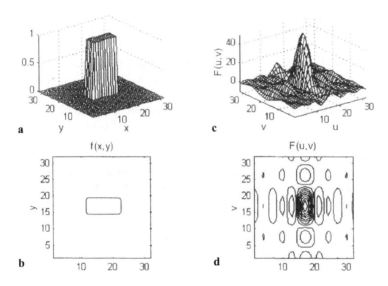

Fig. 8.32. Two-dimensional FOURIER transformation applied to a rectangle function shown in original 3D representation (**a**) and 2D contour plot (**b**) and as FOURIER transforms (**c,d**), (according to DANZER et al. [2001])

reversible from an original data space into an abstract projection for the reason of data processing such as *filtering, cross-* and *autocorrelation, convolution* and *deconvolution* etc.

These and other procedures of signal processing (*smoothing, amplification*) can be used analogously for common signal functions. They are used widely in digital photography to improve the quality of pictures, mainly by signal-to-noise enhancement, resolution improvement, and brightness- and contrast variation.

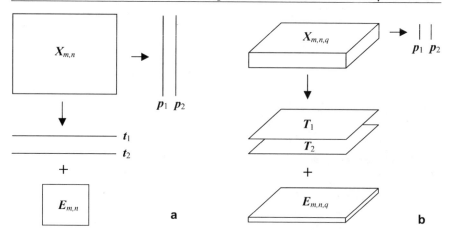

Fig. 8.33. Common PC decomposition of a data matrix $X_{m,n}$ (**a**) and PC decomposition of a data block $X_{m,n,q}$ of a multivariate image (**b**)

Multivariate analytical images may be processed additionally by chemometrical procedures, e.g., by *exploratory data analysis, regression, classification,* and *principal component analysis* (GELADI et al. [1992b]).

The interpretation of a multivariate image is sometimes problematic because the cause for pictorial structures may be complex and cannot be interpreted on the basis of images of single species even if they are processed by filtering etc. In such cases, principal component analysis (PCA) may advantageously be applied. The principle of the PCA is like that of factor analysis which has been mathematically described in Sect. 8.3.4. It is represented schematically in Fig. 8.33.

The data array $X_{m,n,q}$, which may be preprocessed by standardisation and transformation, is decomposed in a set of three-way products of score images, T_k, loading vectors, p_k, and the residual, $E_{m,n,q}$:

$$X = T\,P + E \tag{8.25a}$$

$$x = \sum_{i=1}^{k} T_i p_i + E \quad , \tag{8.25b}$$

cf. Eq. (8.20). The real dimensionality of the data is given by the rank k that separates typical structures from noise. The score matrices T_i have the same geometrical pixel arrangement as the raw images and are, therefore, called *score images*.

Image processing has been applied in wide fields of 2D- and 3D distribution analysis, e.g., in EPMA (WOLKENSTEIN et al. [1997b]), SIMS (RÜDENAUER [1989]; WOLKENSTEIN et al. [1997a]; WICKES et al. [2003]; HUTTER et al. [1996]), AFM and STM (FUCHS et al. [1995]), and XPS (ARTYUSHKOVA and FULGHUM [2002]).

References

Adler B, Schütze P, Will J (1993) *Expert system for interpretation of X-ray diffraction spectra.* Anal Chim Acta 271:287

Ahrens H, Läuter J (1974) *Mehrdimensionale Varianzanalyse.* Akademie-Verlag, Berlin

Akin H, Siemes H (1988) *Praktische Geostatistik.* Springer, Berlin, Heidelberg, New York

Allanic AL, Jezequel JY, Andre JC (1992) *Application of neural networks theory to identify two-dimensional fluorescence spectra.* Anal Chem 64:2618

Allison PD (2002) *Missing data.* Sage, Thousand Oaks, CA

Andrews DF (1972) *Plots of high dimensional data.* Biometrics 28:125

Anker LS, Jurs PC (1992) *Prediction of carbon-13 NMR chemical shifts by artificial neural networks.* Anal Chem 64:1157

Anscombe FJ (1973) *Graphs in statistical analysis.* Am Stat 27:17

Arnold T, Otto M, Wegscheider W (1994) *Interpretation system for automated wavelength dispersive X-ray fluorescence spectrometry.* Talanta 41:1169

Artyushkova K, Fulghum JE (2002) *Multivariate image analysis methods applied to XPS imaging data sets.* Surface Interface Anal 33:185

Ball JW, Jurs PC (1993) *Automated selection of regression models using neural networks for ^{13}C NMR spectral prediction.* Anal Chem 65:505

Barth A (1993), *SpecInfo: an integrated spectroscopic information system.* J Chem Inf Comput Sci 33:52

Benecke C, Grüner T, Kerber A, Laue R, Wieland T (1997) *MOLecular structure GENeration with MOLGEN, new features and future developments.* Fresenius J Anal Chem 359:23

Boger Z, Karpas Z (1994) *Application of neural networks for interpretation of ion mobility and X-ray fluorescence spectra.* Anal Chim Acta 292:243

Borszeki J, Kepes J, Koltay L, Sarudi I (1986a) *Classification of paprika quality using pattern recognition methods based on elemental composition.* Acta Alimentaria 15:93

Borszeki J, Koltay L, Inczedy J, Gegus E (1986b) *Untersuchung der Mineralstoffzusammensetzung von Weinen aus Transdanubien und ihre Klassifikation nach Weingegenden.* Z Lebensmittelunters Forsch 177:15

Bremser W, Grzonka M (1991) *SpecInfo - a multidimensional spectroscopic interpretation system.* Mikrochim Acta 1991/II:483

Brinton WC (1914) *Graphic methods for presenting facts.* The Engineering Magazine Company, New York (reprinted 1980, Arno Press, New York)

Buchanan BG, Feigenbaum EA (1978) *DENDRAL and Meta-DENDRAL: their applications dimension.* Artific Intell 11:5

Buja A, Cook D, Swayne DF (1996) *Interactive high-dimensional data visualization.* J Computat Graph Stat 5:78

Burrough PA (1986) *Principles of Geographical Information Systems for Land Resources Assessment.* Clarendon Press, Oxford

Burrough PA, McDonnell RA (1998) *Principles of Geographical Information Systems.* Oxford University, New York

Canzler D, Hellenbrandt M (1992) *SPECINFO - The spectroscopic information system on STN International.* Fresenius J Anal Chem 344:167

Cherition D, Tarjan RE (1976) *Finding minimum spanning trees.* SIAM J Computing 5:724

Chernoff H (1973) *The use of faces to represent points in k-dimensional space graphically.* J Am Statist Assoc 68:361

Coomans D, Massart DL, Broeckaert I (1981) *Potential methods in pattern recognition. A combination of ALLOC and statistical linear discriminant analysis* . Anal Chim Acta 133:215

Cover TM, Hart PE (1967) *Nearest neighbor pattern classification.* IEEE Transact IT-13:21

Currie LA (1992) *In pursuit of accuracy: nomenclature, assumtions, and standards.* Pure Appl Chem 64:455

Currie LA (1995) IUPAC, Analytical Chemistry Division, Commission on Analytical Nomenclature: *Nomenclature in evaluation of analytical methods including detection and quantification capabilities.* Pure Appl Chem 67:1699

Currie LA (1997) *Detection: international update, and some emerging di-lemmas involving calibration, the blank, and multiple detection decisions.* Chemometrics Intell Lab Syst 37:151

Danzer K (1989) *Robuste Statistik in der analytischen Chemie.* Fresenius Z Anal Chem 335:869

Danzer K, De la Calle D, Thiel G, Reichenbächer M (1999) *Classification of wine samples according to origin and grape varieties on the basis of inorganic and organic trace analysis.* Am Lab 31:26

Danzer K, Florian K, Singer R, Mäurer, F, Abo-Bakr El Nady M, Zimmer K (1987) *Investigation of the origin of archaeological glass artefacts by means of pattern recognition.* Anal Chim Acta 201:289

Danzer K, Hobert H, Fischbacher C, Jagemann K-U (2001) *Chemometrik. Grundlagen und Anwendungen.* Springer, Berlin Heidelberg New York

Danzer K, Singer R, Mäurer F, Florian K, Zimmer K (1984) *Mehrdimensionale Varianz- und Diskriminanzanalyse spektrographischer Daten von Glasperlenfunden.* Fresenius Z Anal Chem 318:517

Danzer K, Wank U, Wienke D (1991) *An expert system for the evaluation and interpretation of interlaboratory comparisons.* Chemom Intell Lab Syst 12:69

De la Calle D, Reichenbächer M, Danzer K, Hurlbeck C, Bartzsch C, Feller K-H (1998) *Analysis of wine bouquet components using headspace solid-phase microextraction-capillary gas chromatography.* J High Resol Chromatogr 21:373

Derde MP, Buydens L, Guns C, Massart DL, Hopke PK (1987) *The use of rule building systems for the classification of analytical data* Anal Chem 59:1868

Derde MP, Massart DL (1986) *UNEQ: a disjoint modelling technique for pattern recognition besed on normal distribution.* Anal Chim Acta 184:33

Derde MP, Massart DL (1988) *Comparison of the performance of the class modelling techniques UNEQ, SIMCA and PRIMA*. Chemom Intell Lab Syst 4:65

Dillon WR, Goldstein M (1984) *Multivariate analysis. Methods and Applications*. Wiley, New York

du Toit SHC, Steyn AWG, Stumpf RH (1986) *Graphical exploratory data analysis*. Springer, Berlin Heidelberg New York

Duewer DL, Kowalski BR (1975) *Forensic data analysis by pattern recognition. Categorization of white bond papers by elemental composition*. Anal Chem 47:526

Ehrlich G, Danzer K (2006) *Nachweisvermögen von Analysenverfahren. Objektive Bewertung und Ergebnisinterpretation*. Springer, Berlin Heidelberg New York

Ehrlich G, Danzer K, Liebich V (1979) *Allgemeine Grundlagen der Verteilungsanalytik*. Tagungsber Techn Hochsch Karl-Marx-Stadt: 2. Tagung Festkörperanalytik 1978. Plenar- und Hauptvorträge 1:69

Einax J, Danzer K (1989) *Die Anwendung der Faktorenanalyse zur Charakterisierung der Quellen staubförmiger Immissionen*. Staub Reinh Luft 49:53

Einax JW, Zwanziger HW, Geiß S (1997) *Chemometrics in Environmental Analysis*. VCH, Weinheim

Elyashberg ME, Martirosian ER, Karasev YuZ, Thiele H, Somberg H (1997) *X-Pert: a user-friendly expert system for molecular structure elucidation by spectral methods*. Anal Chim Acta 337:265

Ferguson IF (1989) *Auger microprobe analysis*. Hilger, Bristol

Fischbacher C, Jagemann K-U, Danzer K, Müller UA, Mertes B (1995) *Non-invasive blood glucose monitoring by chemometrical evaluation of NIR-spectra*. 24th EAS (Eastern Analytical Symposium), Somerset, NJ

Forina M, Armanino C (1982) *Eigenvector projection and simplified non-linear mapping of fatty acid content of Italian olive oils*. Ann Chim [Rome] 72:127

Frank IE (1988) *DASCO: a new classification method*. Chemom Intell Lab Syst 4:215

Frank IE (1989) *Classification models: discriminant analysis, SIMCA, CART*. Chemom Intell Lab Syst 5:247

Frank IE, Friedman JH (1989) *Classification: oldtimers and newcomers*. J Chemom 3:463

Frank IE, Todeschini R (1994) *The data analysis handbook*. Elsevier, Amsterdam

Fresenius CR (1841) *Anleitung zur qualitativen chemischen Analyse*. Vieweg, Braunschweig

Fuchs GM, Prohaska T, Friedbacher G, Hutter H, Grasserbauer M (1995) *Maximum entropy deconvolution of AFM and STM images*. Fresenius J Anal Chem 351:143

Geisler G (1999) *Bestimmung von Spuren- und Ultraspurenelementen in Wein*. Diploma Thesis, Friedrich Schiller Univerity of Jena

Geladi P, Bengtsson E, Esbensen K, Grahn H (1992a) *Image analysis in chemistry. I. Properties of images, greylevel operations, the multivariate image*. Trends Anal Chem 11:41

Geladi P, Grahn H (1996) *Multivariate image analysis*. Wiley, Chichester

Geladi P, Grahn H, Esbensen K, Bengtsson E (1992b) *Image analysis in chemistry. II. Multivariate image analysis.* Trends Anal Chem 11:121

Hampel FR, Roncetti EM, Rousseeuw PJ, Stahel WA (1986) *Robust statistics: the approach based on influence functions.* Wiley, New York

Helsel DR (1990) *Less than obvious: statistical treatment of data below the detection limit.* Environm Sci Technol 24:1756

Hibbert DB (1993) *Genetic algorithms in chemistry.* Chemom Intell Lab Syst 19:277

Horwitz W (1994) *Nomenclature of interlaboratory analytical studies.* Pure Appl Chem 66:1903

Howell DC (2002) *Treatment of missing data.* www.uvm.edu/~dhowell/StatPages/More_Stuff/Missing_Data/Missing.html

Huber PJ (1981) *Robust statistics.* Wiley, New York

Huberty CJ (1994) *Applied discriminant analysis.* Wiley, New York

Hutter H, Brunner C, Nikolov SG, Mittermayr C, Grasserbauer M (1996) *Image surface spectroscopy for two and three dimensional characterization of materials.* Fresenius J Anal Chem 355:585

ISO Guide 33 (1989) *Uses of certified reference materials.* Geneva

ISO/REMCO N 280 (1993) *Proficiency testing of chemical analytical laboratories.* Geneva

IUPAC (1993) Analytical Chemistry Division, Commission on General Aspects of Analytical Chemistry: *The International Harmonized Protocol for the Proficience Testing of (Chemical) Analytical Laboratories.* Pure Appl Chem 65:2123

IUPAC (1994) Analytical Chemistry Division, Commission on General Aspects of Analytical Chemistry: *Nomenclature of interlaboratory analytical studies*, prepared for publication by W Horwitz. Pure Appl Chem 66:1903

IUPAC Orange Book (1997, 2000)
– printed version: *Compendium of analytical nomenclature (definitive rules 1997);* see Inczédy et al. (1997)
– web version (as from 2000): www.iupac.org/publications/analytical_compendium/

Janssens K, van Espen P (1986) *Interpretation system for automated wavelength dispersive X-ray fluorescence spectrometry.* Anal Chim Acta 184:117

Juricskay I, Veress GE (1985) *PRIMA: a new pattern recognition method.* Anal Chim Acta 171:61

Kaiser H (1965) *Zum Problem der Nachweisgrenze.* Fresenius Z Anal Chem 209:1

Kaiser H (1966) *Zur Definition der Nachweisgrenze, der Garantiegrenze und der dabei benutzten Begriffe.* Fresenius Z Anal Chem 216:80

Kaiser H, Specker H (1956) *Bewertung und Vergleich von Analysenverfahren.* Fresenius Z Anal Chem 149:46

Kleiner B, Hartigan JA (1981) *Representing points in many dimensions by trees and castles.* J Am Stat Assoc 76:260

Kohonen T (1982) *Self-organized formation of topologically correct feature maps.* Biol Cybern 43:59

Kohonen T (1984) *Self-organization and associative memory*. Springer, Berlin Heidelberg New York

Kruskal JB (1956) *On the shortest spanning subtree and the traveling salesman problem.* Proc Am Math Soc 7:48

Lapitajs G, Klinkner R: *Marktüberblick LIMS*. http://www.analytik.de/Dokumente/ Artikel/LIMS.html

Lawn RE, Thompson M, Walker RF (1993) *Proficiency testing in analytical chemistry.* Royal Society of Chemistry, London

Li T-H, Lucasius CB, Kateman G (1992) *Optimization of calibration data with the dynamic genetic algorithm.* Anal Chim Acta 268:123

Li-Xian Sun, Danzer K, Thiel G (1997) *Classification of wine samples by means of artificial neural networks and discrimination analytical methods.* Fresenius J Anal Chem 359:143

Limsource: http://www.limsource.com

Little RJA, Rubin DB (1987) *Statistical analysis with mission data.* Wiley, New York

Long GL, Winefordner JD (1983) *Limit of detection. A closer look at the IUPAC definition.* Anal Chem 55:712A

Lucasius CB, Kateman G (1993) *Understanding and using genetic algorithms. Part 1. Concepts, properties and context.* Chemom Intell Lab Syst 19:1

Massart DL, Vandeginste BGM, Deming SN, Michotte Y, Kaufman L (1988) *Chemometrics – A textbook.* Elsevier, Amsterdam

Nagel M, Brenner A, Ostermann R, Henschke K (1996) *Graphische Datenanalyse.* Fischer, Stuttgart

Neitzel V (1996) *To the problem of processing values below the limit of determination.* Vom Wasser 87:223

Neudert R, Penk M (1996) *Enhanced structure elucidation.* J Chem Inf Comput Sci 36:244

Nilsen CL (1996) *Managing the analytical laboratory plain and simple.* CRC, Interpharm Press, Buffalo Grove, IL

Otto M (1998) *Chemometrics. Statistics and computer application in analytical chemistry.* VCH, Weinheim

Prichard E, Green J, Houlgate P, Miller J, Newman E, Phillips G, Rowley A (2001) *Analytical measurement terminology – handbook of terms used in quality assurance of analytical measurement.* LGC, Teddington, Royal Society of Chemistry, Cambridge

Prim RC (1957) *Shortest connection networks and some generalisations.* Bell Syst Techn J 36:1389

Rabiner LR, Gold B (1975) *Theory and application of digital signal processing.* Prentice Hall, New Jersey

Robb EW, Munk ME (1990) *A neural network approach to infrared spectrum interpretation.* Mikrochim Acta [Wien] 1990/I:131

Rousseeuw PJ, Leroy AM (1987) *Robust regression and outlier detection.* Wiley, New York

Rüdenauer FG (1989) *Multidimensional image acquisition and processing in surface analysis.* Fresenius Z Anal Chem 333:308

Rüdenauer FG, Steiger W (1986) *Messung der räumlich 3-dimensionalen Elementvertei-lung in Festkörpern mit Hilfe von SIMS.* Wiss Beitr FSU Jena: 3. Tagung COMPANA '85, Computereinsatz in der Analytik, p 115

Saxberg BEH, Duewer DL, Booker JL, Kowalski BR (1978) *Pattern recognition and blind assay techniques applied to forensic separation of whiskey.* Anal Chim Acta 103:201

Schuur J (1997) *Spektren-Interpretation und -Verwaltung mit SpecInfo.* Nachr Chem Tech Lab 45:401

Sharaf MA, Ilman DL, Kowalski BR (1986) *Chemometrics.* Wiley, New York

Stanley WD (1975) *Digital signal processing.* Reston Publishing, Reston, VA

Stonham TJ, Aleksander I, Camp M, Pike WT, Shaw MA (1975) *Classification of mass spectra using adaptive digital learning networks.* Anal Chem 47:1817

Thiel G, Geisler G, Blechschmidt I, Danzer K (2004) *Determination of trace elements in wines and classification according to their provenance.* Anal Bioanal Chem 378:1630

Thompson M, Wood R (1993) *International harmonized protocol for proficiency testing of (chemical) analytical laboratories.* J AOAC Intern 76:926

Tukey JW (1962) *The future of data analysis.* Ann Math Stat 33:1; 81

Tukey JW (1977) *Exploratory data analysis.* Addison-Wesley, Reading, MA

Tusar M, Rius XF, Zupan J (1991) *Neural networks in chemistry.* Mitt-bl GDCh, Fachgr CIC 19:72

van Leeuwen JA, Buydens LMC, Vandeginste BGM, Kateman G, Schoenmakers PJ, Mul-holland M (1991) *RES, an expert system for the set-up and interpretation of a ruggedness test in HPLC method validation.* Chemom Intell Lab Syst 10:337; 11:37, 161

Wehrens R (1994) *Hybridisation of expert systems in chemometrics.* In: LMC Buydens, WJ Melssen (eds) *Chemometrics. Exploring and exploiting chemical information.* Katholieke Universiteit Nijmegen, p 113

Wehrens R, Lucasius C, Buydens LMC, Kateman G (1993a) *HIPS, a hybrid self-adapting expert system for NMR spectrum interpretation using genetic algorithms.* Anal Chim Acta 277:313

Wehrens R, van Hoof P, Buydens LMC, Kateman G, Vossen M, Mulder WH, Bakker T (1993b) *Sampling of aquatic sediments. The design of a decision-support system and a case study.* Anal Chim Acta 271:11

Weissmantel C, Lenk R, Forker W, Ludloff R, Hoppe J (eds) (1968) *Kleine Enzyklopädie Atom – Struktur der Materie.* Bibliographisches Institut, Leipzig (Tafel 36)

Wickes BT, Kim Y, Castner DG (2003) *Denoising and multivariate analysis of time-of-flight SIMS images.* Surface Interface Anal 35:640

Wienke D, Wank U, Wagner M, Danzer K (1991) *MULTIVAR, PLANEX, INTERLAB – from a collection of algorithms to an expert system: Statistical software written from chemists for chemists.* In: Gmeling J (ed) *Software development in chemistry 5*, Springer, Berlin Heidelberg New York, p 113

Wold S (1976) *Pattern recognition by means of disjoint principal component models.* Pattern Recogn 8:127

Wolkenstein M, Hutter H, Nikolov SG, Grasserbauer M (1997a) *Improvement of SIMS image classification by means of de-noising.* Fresenius J Anal Chem 357:783

Wolkenstein M, Hutter H, Nikolov SG, Schnitz I, Grasserbauer M (1997b) *Comparison of wavelet filtering with well-known techniques for EPMA image de-noising.* J Trace Microprobe Techn 15:33

Wünsch G, Gansen M (1989) *Anwendbarkeit und Anwendung von Expertensystemen in der Analytik.* Fresenius Z Anal Chem 333:607

Wythoff BJ, Levine SP, Tomellini A (1990) *Spectral peak verification and recognition using a multilayered neural network.* Anal Chem 62:2702

Zupan J, Gasteiger J (1993) *Neural networks for chemists. An introduction.* VCH, Weinheim

9 Assessment of Analytical Information

From the end of the 1960s up to the present, analytical chemistry has frequently been defined as the discipline that gains *information* on chemical composition and structure (DANZER et al. [1976, 1977, 1987]; FRESENIUS et al. [1992]); see Sect. 1.1. Consequently, analytical information has been characterized by information theory (DOERFFEL and HILDEBRANDT [1969]; KAISER [1970]; ECKSCHLAGER [1971–1989]; DANZER [1973a,b,c, 1974, 1975a, 1975b, 1978]; MALISSA et al. [1974]; DANZER and ECKSCHLAGER [1978]; ECKSCHLAGER and DANZER [1994]; DANZER and MARX [1979a,b]; DANZER et al. [1987, 1989, 1991]). Before, information theory was been applied to physical measurements (BRILLOUIN [1963]) and in physical chemistry (RACKOW [1963–1969]).

The principle of chemical measurements, as realized in the analytical process (Fig. 2.1), corresponds in its significant steps to the general principle of information processing (Fig. 3.1).

According to classical information theory, founded by SHANNON [1948] (see also SHANNON and WEAVER [1949]), information is eliminated uncertainty about an occurrence or an object, obtained by a message or an experiment. Information is always bound up with signals. They are the carriers of information in the form of definite states or processes of material systems (ECKSCHLAGER and DANZER [1994]; DANZER [2004]); see Sect. 3.1.

9.1
Quantification of Information

Information, I, is the difference between *information entropies*, viz the a priori *information entropy*, H, that characterizes the uncertainty before a message is obtained or an experiment is carried out, and the a posteriori *information entropy*, H', that remains afterwards:

$$I = H - H'$$
(9.1)

In the case of m discrete phenomena x_i (e.g., outputs of an experiment), the information entropy is calculated by means of the respective probabilities $P(x_i)$

$$H\big(P(x)\big) = -\sum_{i=1}^{m} P(x_i) \, \text{lb} \, P(x_i) = \sum_{i=1}^{m} P(x_i) \, \text{lb} \, \frac{1}{P(x_i)}$$
(9.2)

where $\sum P(x_i) = 1$. The binary logarithm as a measure of information has been introduced by HARTLEY [1928]. It holds $\mathrm{lb}\,a = \log_2 a = 3.322 \log_{10} a$, the unit is then $[H] = [I] = \mathrm{bit}$.

On the other hand, the information entropy characterizing the uncertainty of a continuous random quantity with a probability density $p(x)$ is given by

$$H(p) = H\big(p(x)\big) = - \int\limits_{-\infty}^{+\infty} p(x)\,\mathrm{lb}p(x)\,\mathrm{d}x \qquad (9.3)$$

with $\int_{-\infty}^{+\infty} p(x)\mathrm{d}x = 1$.

In the case that the information of chemical results x_i is imparted by signals z_j, then it holds for the entropy (ECKSCHLAGER and DANZER [1994])

$$H\big(P(x_i|\,z_j)\big) = - \sum_{i=1}^{m} P(x_i|\,z_j)\,\mathrm{lb}\,P(x_i|\,z_j) \qquad (9.4)$$

When an unambiguous relationship exists between signal appearance and analytical result then Eq. (9.4) turns into Eq. (9.2).

It is meaningful to differentiate (MEYER-EPPLER [1969]; DANZER et al. [1987]; ECKSCHLAGER and DANZER [1994]) between:

(i) *The specific (partial) information content, I_i, of one definite phenomenon or experimental result*, e.g., the appearance of a given signal and, therefore, the presence of a definite constituent in a sample

$$I_i = H_i + H_i' \qquad (9.5)$$

with the *specific information entropies* $H_i = -\mathrm{lb}P(x_i)$

(ii) *The average information content I_{av} of an experimental result out of a set of m possible*

$$I_{av} = H_{av} + H_{av}' \qquad (9.6)$$

where H_{av} is calculated according to Eq. (9.2) and corresponds to the weighed average of all the specific information entropies H_i characterizing their mathematical expectation

(iii) *The maximum (average) information content I_{max}*

$$I_{max} = H_{max} = \mathrm{lb}m \qquad (9.7)$$

which is relevant if all the m results expected a priori have the same probability $P(x_i) = 1/m$ (in this case, the a posteriori entropy $H_{max}' = \mathrm{lb}1 = 0$).

Table 9.1. Specific (I_+ and I_-) and average (I_{av}) information contents (in bit) and specific plausibilities (Π_+ and Π_-) in cases of various a priori probabilities of qualitative tests

Case	Example	$P(x_+)$	I_+	I_-	I_{av}	Π_+	Π_-
1	General unknown case	0.5	1.00	1.00	1.00	1.00	1.00
2	Mn in steel	0.9	0.15	3.32	0.47	6.67	0.30
3	Silicon in secondary dust	0.99	0.014	6.64	0.078	71.43	0.15

The matter may be illustrated by the example of *qualitative analysis*. As a result of a specific test, it is stated that a constituent searched for is present in the test sample (x_+) or not (x_-) depending on whether a specific signal is detected (z_+) or not (z_-) as represented in Fig. 9.1. In Table 9.1 the different types of information contents are compiled, each for various a priori probabilities.

If the probabilities are equal (case 1), I_+, I_- and I_{av} are the same and I_{av} corresponds to the maximum information content, $I_{av} = I_{max}$. In cases 2 and 3, where the a priori probabilities are different, the expected (more probable) result yields the respective lower information contents whereas the unexpected (less probable) result manifests in high values of the specific information.

In this context, the term *plausibility* which is sometimes used in clinical analytical chemistry, should be considered. Whereas information is formed by the *unexpected*, plausibility is generated even by the *expected*. From this point of view, plausibility has been described inversely to information, viz $\Pi_i = I_i^{-1}$ (DANZER [1983]), but because of the unit, plausibility should be defined by

$$\Pi_i = \frac{lb2}{I_i} \tag{9.8}$$

Qualitative analysis may be carried out by means of visual signals (e.g., spot test reactions) or stationary signals in a given position z_j of a signal function $y = f(z)$, see Sects. 3.2, 3.3 and Fig. 9.1.

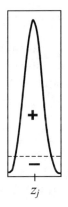

z_j

Fig. 9.1. Signal states in the case of qualitative tests ("+" means signal detected (z_+) and "–" means not detected (z_-))

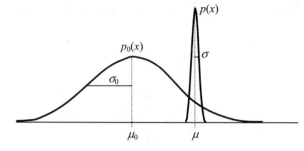

Fig. 9.2. Expectation range (a priori distribution) $p_0(x)$ and distribution of the measured values $p(x)$ in case of normal distributions $N(\mu_0, \sigma_0^2)$, $N(\mu, \sigma^2)$

In the case where a species has to be identified from an instrumental-analytical record like Fig. 3.8 with the help of a definite signal position, then the information content is determined according to KULLBACK's *divergence measure of information* (KULLBACK [1959]; ECKSCHLAGER and STEPANEK [1985]; ECKSCHLAGER and DANZER [1994])

$$I(p, p_0) = H(p, p_0) - H'(p) = \int\limits_{-\infty}^{+\infty} p(x) \, \text{lb} \, \frac{p(x)}{p_0(x)} \, dx \qquad (9.9)$$

In this way, deviations can be characterized between an experimentally found distribution of measured values, $p(x)$, and an a priori distribution $p_0(x)$, e.g., corresponding to an expected normal range of values. There are situations, especially with some spectroscopic methods, in which relations of the signal position, experimentally recorded on the one hand and theoretically expected on the other hand, may contain essential chemical information on the species (chemical shifts).

In case of normally distributed a priori- and a posteriori distributions, $p(x) = N(\mu, \sigma^2)$ and $p_0(x) = N(\mu_0, \sigma_0^2)$, Eq. (9.9) becomes

$$I(p, p_0) = \text{lb} \, \frac{\sigma_0}{\sigma} + k \, \frac{(\mu - \mu_0)^2 + \sigma^2 - \sigma_0^2}{\sigma_0^2} \qquad (9.10)$$

where $k = 1/(2 \ln 2) = 0.72135$. The situation is illustrated by Fig. 9.2.

An example concerning the C=O stretching vibration in IR spectroscopy is given in DANZER et al. [2001], p 51.

In contrast to quantitative analyses, the results of qualitative tests and of identifications cannot be evaluated by means of mathematical statistics. Instead, information theory is a helpful tool to characterize qualitative analyses, in particular in case of multicomponent systems.

9.2
Information Content of Quantitative Analysis

In contrast to qualitative tests, in quantitative analysis more than the two signal states shown in Fig. 9.1 $(+/-)$ are evaluated. Depending on the un-

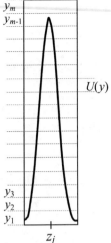

Fig. 9.3. Signal levels in case of quantitative analysis

certainty of the signal intensity, a certain number m of signal levels can be distinguished; see Fig. 9.3.

The information content of quantitative analytical measurements is determined by three items:

(1) The *expectation range* of the results (a priori distribution) $p_0(x)$

(2) The *uncertainty* $U(x)$ of the measured values, characterized by the a posteriori distribution $p(x)$, where $U(x) = U(y)/S$; see Sect. 4.2 and Eq. (4.31)

(3) The *trueness* of the analytical results or their *inaccuracy*, respectively, characterized by the bias $\delta = |x - x_{true}|$; see Sect. 7.1.3.

Classical information theory according to SHANNON [1948] and BRILLOUIN [1963] consider only items (1) and (2); the trueness of information has not been taken into account.

In the case where no bias is relevant ($x = x_{true}$), Eq. (9.9) becomes

$$I(p, p_0) = \text{lb} \, \frac{p(x)}{p_0(x)} \tag{9.11}$$

The a priori distribution, which reflects our assumptions or preliminary information about the analyte in the sample, is frequently considered to be uniform, $U(x_{min}, x_{max})$; their probability density is given by

$$p_0(x) = \begin{cases} \dfrac{1}{x_{max} - x_{min}} & \text{for } x_{min} \leq x \leq x_{max} \\ 0 & \text{otherwise} \end{cases} \tag{9.12}$$

On the other hand, the a posteriori distribution, $p(x)$, mostly can be considered to be a normal distribution, $N(\mu, \sigma^2)$, or, in analytical reality, a t-distribution, $N_t(\overline{x}, s^2)$, with

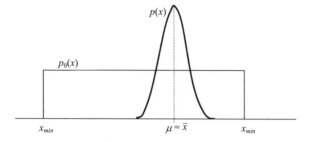

Fig. 9.4. Uniform expectation range between x_{min} and x_{max} and normal-distributed (t-distributed) measured values

$$p(x) = \frac{1}{\sigma\sqrt{2\pi e}} \tag{9.13a}$$

or

$$p(x) = \frac{1}{2\Delta\overline{x}} = \frac{\sqrt{n}}{2s\,t_{1-\alpha,\nu}} \tag{9.13b}$$

respectively. The right-hand side of Eq. (9.13b) represents the reciprocal double-sided confidence interval that stands frequently for the uncertainty of measurement; see Sects. 4.1.2 and 4.2.

With a uniform expectation range $\langle x_{min}, x_{max} \rangle$ and the a posteriori uncertainty according to Eq. (9.13b) as shown in Fig. 9.4, the information content at Eq. (9.11) turns into

$$I(p, p_0) = \mathrm{lb}\,\frac{p(x)}{p_0(x)} = \mathrm{lb}\,\frac{x_{max} - x_{min}}{2\Delta\overline{x}}$$
$$= \mathrm{lb}\,\frac{(x_{max} - x_{min})\sqrt{n}}{2s\,t_{1-\alpha,\nu}} = \mathrm{lb}\,m \tag{9.14}$$

In Eq. (9.14), $m = (x_{max} - x_{min})/(2\Delta\overline{x})$ represents the number of signal levels that can be distinguished significantly within the expectation range (see Fig. 9.3). The same holds for the number of the levels of the chemical quantity, e.g., concentration levels. This is valid on the pre-condition that all the m signal levels are equally probable, viz m is constant over the entire measuring range and, therefore, the standard deviation $s=$const, too.

In some cases, s and with it Δx are changing in dependence on the measurand x. Then the number of significantly distinguishable signal levels (and with it concentration levels) m is given by

$$m = \int_{x_{min}}^{x_{max}} f(x)\mathrm{d}x \tag{9.15}$$

where $f(x) = \Delta x^{-1}$ according to DANZER [1975a]. If $s =$ const, it follows Eq. (9.14). On the other hand, for the frequent case of an approximately constant relative standard deviation $s_{rel} = s/x =$ const, m becomes

$$m = \frac{\sqrt{n}}{2 s_{rel} \, t_{1-\alpha,\nu}} \int\limits_{x_{min}}^{x_{max}} \frac{dx}{x} = \frac{\sqrt{n}}{2 s_{rel} \, t_{1-\alpha,\nu}} \ln \frac{x_{max}}{x_{min}} \qquad (9.16)$$

and the information content in case of constant relative standard deviation

$$I(p, p_0) = \mathrm{lb} \left(\frac{\sqrt{n}}{2 s_{rel} \, t_{1-\alpha,\nu}} \ln \frac{x_{max}}{x_{min}} \right) . \qquad (9.17)$$

If $m = f(x)$ is not explicitly known, Eqs. (9.14) and (9.17) can be used as approximations depending on the size of the expectation range and pre-information on the relative constancy of s or s_{rel}.

In an example it will be shown that the difference between both variants may be small. The determination of an analyte A will be considered with the following data: $x_{min} = 1$ wt%, $x_{max} = 10\%$, $\overline{x}_A = 6.00\%$, $s = 0.05\%$, $s_{rel} = 0.0083$, $n = 10$. It is obtained by:

Eq. (9.14) according to $s = $ const:

$I(p, p_0) = 3.322 \lg[(10 - 1)\sqrt{10}/(2 \cdot 0.05 \cdot 2.26)] = 6.98$ bit

Eq. (9.17) according to $s_{rel} = $ const:

$I(p, p_0) = 3.322 \lg[\sqrt{10}/(2 \cdot 0.0083 \cdot 2.26) \cdot \ln 10] = 7.60$ bit

It is customary in analytical chemistry to examine unknown samples by a screening procedure using a multicomponent method like OES or MS in case of inorganic constituents on the one hand and GC or HPLC in case of organic constituents on the other. In this way, an overview can be obtained on the type of constituents and their approximate contents. In many cases it is necessary to get a deeper insight into the sample composition. For this reason, the one or a few constituents have to determine more precisely.

In such cases, the expectation range corresponds to the uncertainty range of the screening method (to the respective distribution of the values measured by screening) which may be, e.g., a normal distribution, $N(\mu_{scr}, \sigma_{scr}^2)$, or $N_t(\overline{x}_{scr}, s_{scr}^2)$, respectively. The a posteriory distribution is given by the uncertainty of the more precise determination, $N(\mu_{prec}, \sigma_{prec}^2)$, or $N_t(\overline{x}_{prec}, s_{prec}^2)$, respectively. In this case, KULLBACK's measure (Eq. 9.10) have to be applied that reads concretely:

$$I(p, p_0) = \mathrm{lb} \frac{\sigma_{scr}}{\sigma_{prec}} + k \frac{(\mu_{prec} - \mu_{scr})^2 + \sigma_{prec}^2 - \sigma_{scr}^2}{\sigma_{scr}^2} \qquad (9.10')$$

Examples for this proceeding are given in DANZER et al. [2001], p 55.

The situation becomes more complex when aspects of the trueness of analytical results are included in the assessment. Trueness of information cannot be considered neither by the classical SHANNON model nor by KULLBACK's divergence measure if information. Instead, a model that takes account of three distributions, viz the uniform expectation range, $p_0(x)$, the distribution of the measured values, $p(x)$, and that of the true value, $r(x)$, as shown in Fig. 9.5, must be applied.

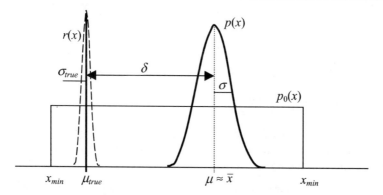

Fig. 9.5. Relationship of a uniform a priori expectation range (between x_{min} and x_{max}), normal-distributed (t-distributed) measured values, and the (conventional) true value x_{true}

The occurrence of a bias δ in the measured results can be dealt with by the KERRIDGE–BONGARD measure of inaccuracy (KERRIDGE [1961]; BONGARD [1970]; ECKSCHLAGER and STEPANEK [1985]; ECKSCHLAGER and DANZER [1994]). With the condition that μ is situated within the range $x_{min} + 3\sigma < \mu < x_{max} - 3\sigma$, the KERRIDGE–BONGARD information is

$$I(r; p, p_0) = \mathrm{lb}\, \frac{x_{max} - x_{min}}{\sigma \sqrt{2\pi e}} - k \left(\frac{\sigma_{true}^2}{\sigma^2} + \frac{(\mu_{true} - \mu)^2}{\sigma^2} - 1 \right) \qquad (9.18)$$

If the true value can be considered to be error-free ($\sigma_{true} \to 0$), $r(x)$ degenerates into a DIRAC impulse $N(\mu_{true}, 0)$. Considering real samples and the bias $\delta = \mu_{true} - x$, the estimate of Eq. (9.18) is given by

$$\hat{I}(r; p, p_0) = \mathrm{lb}\, \frac{(x_{max} - x_{min})\sqrt{n}}{2s\, t_{1-\alpha,\nu}} - k\, \frac{\delta^2 - \sigma^2}{\sigma^2} \qquad (9.19)$$

In the case of a large bias, the second term of Eq. (9.19) can exceed the first one and so the information content formally would become negative. Although this could be interpreted as misinformation, negative information contents are unusual in information theory itself. For this reason, ECKSCHLAGER and STEPANEK [1985] introduced

$$I = \begin{cases} = \hat{I}(r; p, p_0) & \text{if } \hat{I}(r; p, p_0) \geq 0 \\ = 0 & \text{if } \hat{I}(r; p, p_0) < 0 \end{cases} \qquad (9.20)$$

The three-dimensional relationship between I, δ, and σ is represented in Fig. 9.6.

Examples of the evaluation of biased analytical results have been given in DANZER et al. [2001], p 56.

The consideration of both precision and trueness by means of the Kerridge-Bongard model can be generalized as follows:

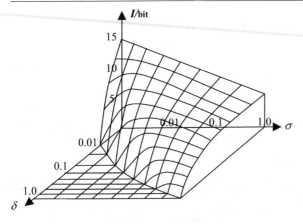

Fig. 9.6. Dependence of the information content on the bias, δ, and the standard deviation of the measured values, σ, as described by the KERRIDGE-BOGARD model (Eq. 9.19) according to DANZER et al. [2001]

(1) Positive values of information content always means a decrease of the a priori uncertainty about the measurand

(2) The information content rises with the increasing width of the expectation range and with the precision of the measured values

(3) A measurement procedure is able to yield information only if the produced results are sufficiently accurate and precise, viz $\delta \to 0$ and $\Delta x < (x_{max} - x_{min})$

(4) Negative information content occurs if a significantly inaccurate measuring result appears. This has more serious consequences the more precise the measured result is (see Fig. 9.6)

(5) The Kerridge-Bongard measure of inaccuracy expresses clearly the consequences of wrong analytical results: inaccurate results do not eliminate uncertainty but misinform the chemist and manifest the initial a priori uncertainty.

The Kerridge-Bongard model of information is of great importance in *quality assurance*, in particular for the assessment of *interlaboratory studies*. Examples of the information-theoretical evaluation of analytical results within the context of interlaboratory comparisons have been given by DANZER et al. [1987, 2001], WIENKE et al. [1991] and DANZER [1993].

9.3
Multicomponent Analysis

Multispecies analyses require two-dimensional analytical information $y = f(x)$, see Sect. 3.4, mostly in the form of spectra and chromatograms. By evaluation of various signals or the entire signal function, simultaneous information on several sample components can be obtained (in the extreme case on all the constituents contained in the sample). The relevant quantity that characterizes multicomponent analyses is the *information amount*,

$M(n)$ being the sum of the information contents of all the n components under consideration:

$$M(n) = \sum_{j=1}^{n} I_j \tag{9.21}$$

For the case that all the n constituents have similar expectation ranges, equally probable signal levels and are estimated with comparable precision, the maximum information amount becomes (DANZER et al. [1987])

$$M(n)_{max} = n \cdot I_{max} = n \cdot \mathrm{lb}m \tag{9.22}$$

In the concrete case of *qualitative tests* on n components, the maximum information amount is $M(n)_{qual} = n$ bit.

As a rule, multicomponent analysis is carried out by instrumental-analytical methods that produce two-dimensional analytical information in form of signal functions. The information amount according to Eqs. (9.21) and (9.22) is based on the condition that the assignment between signals z_j and events x_i is unambiguous and, therefore, $P(x_i|z_j)$ turns into $P(x_i)$; see Eqs. (9.4) and (9.2). Moreover, the signals selected for evaluation must be independent from each other; concerning correlated signals see DUPUIS and DIJKSTRA [1975] and DANZER et al. [1989].

Another matter is the *identification* of a substance, viz the selection of one unknown component out of p which are possible. In case of equal probabilities for all the possibilities, the maximum information content results

$$M(1, p)_{max} = \mathrm{lb}p \tag{9.23}$$

The identification of n constituents out of p possibles in the case of $p \gg n$ is characterized by

$$M(n, p)_{max} = n \cdot \mathrm{lb}p \tag{9.24}$$

In the case where there are different probabilities for the presence of the p possibles, it holds in general that

$$M(n, p) = \sum_{i=1}^{n} \sum_{j=1}^{p} \left(P(x_j) \, \mathrm{lb} \, \frac{1}{P(x_j)} \right)_i \tag{9.25}$$

If the n unknown components have to be found out from a manageable number of p possibles, instead of Eq. (9.24) the following relationship must be applied:

$$M(n, p) = \sum_{k=1}^{n} lb(p-k+1) \tag{9.26}$$

The information amount increases further, if the n identified components additionally have to be determined quantitatively. In the simplest case, the maximum information amount according to Eq. (9.24) increases by the amount $n \cdot \mathrm{lb}m$, given in Eq. (9.22) into

$$M(n, p)_{\max} = n(\mathrm{lb}\,p + \mathrm{lb}\,m) = n \cdot \mathrm{lb}(pm) \tag{9.27}$$

In most unfavourable cases of the chemical practice, not even the number n of the species is known, e.g., in so-called "general unknown" cases in toxicological analytics. Then, at first, it has to be tested by a powerful separation method, e.g. chromatography, how many components are present in the sample before they are identified. For this an information part

$$I_n = \mathrm{lb}\,\frac{p}{n} \tag{9.28}$$

is needed and the total information amount $M(n, p)_{total}$ for the identification and determination of n unknown constituents (out of a set of p possibles) results:

$$M(n, p)_{total} = I_n + M(n, p) + M(n) = \mathrm{lb}\,\frac{p}{n} + n \cdot lb(p \cdot m) \tag{9.29}$$

This information amount is independent of the way in which it is obtained, viz if different techniques are used to recognize the number of constituents (I_n), identify them and quantitatively determine them, or all of this is done by one and the same procedure like GC-MS. Examples are given in ECKSCHLAGER and DANZER [1994], p 56, and DANZER et al. [2001], p 65. From the analytical problem, it has be to distinguish between indentification of n unknown constituents and the analysis of n known components as characterized by Eq. (9.22).

For the simultaneous analysis of n components, at least $N \geq n$ useful signals must be available. Instrumental-analytical methods produce as a rule two-dimensional information $y = f(z)$, e.g., in form of spectra, chromatograms etc., as schematically shown in Fig. 7.11.

The potential amount of signals, viz the number of signals that can maximally be observed with no disturbance, is characterized by the *analytical resolving power* (see Sect. 7.6) $R_z = N$ that is given by the boundaries of the registration range $z_{\min}$ and $z_{\max}$ as well as the signal half width Δz; see Eqs. (7.53) to (7.55) and DOERFFEL and HILDEBRANDT [1969], KAISER [1970] and DANZER [1975a]. According to them, two neighbouring signals can be separately detected if their distance is at least Δz.

The potential information amount M_{pot} of an analytical method is directly characterized by their signal resolution power (analytical resolution power) $R_z = N$

$$M_{pot} = N \cdot I_{\max} = N \cdot \mathrm{lb}\,m \tag{9.30}$$

Table 7.6 gives an overview about the analytical resolving power of various analytical techniques.

Whereas some instrumental methods produce just one signal per component, $N = n$, e.g., chromatographic methods, some other methods such as atomic spectroscopy generate much more signals as required, $N \gg n$.

The surplus of signals, $N - n$, leads to *redundancy*, R_M,

$$R_M = M_{pot} - M(n) = (N - n) \cdot \mathrm{lb}\,m \tag{9.31a}$$

frequently expressed in the form of *relative redundancy*, r_M,

$$r_M = \frac{R_M}{M_{pot}} = 1 - \frac{n}{N} \quad . \tag{9.31b}$$

In more detail, the problem of redundancy is described in DANZER et al. [1989] and ECKSCHLAGER and DANZER [1994], where, in the example of qualitative evaluation in optical emission spectrography, the contrast between "*helpful*" and "*blank*" redundancy is illustrated. In OES, for about 80 elements there exist hundreds of thousands of spectral lines, but the spectrochemists focus their interest only on relatively few of them: the so-called main detection lines (prominent lines) according to GERLACH and SCHWEIZER [1930], GERLACH and GERLACH [1936] and GERLACH and RIEDL [1939]. These prominent lines are a set of lines associated with promoting redundancy, whereas the other lines fall into the category of blank redundancy.

On the other hand, not only the enormous number of signals in multicomponent methods but also the large number of species that can be detected in highly resolved spectra and chromatograms, respectively, influence the information amount. Therefore, MATHERNY and ECKSCHLAGER [1996] proposed the introduction of so-called *relevancy coefficients*, k, into the system of information-theoretical assessment. In analytical practice, the coefficients k can be considered as being *weight factors* of the information contents of the respective species with which Eq. (9.21) becomes

$$M(n)_w = \sum_{j=1}^{n} k_j \cdot I_j \tag{9.32}$$

The values of these relevancy coefficients can be fixed depending on the practical chemical problem using static or dynamic models (MATHERNY and ECKSCHLAGER [1996]; ECKSCHLAGER and STEPANEK [1985]; ECKSCHLAGER and DANZER [1994]).

9.4
Process and Image Analysis

In specialized studies, chemical measurements achieve particular importance by their temporal and spatial resolution. In process control, the momentary value is interesting only in such exceptional situations in which a warning or control limit is passed. Much more information is contained in the time function of an analytical system, $y = f(t)$, or $y = f(z, t)$, in case of multicomponent systems as shown in Fig. 3.11.

The information amount of *process analyses* (under which term all time-dependent studies from chemical process control up to dynamic and kinetic studies are summarized) increases by the factor of *time resolving power*

$$R_t = \frac{t}{\Delta t} \tag{9.33}$$

compared with that of static chemical measurements. The information amount of temporally resolved measurements is given by

$$M_t = R_t \cdot I_{max} = R_t \cdot lbm \tag{9.34}$$

in case of single component analysis (DANZER [1978]) and

$$M(n)_t = R_t \cdot M(n) = R_t \cdot n \cdot lbm \tag{9.35}$$

for multicomponent analysis.

Similar is the situation in the case of *geometrically resolved information* (*image information* in the most general sense). The information amount increases in the following cases (see also Table 3.1):

- *Lateral profiles*, $y = f(l_x)$, viz line scans along a line l_x across the sample with y being a intensity measure or content; see, e.g., Fig. 3.11b
- *Concentration profiles*, $y = f(l_x, l_y)$, on a sample surface $l_x \times l_y$ (*surface scan*; see Fig. 3.12a,b)
- *Elemental images*, $z = f(l_x, l_y)$, with z being a component-specific signal; see Fig. 3.12c
- Black and white *image information*, $b = f(l_x, l_y)$, where b is the number of grey levels (e.g. of photos)
- Coloured *image information*, $B = f(\varphi, l_x, l_y)$, where B is the number of intensity levels of φ colours

Compared with that of common analyses which determine an average composition of the sample, the information amount of spatially resolved analyses increases by the factor of *geometrical resolving power* R_A that may be in the concrete case

- *Lateral resolving power*:

$$R_{lateral} = \frac{l_i}{\Delta l_i} \tag{9.36a}$$

where l_i is the entire lateral distance scanned on the sample, Δl the smallest resolvable lateral distance, given by the diameter of the measuring point (i is the index of the spatial direction, x, y, or $z = $ depth).

- *Surface resolving power*:

$$R_{surface} = \frac{a}{\Delta a} \tag{9.36b}$$

with a being the plane (area) of the entire sample under study, Δa the smallest resolvable element of plane (*plane resolving limit*, "*pixel*"; see RÜDENAUER [1989]). Areal resolving power plays an important role for all planar microprobe techniques like Electron Probe Micro Analysis (EPMA), Scanning Auger Microprobe (SAM) and 2D Secondary Ion Mass Spectrometry (SIMS); see GRASSERBAUER [1985], GRASSERBAUER et al. [1986a,b]; RÜDENAUER [1989] and HUTTER et al. [1996].

- *Volume resolving power:*

$$R_{volume} = \frac{v}{\Delta v} \tag{9.36c}$$

where v is the sample volume under investigation, $\Delta v = \Delta l_x \, \Delta l_y \, \Delta l_z$ $\approx (\Delta l_i)^3$ smallest resolvable volume element (*volume resolving limit*, "*voxel*"; see RÜDENAUER [1989]). A well-known method that yields three-dimensional analytical information is the ion microprobe (SIMS).

In the special case of Eq. (9.36a) where $\Delta l_i = l_z$ is fixed perpendicularly to the surface of the sample, the depth resolving power $R_{depth} = l_z/\Delta l_z$ results which becomes important for depth profiles and thin film analysis.

For distribution analysis and analytical images the potential information amount increases by the factor A in case that all the distinguishable points (all the pixels or voxels, respectively) are analyzed. In the ideal case, it results in

$$M_A = R_A \cdot I_{max} = R_A \cdot \text{lbm} \tag{9.37}$$

in case of single component analysis and

$$M(N)_{R_A} = R_A \cdot M(N) = R_A \cdot N \cdot \text{lbm} \tag{9.38}$$

for multicomponent analysis (DANZER [1974]; DANZER et al. [1991]; ECKSCHLAGER and DANZER [1994]) with a distribution-analytical method with the potential analytical resolving power N.

The relationships at Eqs. (9.37) and (9.38) do not take into account a possible loss of precision with increasing geometrical resolving power as well as errors in adjusting the measuring points. Furthermore, the independence of all the R_A measuring points must be presupposed for the estimation of the ideal information amount.

Another limiting case is given by the real information amount of a given distribution-analytical result or problem, respectively,

$$M(n)_a = a \cdot M(n) = a \cdot n \cdot \text{lbm} \tag{9.39}$$

that will be determined by the practically used geometrical resolving power, $R_{real} = a$. Therefore, the information amount is a times greater than that of average analysis, $M(n)$. The number a measuring points, e.g., spark spots or laser craters in Optical Emission Spectroscopy or Mass Spectrometry, are determined by a given problem, e.g., testing of homogeneity (see DANZER et al. [1991]; ECKSCHLAGER and DANZER [1994]). However, the real information amount, $M(n)_a$, is always lower than the ideal one, $M(N)_{R_A}$, because $a < R_A$ and $n < N$. Practical limitations of distribution analysis as well as examples that represent real distribution-analytical problems like multi-phase and homogeneity studies are given in more detail by DANZER et al. [1991]. There may also be found a comparison of detailed expressions of potential and practical information amounts.

Distribution analysis in atomic dimensions becomes structure analysis. But because of its specific methodology, it makes sense to consider structure analysis as a separate field of analytical chemistry; see Sect. 1.2. Therefore, the information-theoretical fundamentals of structure analysis are different from that of element analysis and have been represented by DANZER and MARX [1979a,b].

Both time- and position-dependent concentration functions can be dealt with by the theory of stochastic processes (BOHACEK [1977]). Time functions playing a role in process analysis can be assessed not only by means of information amount $M(n)_t$ but also – sometimes in a more effective way – by means of the *information flow*, J, which is generally given by

$$J = \frac{dM}{dt} \tag{9.40}$$

Based on the information flow, a number of information-theoretical performance quantities can be derived, and some important ones are compiled in Table 9.2. The *information performance* of analytical methods can be related to the information requirement of an given analytical problem. The resulting measures, *information efficiency* and *information profitability*, may be used to assess economical aspects of analytical chemistry.

The partial efficiency coefficients are defined in detail as follow:

- $e_m = lbm^*/lbm$
  Precision efficiency coefficient (m is the number of signal or concentration levels distinguishable by the method and m^* the number of levels needed to solve the given problem)

- $e_n = n/N$
  Signal efficiency coefficient ($N = R_Z$ analytical resolving power according to Eqs. (7.53) to (7.55) and Eq. (9.30), n number of components to be analysed)

- $e_a = a/R_A$
  Spatial efficiency coefficient (R_A spatial resolving power; see Eqs. (9.36a–c), a number of spots needed to be investigated)

- $e_\vartheta = \vartheta/R_t$
  Time resolution efficiency coefficient (R_t time resolving power, see Eq. (9.33), ϑ time resolution needed)

- $e_t = t_{ana}/t_{req}$
  Time efficiency coefficient (t_{req} time required for analysis, t_{ana} real analysis time)

A peripheral condition of the efficiency coefficients is that they must be less than 1, otherwise the analytical problem cannot be solved. Therefore, e_i is relevant for $e_i \leq 1$ and $e_i = 0$ is assigned for $e_i > 0$. Consequently, the efficiency E according to Table 9.2 becomes 0 if at least one of the coefficients e_i does not fulfil the condition and, therefore the problem is insoluble (DANZER and ECKSCHLAGER [1978]; ECKSCHLAGER and DANZER [1994]).

Table 9.2. Information-theoretical performance parameters, according to DANZER and ECKSCHLAGER [1978] and ECKSCHLAGER and DANZER [1994]

Designation	Quantity	Unit	Application in analytical process	Note
Information density (storage density)	$D_a = M_a/a$ $D_v = M_v/v$	bit/cm^2 bit/cm^3	Static detectors, e.g., photographic plate	a, v spatial storage unit
Information capacity (channel capacity)	$C = M_{pot}/t$	bit/s	Dynamic detectors, e.g., photomultiplier	t time
Information flow (momentary information performance)	$J = \mathrm{d}M_{pot}/\mathrm{d}t$	bit/s	Load of spectrometer channels, Chromatography	
Information performance	$P = M_{pot}/t_{ana}$	bit/s bit/h	Information power of analytical procedures	t_{ana} time of analysis
Information efficiency	$E = \prod_{i=1}^{k} e_i$	1	Efficiency test	e_i partial efficiency coefficients
Information profitability	$\Phi = E \cdot M_{pot}/\gamma$	bit/s/cu bit/h/cu	Performance/cost relation	γ cost of an analysis

References

Bohacek, P (1977) *Chemical inhomogeneity of materials and its determination. Compact materials.* Coll Czechoslov Chem Commun 4:2982

Bongard MM (1970) *Pattern recognition.* Spartan Books, New York

Brillouin L (1963) *Science and information theory.* Academic Press, New York

Danzer K (1973a) *Zu einigen informationstheoretischen Aspekten der Analytik.* Z Chem 13:20

Danzer K (1973b) *Die Informationsmenge als Kenngröße zweidimensionaler analytischer Aussagen.* Z Chem 13:69

Danzer K (1973c) *Ermittlung der Informationsmenge qualitaiver Analysengänge.* Z Chem 13:229

Danzer K (1974) *Informationstheoretische Charakterisierung von Verteilungsanalysen.* Z Chem 14:73

Danzer K (1975a) *Zur Ermittlung der Informationsmenge bei spektrochemischen Analysenverfahren.* Z Chem 15:158

Danzer K (1975b) *Kennzeichnung des Leistungsvermögens und des rentablen Einsatzes von Analysenverfahren mit Hilfe der Informationstheorie.* Z Chem 15:326

Danzer K (1978) *Zur Ermittlung der Informationsmenge und des Informationsflusses in der Prozeßanalytik.* Z Chem 18:104

Danzer K (1983) *Möglichkeiten der informationstheoretischen Bewertung von Analysenverfahren und -ergebnissen.* In Müller RK (ed): *Instrumentelle Analytik in der Toxikologie.* Universität Leipzig, p 26

Danzer K (1993) *Information content of an analytical result and its assessment by interlaboratory studies.* IDF Special Issue 9302: *Analytical quality assurance and good laboratory practice in dairy laboratories.* IDF, Brussels

Danzer K (2004) *A closer look at analytical signals.* Anal Bioanal Chem 380:376

Danzer K, Eckschlager K (1978) *Information efficiency of analytical methods.* Talanta 25:725

Danzer K, Marx G (1979a) *Zu einigen informationstheoretischen Grundlagen der Strukturanalytik, I. Ermittlung der Informationsmenge qualitativer Strukturaussagen.* Chem Analit [Warsaw] 24:33

Danzer K, Marx G (1979b) *Zu einigen informationstheoretischen Grundlagen der Strukturanalytik, II. Die Informationsmenge quantitativer Strukturinformationen und Analysenverfahren.* Chem Analit [Warsaw] 24:43

Danzer K, Than E, Molch D (1976, 1977, 1987) *Analytik. Systematischer Überblick.* Akademische Verlagsgesellschaft Geest & Portig, Leipzig, 1976; Wissenschaftliche Verlagsgesellschaft, Stuttgart, 1977 (2nd edn 1987)

Danzer K, Eckschlager K, Wienke D (1987) *Informationstheorie in der Analytik, I. Grundlagen und Anwendungen auf Einkomponentenanalysen.* Fresenius Z Anal Chem 327:312

Danzer K, Eckschlager K, Matherny M (1989) *Informationstheorie in der Analytik, II. Mehrkomponentenanalyse.* Fresenius Z Anal Chem 334:1

Danzer K, Schubert M, Liebich V (1991) *Information theory in analytical chemistry, III. Distribution-analytical investigations.* Fresenius J Anal Chem 341:511

Danzer K, Hobert H, Fischbacher C, Jagemann K-U (2001) *Chemometrik. Grundlagen und Anwendungen.* Springer, Berlin Heidelberg New York

Doerffel K, Hildebrandt W (1969) *Beurteilung von Analysenverfahren unter Einsatz der Informationstheorie.* Wiss Z Techn Hochsch Leuna-Merseburg 11:30

Dupuis F, Dijkstra A (1975) *Application of information theory to analytical chemistry. Identification by retrieval of gas chromatographic retention indices.* Anal Chem 47:379

Eckschlager K (1971-1989) *Theory of information as applied to analytical chemistry.* Collect Czech Chem Commun 36: 3016; 37: 137, 1486; 38: 1330; 39: 1426, 3076; 40: 3627; 41: 1875, 2527; 42: 225, 1935; 43: 231; 44: 2373; 45: 2516; 46: 478; 47: 1195, 1580; 49: 2342; 50: 1359; 53: 1647, 3021; 54: 1770, 3031

Eckschlager K, Danzer K (1994) *Information theory in analytical chemistry.* Wiley, New York

Eckschlager K, Stepanek V (1985) *Analytical measurement and information.* Letchworth, Res Studies Press

Fresenius W, Malissa H Sr (eds), Grasserbauer M (coord), Cammann K, Valcarcel M, Zuckerman AM, Nan Z, Koch KH, Perez-Bustamante JA, Ortner, HM, Danzer K, Green

JD, Stulik K, Zyka J, Kuznetsov VI (1992) *Analytical chemistry – today's definition and interpretation.* Fresenius J Anal Chem 343:809–835

Gerlach W, Gerlach W (1936) *Die chemische Emissionsspektralanalyse. II. Anwendungen in Medizin, Chemie und Mineralogie.* Voss, Leipzig

Gerlach W, Riedl E (1939) *Die chemische Emissionsspektralanalyse. III. Tabellen zur quantitativen Analyse.* Voss, Leipzig

Gerlach W, Schweizer E (1930) *Die chemische Emissionsspektralanalyse. I. Grundlagen und Methoden.* Voss, Leipzig

Grasserbauer M (1985) *Distribution analysis in materials research.* Fresenius Z Anal Chem 322:105

Grasserbauer M, Dudek HJ, Ebel MF (1986a) *Angewandte Oberflächenanalyse.* Springer, Berlin Heidelberg New York

Grasserbauer M, Stingeder G, Pötzl H, Guerrero E (1986b) Fresenius Z Anal Chem 323:421

Hartley RVL (1928) *Transmission of information.* Bell Syst Techn J 7:535

Hutter H, Brunner C, Nikolov S, Mittermayer C, Grasserbauer M (1996) *Imaging surface spectroscopy for two- and three-dimensional characterization of materials.* Fresenius J Anal Chem 355:585

Kaiser H (1970) *Quantitation in elemental analysis.* Anal Chem 42(2):24A, 42(4):26A

Kerridge DF (1961) *Inaccuaracy and inference.* J R Statist Soc B23:184

Kullback S (1959) *Information theory and statistics.* Wiley, New York

Malissa H, Rendl J, Clerc JT, Gottschalk G, Kaiser R, Schwarz-Bergkampf E, Spitzy H, Werder D, Zettler H (1974) Arbeitskreis "Automation in der Analyse", *Informationstheorie in der Analytik.* Fresenius Z Anal Chem 272:1

Matherny M, Eckschlager K (1996) *Determination of relevancy and redundancy of the parameters of information theory.* Chemom Intell Lab Syst 32:67

Meyer-Eppler W (1969) *Grundlagen und Anwendungen der Informationstheorie.* Springer, Berlin Heidelberg New York

Rackow B (1963–1969) *(30 papers on information-theoretical aspects of physicochemistry, e.g. introduction of the quantity "molbit", relationship between information and thermodynamical entropy, etc)* (in German). Z Chem 3: 268, 316, 437, 477; 4: 36, 72, 73, 109, 155, 196, 236, 275, 311; 5: 67, 116, 159, 195, 238, 278, 434; 7: 398, 444, 472; 8: 33, 117, 157, 434; 9: 318; 11: 398, 436

Rüdenauer FG (1989) *Multidimensional image acquisition and processing in surface analysis.* Fresenius J Anal Chem 333:308

Shannon CE (1948) *A mathematical theory of communication.* Bell Syst Techn J 27:379, 623

Shannon CE, Weaver W (1949) *The mathematical theory of communication.* Univ Illinois Press, Urbana

Wienke D, Wank U, Wagner M, Danzer K (1991) *MULTIVAR, PLANEX, INTERLAB – From a collection of algorithms to an expert system: statistics software written from chemists for chemists.* In: Gmeling J (ed) Software development in chemistry, vol 5, p 113

Glossary of Analytical Terms

Globalization in science, technology and economy causes, amongst other things, problems in derivation, adaptation and acceptance of technical terms in general, and in Analytical Chemistry, too. Both vague and contradictory definitions have been developed and established over the years. Therefore, in the last decade the efforts have been increased to harmonize the use of analytical terms. In particular international organisations like ISO, IUPAC render outstanding services to the fixing and harmonization of essential analytical terms. But also additional activities have been done in this field, e.g., by the Royal Society of Chemistry, the EURACHEM Education and Training Working Group and the Federation of European Chemical Societies (FECS), Division on Analytical Chemistry (DAC) as well as by publications in form of books (e.g., PRICHARD et al. [2001]) and other publications, e.g. the series "*Glossary of Analytical Terms*" (*GAT*) in the journal *Accreditation and Quality Assurance*, see GAT I to X [1996-1998], where multilingual terms are given, and HOLCOMBE [1999/2000].

In the following, definitions of essential analytical terms are compiled, if possible on the basis of international agreements. Attached are sparse references and cross-references. The symbols, being used here, means: $\rightarrow$ "see also"(cross-references to terms, additional references as well as paragraphs, chapters, equations and figures of this book), and $\times$ is a warning notice.

Accuracy

"Closeness of agreement between the result of a measurement and a true value of the measurand".	ISO 3534-1 [1993] $\rightarrow$ Sect. 7.1 $\rightarrow$ *Trueness* $\times$ Do not confuse with *Precision*

Analysis (of a sample)

Investigation of a *sample* to identify and/or determine (an) analyte(s) or assay a material.	$\rightarrow$ Quotation from ANAL CHEM [1975] at the end of the Glossary

Analyte

"The chemical entity being investigated (qualitatively or quantitatively)".

According to PRITCHARD et al. [2001]

Analytical function (*evaluation function*)

Inverse of the calibration function, $x = f^{-1}(y)$, describing the dependence of the analytical values from the measured values.

Analytical method

"Logical sequence of operations, described generally, used in the performance of measurements", e.g., the links of a given analytical technique with particular excitation and detection.

ISO 3524-1 [1993]
$\longrightarrow$ Fig. 7.1

Analytical procedure

"Set of operations, described specifically, used in the performance of particular analytical measurements according to a given method".

ISO 3524-1 [1993]
PRITCHARD et al. [2001]
$\longrightarrow$ Fig. 7.1

Analytical process

Logic sequence of objects linked by general analytical standard operations.

$\longrightarrow$ Fig. 2.1

Analytical result

Analytical value attributed to a measurand, obtained by measurement and completed by information on the uncertainty of measurement.

According to ISO 3524-1 [1993]
$\longrightarrow$ 8.1

Analytical quantity

"Particular quantity subject to analytical measurement".

According to ISO 3524-1 [1993]
$\longrightarrow$ *Measurand*

Analytical technique

"Generic analytical application of a scientific principle".

According to PRITCHARD et al. [2001]
$\longrightarrow$ Fig. 7.1

Assay

Determination of how much of a sample is the material indicated by the name.

→ Quotation from ANAL CHEM [1975] at the end of the Glossary
→ e.g. analysis of ores

Analytical value

Magnitude of an analytical quantity, x, measured at test samples on the one hand and given for reference samples used for calibration on the other hand.

Background (instrumental background, background signal)

Instrumental background is the null signal, obtained in the absence of any analyte- or interference-derived signal.

→ IUPAC [1995]; CURRIE [1999]
→ Background may be set to zero, on the average, for certain instruments

Baseline

"Summation of the instrumental background plus signals in the analyte (peak) region of interest due to interfering species".

IUPAC [1995]; CURRIE [1999]

Bias

"The difference between the expectation of the test results and an accepted reference value".

According to GAT VIII [1997]

Blank

A value, y_B, obtained my measuring a blank sample (in calibration, the intercept of the calibration curve is considered to be equal to the blank).
Blanks may be differentiated into *instrumental blank* (*background* and *baseline*, respectively) and *chemical blank* (*analyte blank*).

→ IUPAC [1995]; CURRIE [1999]
→ *Background*
→ *Baseline*
→ *Chemical blank*

Blank measurement

Procedure by which a measured value is obtained with a sample in that the analyte of interest is intentionally absent.

According to PRITCHARD et al. [2001]; TAYLOR [1987]; SHARAF et al. [1986]
$\longrightarrow$ Blank sample

Blank sample

A sample whose analyte concentration is below the limit of decision of the analytical procedure being used.

According to PRITCHARD et al. [2001]
$\longrightarrow$ Blank measurement

Calibration

Set of operations that establish, under specified conditions, the relationship between values of quantities indicated by a measuring system and the corresponding values of quantities represented by a material both in form of reference materials and samples. In a wider sense, calibration represents a set of operations that establish relationships between quantities in the sample domain with quantities in the signal domain, viz $y = f(x)$ and $z = f(Q)$.

$\longrightarrow$ ISO 3524-1 [1993]
$\longrightarrow$ GAT IV [1996]
$\longrightarrow$ PRITCHARD et al. [2001]
$\longrightarrow$ IUPAC [1998]
$\longrightarrow$ Sect. 6.1
$\longrightarrow$ Sample domain
$\longrightarrow$ Signal domain

Calibration function

Equation for the estimation of the values of a measuring quantity from given values of a analytical quantity. The calibration function may be known a priori by natural laws or estimated experimentally by means of calibration samples.
The calibration function represents that segment of the response function that is chosen for estimating the analytical value of an unknown sample.

$\longrightarrow$ PRITCHARD et al. [2001]
$\longrightarrow$ IUPAC [1998]
$\longrightarrow$ SHARAF et al. [1986]
$\longrightarrow$ Sensitivity
$\longrightarrow$ Response function

Calibration samples

Set of samples characterized by accurate and precise values of the measurand. In a concrete case, calibration samples may be portions of (certified) reference materials, in-house reference materials (laboratory standard samples), or spiked samples, and, in addition, blank samples.

$\rightarrow$ PRITCHARD et al. [2001]
$\rightarrow$ *Reference material*
$\rightarrow$ *Certified reference material*

Certified reference material (CRM)

"A reference material, accompanied by a certificate, one or more of whose property values are certified by a procedure which establishes traceability to an accurate realization of the unit in which the property values are expressed, and for which each certified value is accompanied by an uncertainty at a stated level of confidence".

ISO 3524-1 [1993]
$\rightarrow$ GAT IV [1996]
$\rightarrow$ PRITCHARD et al. [2001]

Chemical blank (*analyte blank*)

"Blank which arises from contamination from the reagents, sampling procedure, or sample preparation steps which correspond to the very analyte being sought".

IUPAC [1995], CURRIE [1999]

Coefficient of variation: The term is not recommended by IUPAC;
$\rightarrow$ ***Relative standard deviation***

Concentration domain $\rightarrow$ *Sample domain*

One of the dimensions of the sample domain.

Confidence interval (*CI*)

Statistical interval, e.g., of a mean, $\overline{y}$, $cnf(\overline{y}) = \overline{y} \pm \Delta\overline{y}_{cnf}$, which express the uncertainty of measured values. *CI*s are applied for significance tests and to establish quantities for limit values (*CV*).

$\rightarrow$ Sect. 7.5
$\rightarrow$ *Critical value*
$\rightarrow$ *Prediction interval*

Conventional true value

"Value attributed to a particular quantity and accepted, sometimes by convention, as having an uncertainty appropriate for a given purpose".

ISO 3534-1 [1993]
$\longrightarrow$ *True value*

Correlation

Stochastic relationship between random variables in such a way that one depends on the other. The degree of relationship may be estimated by the *correlation coefficient*.

Correlation coefficient

The correlation coefficient, r_{xy}, is given by the covariance of two random variables x and y, $\operatorname{cov}(x, y) = s_{xy}$, divided by the standard deviations s_x and s_y, see Eq. (6.3). The correlation coefficient becomes $r_{xy} = 0$ if there is no relationship between x and y, and $r_{xy} = \pm1$ if there exist a stringent deterministic dependence.

$\longrightarrow$ Sect. 6.1.3
$\times$ The correlation coefficient is not of any relevance in calibration, as a rule. This is because only the measured value is a random variable and, in contrast, the analytical value is a fixed value and not selected randomly

Correlation matrix

Matrix formed by a set of correlation coefficients related to m variables in multivariate data sets, $\boldsymbol{R} = (r_{x_i,x_j})$. It is relevant in multicomponent analysis.

$\longrightarrow$ Eqs. (6.4) and (8.14)

Critical value (CV)

Limit in the signal domain, estimated from the average blank plus its uncertainty, generally according to $y_c = \overline{y}_{BL} + U(\overline{y}_{BL})$, in analytical chemistry frequently according to $y_c = \overline{y}_{BL} + 3s_{BL}$. If the critical value is exceeded, the respective analyte is reliably detected (except for a remaining risk of error α). Therefore, the CV stands for the *guarantee of presence* of an analyte.

EHRLICH and DANZER [2006]; CURRIE [1999]
$\longrightarrow$ Sect. 7.5
$\longrightarrow$ Fig. 7.8
$\longrightarrow$ *Decision limit*

Cross sensitivity (partial sensitivity)

Dependence of the measured value (signal intensity), y_A, from other constituents than the analyte A, present in the measuring sample, quantitatively expressed by the respective partial differential quotient.

→ KAISER [1972]; DANZER [2001]
→ Sect. 7.2
→ Eq. (3.11)
→ Sensitivity
→ Sensitivity matrix
→ Total sensitivity

Determination

Analysis of a sample to estimate quantitatively the amount (content, concentration) of (an) analyte(s).

→ Quotation from ANAL CHEM [1975] at the end of the Glossary

Evaluation function (analytical function)

Inverse of the calibration function, $x = f^{-1}(y)$, describing the dependence of the analytical values from the measured values, being so the basis of analytical evaluation.

Homogeneity

A qualitative term used to describe that the analyte is uniformly distributed through the sample. The degree of homogeneity may also be characterized quantitatively as a result of a statistical test.

→ PRITCHARD et al. [2001]
→ Inhomogeneity
→ Sect. 2.1
→ Eq. (2.9)

Hyphenated techniques

Coupling of two (or more) separate analytical techniques via appropriate interfaces and computer with the goal to obtain faster a higher amount of information on the subject under investigation.

→ HIRSCHFELD [1980]
→ KELLNER et al. [1998]
→ By hyphenating analytical methods, the dimension of analytical information will be increased (usually by one)
→ Sect. 3.4

Identification

Recognizing of (an) unknown constituent(s) in an analytical test sample. In contrast, by qualitative analysis it is tested whether (a) known constituent(s) are present or absent.

→ Sect. 9.3

Imprecision

A quantitative term to describe the (lack of) "precision" of an analytical procedure (e.g. by *standard deviation*).

$\longrightarrow$ IUPAC [1995]; CURRIE [1999]
$\longrightarrow$ *Precision*
$\longrightarrow$ *Imprecision of analytical results*, see Sect. 7.1
$\longrightarrow$ *Standard deviation*

Inaccuracy

A quantitative term to describe the (lack of) accuracy of an analytical procedure which comprises the imprecision *and* the bias.

$\longrightarrow$ IUPAC [1995]; CURRIE [1999]
$\longrightarrow$ *Accuracy*
X   *Inaccuracy* should not be confused with *uncertainty*, see IUPAC [1994a]

Inhomogeneity

"Term used to describe situations where the analyte is unevenly distributed through the sample matrix".
The degree of inhomogeneity may be characterized quantitatively by Eq. (2.9) the value of which becomes negative with the transition from homogeneity to inhomogeneity.

$\longrightarrow$ PRITCHARD et al. [2001]
$\longrightarrow$ Sect. 2.1
X   The term *inhomogeneity* should not be confused with *heterogeneity*

Interlaboratory study

"A study in which several laboratories measure a quantity in one or more identical portions of homogeneous, stable materials under documented conditions, the results of which are compiled into a single report".
According to the evaluation types, it is differentiated between:
(1) Method-performance studies.
(2) Laboratory-performance studies.
(3) Material-certification studies.

IUPAC [1994b]
$\longrightarrow$ A minimum of five laboratories should be used to provide meaningful statistical conclusions from interlaboratory studies
$\longrightarrow$ Sect. 8.2.4

Limit of decision ("3σ-limit of detection")

The analytical value (e.g. the concentration) that corresponds to the critical value. The limit of decision is of minor importance in analytical chemistry because the detection at this level of concentration succeeds only in 50% of all cases.

EHRLICH and DANZER [2006]; IUPAC [1995]; CURRIE [1999]
→ Sect. 7.5
→ Fig. 7.7
→ *Critical value*
→ *Detection limit*
✗ The decision limit should not be used as a performance characteristic of analytical methods and also not as a limit of guarantee of an analyte

Limit of detection (LD)

The analytical value, x_{LD}, that always produce a signal which can be distinguished from the blank (except for a remaining risk of error β).
LD is the limit in the sample domain (analyte domain). It characterizes analytical procedures, in particular with regard to the limit concentration that can be detected. Therefore, the LD stands for the *guarantee of absence* of an analyte.

EHRLICH and DANZER [2006]; IUPAC [1995]; CURRIE [1999]
→ Sect. 7.5
→ Fig. 7.7
→ *Critical value*

Limit of determination → *Limit of quantitation*

Limit of quantitation (LQ)

An analytical value, x_{LQ}, above which quantitative determinations are possible with a given minimum precision. The condition on precision must be declared in each case. For a given precision $k = x_{LQ}/\Delta x_{LQ}$, the limit of quantification can be estimated by Eqs. (7.48) and (7.49).

EHRLICH and DANZER [2006]; IUPAC [1995]; CURRIE [1999]
→ *Precision*
→ For factual reasons, the limit of quantification cannot be lower than the limit of detection
→ The declaration of precision must always be given because it is an inherent component of LQ

Linear dynamic range

The range of concentration in which the response varies linearly with the analyte concentration.

→ SHARAF et al. [1986]

Linearity

Ability of an analytical method to give a response which depends linearly on the analyte concentration.

→ PRITCHARD et al. [2001]
→ SHARAF et al. [1986]

Matrix

All of the constituents of a sample except the analyte. The matrix is the carrier of the analyte.

→ IUPAC ORANGE BOOK
→ [1997, 2000]
→ Analyte

Matrix effect

Influence of one or more matrix constituent(s) on the analyte under study. Matrix influences may affect the analyte signal directly by interferences or indirectly by signal depression or amplification.

→ Sect. 3.5
→ Eqs. (3.12)–(3.14); (3.16); (3.17)

Measurand

"Particular quantity subject to measurement".

ISO 3524-1 [1993]

Measured result

Measured value, obtained by measurement and completed by information on the uncertainty of measurement.

→ Sect. 8.1
→ Measured value
→ Uncertainty

Measured value

"Outcome of an analytical measurement" or "value attributed to a measurand". A measured value is a "Magnitude of a measuring quantity generally expressed as a unit of measurement multiplied by a number".

ISO 3524-1 [1993]
IUPAC [1995]; CURRIE [1999]
→ Measuring quantity

Measuring quantity

"Attribute of a phenomenon ... that may be distinguished qualitatively and determined quantitatively".

ISO 3524-1 [1993]

Measuring sample

Sample that is directly introduced into analytical measurement. A measured sample is created from a test sample by conversion into a measurable form by means of a procedure of sample preparation.

→ Sect. 2.2

Metrology

"Science of measurement".

ISO 3524-1 [1993]

Monitoring

Continuous or repeated observation, measurement, and evaluation of a process in a certain field of application (e.g., environmental surveillance, health checking, foodstuff inspection, quality assurance in manufacturing), according to given schedules in space and time.

Multicomponent analysis (multispecies analysis)

Simultaneous determination of several analytes (species) by means of a multicomponent sensing technique or hyphenated techniques.

→ IUPAC [1995]; CURRIE [1999]
→ IUPAC [2004]

Noise

Fluctuations of the baseline- or background record of an (analytical) instrument. Noise do not provide meaningful information, on the contrary, it degrades the quality of signals and, therefore their detectability.

→ Background
→ Baseline
→ Signal-to-noise ratio, R/N, is a measure of the quality of signals
→ Sect. 7.5
→ Figs. 7.9B and 7.10

Population

"Finite or infinite set of individuals (objects, items). A population implicitly contains all the useful information for calculating the true values of the population parameters", e.g., the mean μ and the standard deviation σ.

FRANK and TODESCHINI [1994]
→ Sample (in the statistical sense)

Precision

"The closeness of agreement between in-dependent test results obtained under stipulated conditions".
"The precision of a set of results of measurements may be quantified as a standard deviation".

ISO 3524-1 [1993]; GAT II [1996]
$\longrightarrow$ KAISER and SPECKER [1956]
$\times$ In fact, standard deviation characterizes *imprecision*
$\longrightarrow$ Sect. 7.1.1; Eqs. (7.8) and (7.9)

Prediction interval (PI)

Statistical interval, e.g., of a mean, $\overline{x}$, $prd(\overline{x}) = \overline{x} \pm \Delta\overline{x}_{prd}$, that express the uncertainty of analytical values which are predicted on the basis of experimental calibration. *PIs* are applied for significance tests and to establish quantities for limit values (*LD, LQ*).

$\longrightarrow$ Sect. 7.5
$\longrightarrow$ *Limit of detection*
$\longrightarrow$ *Limit of quantification*
$\longrightarrow$ *Confidence interval*

Proficiency test

"Study of laboratory performance by means of ongoing interlaboratory test comparisons".

ISO GUIDE 33 [1989]
PRITCHARD et al. [2001]
$\longrightarrow$ *Interlaboratory study*

Qualitative analysis

Testing whether (a) known constituent(s) are present or absent in test samples.
In contrast, identification means recognizing of (an) unknown constituent(s) in a test sample.

$\longrightarrow$ Sect. 9.3

Quantitative analysis

Determination of the amount(s) of (an) analyte(s) in a test sample.

$\longrightarrow$ *Assay*

Random variable

A quantity that appears in a random experiment. Random variables relate events into a set of values.

$\longrightarrow$ SACHS [1992]
$\longrightarrow$ FRANK and TODESCHINI [1994]

Range (in the analytical sense)

"The interval between the upper and the
lower concentration of the analyte in the
sample for which it has been determined
that the method is applicable".

PRITCHARD et al. [2001]

Range (in the statistical sense)

Difference between the greatest and the
smallest values of a series of measurements.

→ PRITCHARD et al. [2001]
→ Sect. 4.3.2

Recalibration

Updating of a calibration model in the
case that details of the analytical procedure are changed.

→ *Standard operating procedure*

Reference material

"A material or substance one or more
of whose property values are sufficiently
homogeneous and well established to be
used for the calibration of an apparatus,
the assessment of a measurement method
or for assigning values to materials".

ISO 3524-1 [1993]
→ GAT IV [1996]
→ PRITCHARD et al. [2001]

Regression

Statistical method to model a mathematical equation that describes the relationship between random variables (usually
x and y). The goal of regression analysis
is both modelling and predicting.

→ SACHS [1992]
→ FRANK and TODESCHINI
[1994]
→ *Regression model*
→ Sect. 6.1.3

Regression coefficients (*regression parameter*)

Coefficients of the predictors in a regression model, e.g., a_x and b_x or a_y and b_y,
respectively, in linear regression models.

→ Eqs. (6.8) to (6.10)

Regression model

Mathematical model that describes the relationship between random variables (usually x and y) by means of regression coefficients and their uncertainties as well as uncertainties of model and the prediction.

In linear regression there are two different models:

that of the prediction of y from x

$$\hat{y} = a_x + b_x x \qquad (6.6)$$

and that of the prediction of x from y

$$\hat{x} = a_y + b_y y \qquad (6.7)$$

$\rightarrow$ SACHS [1992]
$\rightarrow$ FRANK and TODESCHINI [1994]
$\rightarrow$ Sect. 6.1.3
$\times$ Eq. (6.7) is not the inverse of Eq. (6.6), viz

$$\hat{x} = a_y + b_y y \neq \frac{\hat{y} - a_x}{b_x}$$

though Eq. (6.7) approximates to Eq. (6.6) in the same degree as the correlation coefficient r_{xy} approximates to 1

Relative standard deviation (RSD)

Standard deviation expressed as a fraction of the mean $s_{rel} = s/\overline{x}$. RSD is a dimensionless quantity; sometimes it is multiplied by 100 and expressed as a percentage.

$\times$ The use of the term "*coefficient of variation*" ("*variation coefficient*") is not recommended by IUPAC
$\rightarrow$ IUPAC Orange Book [1997, 2000]

Reliability

A qualitative term that covers precision and accuracy as well as robustness (ruggedness).

Repeatability (of results of measurements)

"Closeness of the agreement between the results of successive measurements of the same measurand carried out under the same conditions of measurement" (Precision under repeatability conditions). Repeatability may be expressed quantitatively in terms of suitable dispersion characteristics.

ISO 3524-1 [1993]
$\rightarrow$ GAT I [1996]
$\rightarrow$ Sect. 7.1.3
$\rightarrow$ Repeatability conditions include the same measurement procedure, the same observer, the same measuring instrument, used under the same conditions, the same location, and repetition over a short period of time
$\rightarrow$ *Repeatability standard deviation*
$\rightarrow$ *Repeatability interval*

Repeatability standard deviation (s_{repeat})

Experimental standard deviation obtained from a series of n measurements under repeatability conditions.

$\longrightarrow$ PRITCHARD et al. [2001]
$\longrightarrow$ The number of measurements should be about $n = 10$

Repeatability interval (repeatability limit)

A confidence interval representing the maximum permitted difference between two single test results under repeatability conditions:

$$r = t_{1-\alpha,\nu}\sqrt{2} \cdot s_{repeat}$$

$\longrightarrow$ PRITCHARD et al. [2001]
In the given formula, $t_{1-\alpha,\nu}$ is the quantile of the respective t-distribution (the degrees of freedom ν relates to the number of replicates by which s_{repeat} has been estimated)

Reproducibility (of results of measurements)

"Closeness of the agreement between the results of measurements of the same measurand carried out under changed conditions of measurement" (Precision under reproducibility conditions).
Reproducibility may be expressed quantitatively in terms of suitable dispersion characteristics.

ISO 3524-1 [1993]
$\longrightarrow$ GAT I [1996]
$\longrightarrow$ Sect. 7.1.3
$\longrightarrow$ Reproducibility conditions are characterized by changing conditions such as: observer, measuring instrument, conditions of use, location, time, but applying the same method
$\longrightarrow$ Reproducibility standard deviation
$\longrightarrow$ Reproducibility interval

Reproducibility standard deviation (s_{repro})

Experimental standard deviation obtained from a series of measurements under reproducibility conditions.

$\longrightarrow$ PRITCHARD et al. [2001]
$\longrightarrow$ The number of measurements should be sufficiently large to estimate a representative reproducibility standard deviation

Reproducibility interval (reproducibility limit)

A confidence interval representing the maximum permitted difference between two single test results under reproducibility conditions:

$$R = t_{1-\alpha,\nu}\sqrt{2} \cdot s_{repro}$$

$\longrightarrow$ PRITCHARD et al. [2001]
$\longrightarrow$ In the given formula, $t_{1-\alpha,\nu}$ is the quantile of the t-distribution (the degrees of freedom ν relates to the number of replicates by which s_{repro} has been estimated)

Resolution

Process by which a composite signal is split up into individual forms. The resolution can be related to:
(i) Signal overlappings and fine structure (z-scale)
(ii) Signals in close succession in time and space

$\longrightarrow$ Sect. 6.4.1
$\longrightarrow$ Sect. 7.6

Resolution limit

The smallest difference Δz at which two adjacent signals can be separately observed, i.e., their overlap does not exceed a threshold of 50% of the individual profiles.

$\longrightarrow$ SHARAF et al. [1986]
$\longrightarrow$ In case (ii) of resolution of the resolution problem, Δt and Δl are the crucial parameters
$\longrightarrow$ Sect. 7.6

Resolution power

Ability of an analytical procedure to detect signals of small differences as separate signals. Resolution power is inversely proportional to resolution limit, e.g., $R = z/\Delta z$

$\longrightarrow$ SHARAF et al. [1986]
$\longrightarrow$ Sect. 7.6

Response

Output of an analytical system as a reaction to a certain stimulus.

$\longrightarrow$ Stimulus
$\longrightarrow$ The output may be an observable or measurable effect

Response function

Relationship between the response of the analytical system and the amount of analyte. The overall response function is frequently nonlinear.

$\longrightarrow$ SHARAF et al. [1986]
$\longrightarrow$ Calibration function

Response variable (dependent variable)

$\longrightarrow$ *Measuring quantity, measured value*

Robustness

Property of an analytical procedure that indicates insensitivity against changes of known operational parameters on the results of the method and hence its suitability for its defined purpose.

$\longrightarrow$ BURNS et al. [2005]
$\longrightarrow$ ICH [1996]
$\longrightarrow$ Robustness may be quantified by means of quantities characterizing signal effects
$\longrightarrow$ Eq. (7.31)

Round robin test $\longrightarrow$ interlaboratory study

Ruggedness

Property of an analytical procedure that indicates insensitivity against changes of known operational variables and in addition any variations (not discovered in intra-laboratory experiments) which may be revealed by inter-laboratory studies.

$\longrightarrow$ BURNS et al. [2005]
$\longrightarrow$ Ruggedness may be quantified by means of quantities characterizing signal effects
$\longrightarrow$ Eq. (7.33)

Sample (in the analytical sense)

Portion of the object under study (the material submitted for analysis).
A sample consists of the analyte and the matrix.

$\longrightarrow$ PRITCHARD et al. [2001]
$\longrightarrow$ There are various types of samples within given sampling schemes, e.g., bulk samples > primary samples > gross samples > subsamples > test samples > measuring samples
$\longrightarrow$ Fig. 2.4

Sample (in the statistical sense)

"Subset of a population that is collected in order to estimate the properties of the underlying population", e.g., the sample parameters mean $\bar{x}$ and standard deviation s. In the ideal case of representative sampling, the sample parameter fit the parameter of the population μ and σ, respectively.

FRANK and TODESCHINI [1994]

Sample domain (analyte domain)

Field of analytical operation that is characterized by samples' properties such as type of analytes, Q, and their amount, x_Q. The transition to signal domain is done by calibration and analytical measurement.

$\longrightarrow$ Fig. 2.12
$\longrightarrow$ Signal domain

Sampling

"Sequence of selective and non-selective operations ending with the selection of one or several test portions submitted to the analytical process in their entirety. Their physical properties (maximum particle size, mass, etc) are specified in the analytical procedure." Sampling covers sampling (in the narrow sense) and sample reduction.

$\longrightarrow$ Gy [1992]
$\longrightarrow$ Sect. 2.1
$\longrightarrow$ Fig. 2.4

Screening

Testing of (a large number of) objects in order to identify those with particular characteristics.

$\longrightarrow$ Sect. 1.2
$\longrightarrow$ Fig. 1.5

Selectivity

The extent to which n given analytes can be measured simultaneously by (a least) n sensors (detecting channels) without interferences by other components and, therefore, can be detected and determined independently and undisturbedly.

$\longrightarrow$ KAISER [1972]; DANZER [2001]
$\longrightarrow$ Sect. 7.3
$\longrightarrow$ Eq. (7.24)
$\times$ Selectivity should not be merged with specificity: selectivity relates to multicomponent analysis and specificity to single component analysis

Sensitivity

"Change in the response of a measuring instrument divided by the corresponding change in the stimulus". In analytical measurements is this, in fact, the differential quotient of the measured value to the analytical value.

ISO 3435-1 [1993]; GAT VII [1997]
$\longrightarrow$ Sect. 7.2
$\longrightarrow$ Eq. (7.12)
$\times$ Sensitivity should not be confused with limit of detection

Sensitivity matrix (*Matrix of partial sensitivities*)

Matrix that contains all the sensitivities and cross sensitivities of a multicomponent (multidetector) analytical system.

$\longrightarrow$ KAISER [1972]; DANZER [2001]
$\longrightarrow$ Sect. 7.2
$\longrightarrow$ Eq. (7.17)
$\longrightarrow$ *Cross sensitivity*

Signal

"Response of a device (usually an instrument or a module of an instrument) to certain stimuli". A signal is characterized by at least three parameters: position, intensity, and width (symmetry, shape).

SHARAF et al. [1986]
$\longrightarrow$ Sect. 3.3
$\longrightarrow$ Fig. 3.6

Signal domain (*response domain*)

Field of analytical operation that is characterized by signal properties such as signal position, z, and signal intensity, y_z. The transition to sample domain is done by analytical evaluation (signal decoding).

$\longrightarrow$ Fig. 2.12
$\longrightarrow$ *Sample domain*

Signal function

Record of signal intensity in dependence of the signal position over a certain range of the z-scale: $y = f(z)$.

Signal-to-noise ratio

Measure of the precision of signal measurement, expressed mostly by the ratio of the net signal value to a noise parameter (standard deviation or peak-to-peak distance).

$\longrightarrow$ Sect. 7.1, Fig. 7.2
$\longrightarrow$ Eqs. (7.1)–(7.6)

Specificity

The extent to which *one individual* analyte can be measured undisturbedly in a real sample by a specific reagent, a particular sensor or a comparable specific measuring system.

$\longrightarrow$ KAISER [1972]; DANZER [2001]
$\longrightarrow$ Sect. 7.3
$\longrightarrow$ Eq. (7.26)
$\longrightarrow$ *Selectivity*
$\times$ *Specificity* should not be merged with *selectivity*

Specimen

Fraction of a lot (batch sample) taken without respecting the rules for sampling correctness or under unknown conditions.

Gy [1992]

Standard deviation (SD)

Dispersion parameter for the distribution of measured values, s_y, or analytical results, s_x, for a given sample or the population, σ_y and σ_x. The SD is the square root of the variance.

→ IUPAC [1995]; CURRIE [1999]
→ SACHS [1992]
→ DIXON and MASSEY [1969]
→ Sect. 4.1.2
→ Eqs. (4.12)–(4.14)

Standard error

The term "standard error" is not explicitly introduced. It is used sometimes (a) synonymously for standard deviation and (b) for the residual standard deviation in modelling and calibration.

→ SACHS [1992]
→ FRANK and TODESCHINI [1994]
✗ The term standard error should be avoided

Stimulus

Property of an analytical system to produce a response of an observation- or measuring system. Rousing effect of an analyte that can be characterized qualitatively and quantitatively.

Standard operating procedure (SOP)

"A set of written instructions that document a routine or repetitive activity followed by an organization".

EPA [2001]
→ PRITCHARD et al. [2001]
→ Fig. 7.1

Test

Process of analyzing the sample to recognize (an) analyte(s) and/or determine the amount(s) of (an) analyte(s).

→ PRITCHARD et al. [2001]
→ Quotation from ANAL CHEM [1975] at the end of the Glossary

Total sensitivity (total multicomponent sensitivity)

Sensitivity of a multicomponent analysis. In the simplest case it is given by the determinant of the sensitivity matrix.

→ SHARAF et al. [1986]
→ MASSART et al. [1988]
→ Sect. 7.2
→ Eqs. (7.18)–(7.20)

Traceability

"The property of a result of measurement whereby it can be related to appropriate standards, generally international or national standards, through an unbroken chain of comparisons".

GAT I [1996]
→ ISO 3435-1 [1993]
→ All the standards used should have stated uncertainties

Trackability

"The property of a result of a measurement whereby the result can be uniquely related to the sample".

GAT I [1996]

True value

"Value consistent with the definition of a given particular quantity" and "value which characterizes a quantity perfectly defined in the conditions which exist when that quantity is considered".

ISO 3534-1 [1993]
GAT III [1996]
→ Conventional true value
→ Sect. 7.1

Trueness

"Closeness of agreement between the average value obtained from a large series of test results and an accepted reference value".
Trueness has been referred to as "accuracy of the mean".

IUPAC ORANGE BOOK [1997, 2000]
→ CODEX ALIMENTARIUS COMMISSION [1997]
→ Sect. 7.1.3

Uncertainty of measurement

"Parameter, associated with the result of a measurement, that characterizes the dispersion of the values that could reasonably be attributed to the measurand". The uncertainty should combine both statistical and non-statistical contributions to the variation of the measured values which may occur in all steps of the analytical process.

ISO 3524-1 [1993]
EURACHEM [1995]
GAT I [1996]
$\longrightarrow$ The uncertainty of measurement may be expressed by the combined or extended uncertainty, $u(y)$ or $U(y)$, respectively
$\longrightarrow$ Sect. 4.2
$\longrightarrow$ Eqs. (4.25), (4.26) and (4.29) to (4.32)

Uncertainty of an analytical result

Interval, e.g., of a mean, $U(\overline{x})$, that express the uncertainty of analytical values considering statistical and non-statistical variations within the measurement process plus uncertainties of experimental calibration.

$\longrightarrow$ Sect. 4.2
$\longrightarrow$ Eq. (4.32)

Validation (of an analytical method)

"Process by which it is established, by laboratory studies, that the performance characteristics of the method meet the requirements for the intended analytical applications".

USP XXII < 1225 > [1990]
WEGSCHEIDER [1996]
EURACHEM [1998]
$\longrightarrow$ Typical performance characteristics that should be considered in the validation are: *precision, accuracy, limit of detection, limit of quantitation, selectivity, range, linearity, robustness, ruggedness*

Variable

"Characteristic of an object that may take on any value from a specified set".

FRANK and TODESCHINI [1994]
$\longrightarrow$ There are several types of variables, e.g., *categorical, dependent* and *independent, experimental* and *theoretical, manifest* and *latent, random, standardized variables*

Variance

Dispersion parameter for the distribution of measured values, s_y^2, or analytical results, s_x^2, for a given sample or the population, σ_y^2 and σ_x^2. Statistically defined as the second moment about the mean.	$\longrightarrow$ IUPAC [1995]; CURRIE [1999] $\longrightarrow$ SACHS [1992] $\longrightarrow$ DIXON and MASSEY [1969] $\longrightarrow$ Sect. 4.1.2 $\longrightarrow$ Eqs. (4.8); (4.10)

Working range $\longrightarrow$ *Range* (in the analytical sense)

Analysis, Identification, Determination, and Assay (quoted from ANAL CHEM [1975])

"While most chemists probably realize the difference between the terms *analyze*, *identify*, and *determine*, they are frequently careless when using them. Most frequently the term *analysis* is used when *determination* is meant.

A study of the nomenclature problem indicates that only samples are *analyzed*; elements, ions, and compounds are *identified* or *determined*. The difficulty occurs when the sample is nominally an element or compound (of unknown purity). 'Analysis of ...' (an element or compound) must be understood to mean the identification or determination of impurities. When the intent is to determine how much of such a sample is the material indicated by the name, *assay* is the proper word."

References

Anal Chem (1975), published in several issues amongst the Instructions for Authors, e.g., 47:2527

Burns DT, Danzer K, Townshend A (2005) IUPAC, Analytical Chemistry Division. *Use of the terms "Robust" and "Rugged" and the associated characteristics of "Robustness" and "Ruggedness" in descriptions of analytical procedures.* Draft 2005

Codex Alimentarius Commission (1997) ALINORM 97/23A, Appendix III: *Analytical terminology for codex use*

Currie LA (1999) *Nomenclature in evaluation of analytical methods including detection and quantification capabilities (IUPAC Recommendations 1995).* Anal Chim Acta 391:105

Danzer, K (2001) *Selectivity and specificity in analytical chemistry. General considerations and attempt of a definition and quantification.* Fresenius J Anal Chem 369:397

Dixon WJ, Massey, FJ (1969) *Introduction to statistical analysis.* McGraw-Hill, New York

Ehrlich, G, Danzer, K (2006) *Nachweisvermögen von Analysenverfahren. Objektive Bewertung und Ergebnisinterpretation.* Springer, Berlin Heidelberg New York

EPA (2001) Environmental Protection Agency. *Requirements for quality assurance project plans for environmental data operations* (EPA QA R/5)

EURACHEM (1995) *Quantifying uncertainty in analytical measurement.* Teddington

EURACHEM (1998) *The fitness for purpose of analytical methods.* Teddington

Frank IE, Todeschini R (1994) *The data analysis handbook.* Elsevier, Amsterdam

GAT: *Glossary of analytical terms.* Series in the journal "Accreditation and Quality Assurance"

I. (1996) 1:41 (Fleming J, Neidhart B, Tausch C, Wegscheider W)

II. (1996) 1:87 (Fleming J, Albus H, Neidhart B, Wegscheider W)

III. (1996) 1:135 (Fleming J, Neidhart B, Albus H, Wegscheider W)

IV. (1996) 1:190 (Fleming J, Albus H, Neidhart B, Wegscheider W)

V. (1996) 1:233 (Fleming J, Albus H, Neidhart B, Wegscheider W)

VI. (1996) 1:277 (Fleming J, Albus H, Neidhart B, Wegscheider W)

VII. (1997) 2:51 (Fleming J, Albus H, Neidhart B, Wegscheider W)

VIII. (1997) 2:160 (Fleming J, Albus H, Neidhart B, Wegscheider W)

IX. (1997) 2:348 (Prichard E, Albus H, Neidhart B, Wegscheider W)

X. (1998) 3:171 (Prichard E, Albus H, Neidhart B, Wegscheider W)

Gy PM (1992) *Sampling of heterogeneous and dynamic material systems: theories of heterogeneity, sampling and homogenizing.* Elsevier, Amsterdam

Hirschfeld T (1980) *The hy-phen-ated methods.* Anal Chem 52(2):297A

Holcombe D (1999/2000) *Alphabetical index of defined terms and where they can be found.* Accr Qual Assur *Part I: A–F* 4:525; *Part II: G–Q* 5:77; *Part III: R–Z* 5:159

ICH (1996) ICH Topic Q2B (ICH Harmonised Tripartite Guideline). *Validation of analytical methods: methodology* (CPMP/ICH/281/95), ICH, London

ISO 3534-1 (1993) International Organization for Standardization (BIPM, IEC, IFCC, ISO, IUPAC, IUPAP, OIML), *International vocabulary of basic and general terms in metrology.* Geneva

ISO Guide 33 (1989) International Organization for Standardization, *Use of certified reference materials.* Genova

IUPAC (1994a) Analytical Chemistry Division, Commission on Analytical Nomenclature: *Recommendations for the presentation of results of chemical analysis.* Prepared for publication by LA Currie and G Svehla. Pure Appl Chem 66:595

IUPAC (1994b) Analytical Chemistry Division, Commission on Analytical Nomenclature, Commission on General Aspects of Analytical Chemistry: *Nomenclature of interlaboratory analytical studies.* Prepared for publication by W Horwitz. Pure Appl Chem 66:1903

IUPAC (1995) Analytical Chemistry Division, Commission on General Aspects of Analytical Chemistry: *Nomenclature in evaluation of analytical methods including detection and quantification capabilities (IUPAC Recommendations 1995).* Prepared for publication by LA Currie. Pure Appl Chem 67:1699

IUPAC (1998) Analytical Chemistry Division, Commission on General Aspects of Analytical Chemistry: *Guidelines for calibration in analytical chemistry*. Prepared for publication by K Danzer, LA Currie. Pure Appl Chem 70:993

IUPAC (2004) Analytical Chemistry Division: *Guidelines for calibration in analytical chemistry. Part 2. Multispecies calibration (IUPAC Technical Report)*. Prepared for publication by K Danzer, M Otto, LA Currie. Pure Appl Chem 76:1215

IUPAC Orange Book (1997, 2000)
– printed version: *Compendium of Analytical Nomenclature (Definitive Rules 1997)*, 3rd edn. Edited by J Inczédy, JT Lengyiel, AM Ure, A Geleneser, A Hulanicki. Blackwell, Oxford, 1997
– web version (from 2000 on): www.iupac.org/publications/analytical_compendium/

Kaiser H, Specker H (1956) *Bewertung und Vergleich von Analysenmethoden*. Fresenius Z Anal Chem 149:46

Kaiser H (1972) *Zur Definition von Selektivität, Spezifität und Empfindlichkeit von Analysenverfahren*. Fresenius Z Anal Chem 260:252

Kellner R, Mermet J-M, Otto M, Widmer HM (Eds) (1998) *Analytical chemistry. The approved text to the FECS curriculum analytical chemistry*. Wiley-VCH, Weinheim, New York

Massart DL, Vandeginste BGM, Deming SN, Michotte Y, Kaufman L (1988) *Chemometrics: a textbook*. Elsevier, Amsterdam

Prichard, E, Green, J, Houlgate, P, Miller, J, Newman, E, Phillips, G, Rowley, A (2001) *Analytical measurement terminology – handbook of terms used in quality assurance of analytical measurement*. LGC, Teddington, Royal Society of Chemistry, Cambridge

Sachs L (1992) *Angewandte Statistik 7th edn*. Springer, Berlin Heidelberg New York

Sharaf MA, Illman DL, Kowalski BR (1986) *Chemometrics*. Wiley, New York

Taylor JK (1987) *Quality assurance of chemical measurements*. Lewis Publishers, Chelsea, MI

USP XXII < 1225 > (1990), United States Pharmacopoeia, *Validation of compendial methods*

Wegscheider W (1996) *Validation of analytical methods*. Chap 6 in: Günzler H (ed) *Accreditation and quality assurance in analytical chemistry*. Springer, Berlin Heidelberg New York

Index